W9-CRL-664

Heteroatom Chemistry

Heteroatom Chemistry

Eric Block, Editor

Department of Chemistry
State University of New York at Albany
Albany, NY 12222

03790745

CHEMISTRY

Eric Block
Department of Chemistry
State University of New York at Albany
Albany, NY 12222

Library of Congress Cataloging-in-Publication Data

International Conference on Heteroatom Chemistry (2nd: 1989: Albany.
   N.Y.)
      Heteroatom chemistry/edited by Eric Block.
      p.   cm.
      ISBN 0-89573-743-4
      1. Heterocyclic chemistry—Congresses.   I. Block, Eric.
   II. Title.
   QD399.I54 1989
   547′.59—dc20   90-11961
                                                             CIP

British Library Cataloguing in Publication Data

International Conference on Heteroatom chemistry. 2nd (1989)
   Heteroatom chemistry
   1. Chemical elements & chemical compounds
   I. Title.   II. Block, Eric
   540

   ISBN 0-89573-743-4

Printed in the United States of America.
ISBN 0-89573-743-4 VCH Publishers
ISBN 3-527-27858-3 VCH Verlagsgesellschaft

Print History:
10 9 8 7 6 5 4 3 2 1

Published jointly by:

VCH Publishers, Inc.	VCH Verlagsgesellschaft mbH	VCH Publishers (UK) Ltd.
220 East 23rd Street	P.O. Box 10 11 61	8 Wellington Court
Suite 909	D-6940 Weinheim	Cambridge CB1 1HW
New York, New York 10010	Federal Republic of Germany	United Kingdom

To my wife Judith
and children David and Melinda

Contents

Preface

Heteroatom chemistry allows a broadening of the boundaries of organic chemistry. For example, until recently, compounds with multiple bonding between carbon and non–first-row elements were viewed as chemical oddities, synthetic reagents employing heavier heteroatoms were considered to be laboratory curiosities, and physical organic studies rarely focused on the influence of heavier heteroatoms. Current research in organosilicon, organosulfur, and organophosphorus chemistry has uncovered numerous compounds with $C=Si$, $Si=Si$, $C=S$, and $C=P$ double bonds and $C\equiv P$ and $C\equiv S$ triple bonds, including compounds of types unimaginable a few years ago. Useful new regiospecific or stereospecific reagents for organic synthesis employing such previously uncommon heteroatoms as arsenic, selenium, tellurium, and bismuth are being reported daily. Determination of rates of solvolysis involving tin and silicon as neighboring groups show the most dramatic rate acceleration effects ever seen.

To bring together researchers studying the organic chemistry of such heteroatoms as boron, sulfur, selenium, tellurium, silicon, tin, phosphorus, arsenic, antimony, and bismuth, and to highlight recent trends in this broad and active field, a series of international conferences on heteroatom chemistry (ICHAC) was established, the first being in Kobe, Japan in 1987 and the second in Albany, New York, in July, 1989. This book contains the texts of 20 of the invited lectures at ICHAC-2. As such it constitutes the first compilation of recent research on the organic chemistry of the principal heteroatoms other than nitrogen and oxygen. Where possible the focus is on what happens at the *heteroatom* rather than on the chemistry of molecules in which the presence of heteroatoms is incidental. Discussions of mechanism and bonding appear side by side with specific applications of heteroatoms in synthesis. It is hardly necessary to mention that the field of heteroatom-directed synthesis is one of the most vigorously growing areas of research. When used together with the many excellent available books that provide an introduction to the chemistry of each of the specific heteroatoms, this volume summarizes for both the student and seasoned investigator current areas of research activity involving each heteroatom along with up-to-date references.

The emphasis in several chapters is on the synthesis of heteroatomic compounds with unusual bonding. Adrian Brook introduces us to the unusual world of compounds with $C=Si$ bonds (silenes) that can in turn serve as starting materials for synthesis of a host of novel organosilicon heterocyclic and acyclic compounds. His chapter provides us with numerous challenging mechanistic puzzles. Bulky alkylsilanes are also of value in the stabilization of various boron compounds, as demonstrated by Michael Lappert and co-workers, and in adding steric hindrance to thiolate ligands, as demonstrated by Kaluo Tang and Youqi Tang. As shown by Manfred Regitz, phosphaalkynes (RC≡P) constitute a valuable entry to a variety of fascinating trivalent phosphorus compounds, such as phosphacubanes, Dewar phosphabenzenes, nonplanar heterobenzenes, 1H-phosphirenes, phosphacyclobutadienes, and metal complexes of phosphaalkynes. Wataru Ando and Norihiro Tokitoh look at the photochemistry and pyrolysis of 1,2,3-selenadiazoles and 1,2,3-thiadiazoles in the presence of unsaturated compounds and elemental sulfur. Thioketenes, selenoketenes, 1,2-diselenones, selenirenes, and (isolable) seleniranes and novel cyclic polysulfides–polyselenosulfides are generated. A structural study by Konrad Seppelt clarifies the molecular basis for the color of aryl bismuth(V).

Three chapters deal with complexes from different perspectives. The chapter by Anthony Arduengo and David Dixon examines bonding trends in complexes of the pnictogens, with emphasis on phosphorus and arsenic as well as comparison with related organosulfur heterocycles. The focus of the chapter by Henry Kuivila is on the development of multidentate Lewis acids, including cyclic systems, involving boron, tin, and mercury, for example, as catalysts. In his chapter, Hideki Sakurai focuses on macrocyclic compounds containing both carbon–carbon triple bonds and silicon. These compounds undergo metal-promoted intramolecular cyclization processes leading to unusual cyclic structures and metal complexes.

A major portion of this volume deals with new synthetic methodology involving heavier heteroatoms. Uses of organoarsenic ylides in organic synthesis is the subject of a chapter by Yao-Zeng Huang and co-workers. The use of aryl bismuth(V) and lead(VI) reagents in the arylation of a wide range of nucleophilic organic molecules [indoles, phenols, ketones, nitroalkanes, 4-hydroxycoumarins (C-phenylation), amines (N-phenylation), glycols, and sulfinic acid salts (O-phenylation)] and mechanistic studies of these ligand-coupling reactions are covered by Sir Derek H. R. Barton. Other examples of ligand coupling, involving trivalent compounds of sulfur, selenium, and tellurium, are presented by Naomichi Furukawa.

The thesis of the chapter by Alain Krief and co-workers is that organoselenium compounds are quite useful in the synthesis of molecules containing adjacent quaternary centers. When sulfur and selenium combine forces in selenosulfones, $RSeSO_2R'$, they make available a useful procedure for organic synthesis known as selenosulfonation. Thomas Back and co-workers illustrate the use of these reagents under both free radical and electrophilic addition conditions for regioselective and stereoselective synthesis of sulfonyl-substituted allylic alcohols, 1,2-dienes, alkynes, and allenes, as well as sulfur-free marine sterols and a variety of other products. The removal of sulfur when done under Theodore Cohen's conditions of reductive elimination provides an extremely useful method for generating organolithium compounds not easily prepared by other methods, including α-lithiothers and hydrocarbon allyl anions. Synthetic methods based on reductive cleavage of ethers coupled with directing effects associated with cerium π-allyl complexes round out his chapter.

Ottorino De Lucchi and co-workers illustrate a synthetically valuable new procedure involving a chiral acetylene equivalent in Diels–Alder reactions and a chiral carbonyl anion equivalent. The first procedure involves the use of 1,1'-binaphthalene-2,2'-bis(sulfonyl)ethylene, whereas the second procedure involves carbanions from dinaphtho[2,1-d:1',1'-f][1,3]-dithiepine and its S-oxides. Henri Kagan reports efficient new routes to chiral sulfoxides and then illustrates diverse applications of these chiral compounds as auxiliaries in asymmetric synthesis. A complementary chapter by Atsuyoshi Ohno reviews the controversial subject of the stereochemistry of deprotonation and electrophilic substitution at the α carbon of chiral sulfoxides.

The entire field of heteroatom chemistry has benefited through application of theoretical calculations to important problems. Yitzhak Apeloig in his chapter illustrates the application of modern theoretical methods to such areas of organosilicon chemistry as multiply bonded silicon compounds ($C = Si$ and $Si = Si$), silicenium ions, the β-silicon carbocation hyperconjugative effect, and cationic rearrangements involving silicon, providing a suitable theoretical backdrop for the chapters by Brook and Lambert. Physical organic chemistry provides a sound theoretical insight into organic reactions. Joseph Lambert and co-workers examine the physical basis for the β effect of the group IVB elements silicon, germanium, tin, and lead and come up with evidence for extraordinary acceleration by the heavier of these elements for ionization on β carbons, particularly when antiperiplanar geometry is possible.

ICHAC-2, held at the State University of New York at Albany, July 17–21, 1989, and sponsored by the Organic Divisions of the American Chemical Society and the Canadian Society for Chemistry, was generously supported by the following corporations, agencies and foundations, to whom I express my sincere gratitude: American Chemical Society Petroleum Research Fund; American Cyanamid Company; Amoco Performance Products, Inc.; Bristol-Myers Company; Chemistry Department–SUNYA; CIBA-GEIGY; Dow Corning Corporation; Eastman Kodak; E. I. Du Pont De Nemours & Company; General Electric Corporate R&D Center; General Electric Silicone Division; Hoechst Celanese; Hoechst-Roussel Pharmaceuticals, Inc.; Hüls America, Inc.–Petrarch Systems; Janssen Research Foundation; Lubrizol Corporation; Merck Frosst Canada, Inc.; Merck Sharp & Dohme Research Laboratories; Monsanto Company; National Science Foundation; PPG Industries, Inc.; Pfizer, Inc.; Research Foundation–SUNY; Sandoz Research Institute; Schering Research; SmithKline Beckman; Société Nationale Elf Aquitaine; State University of New York at Albany; Sterling-Winthrop Research; Syntex Research; Upjohn Company; VCH Publishers; and Wakunaga of America Company. I would also like to acknowledge the excellent editorial efforts of Mark Sacher, and the considerable advice and assistance of the members of the International Advisory Board of ICHAC-2, D. H. R. Barton, S. Gronowitz, C. R. Johnson, J. F. King, J. C. Martin, J. Michalski, G. Modena, S. Oae, B. Zwanenburg, and X.-Y. Li, and the members of the Organizing Committee, J. J. Eisch, R. S. Glass, B. Kavanaugh, E. McLaren, P. Toscano, J. J. Tufariello, and J. Welch. I would especially like to express my gratitude to Ellen Faust, Patricia Kelly, and my wife, Judith Block, for countless significant contributions to the success of the conference.

Eric Block

Albany, New York
January 1990

New Chemistry of Fused 1,2,3-Selenadiazoles and 1,2,3-Thiadiazoles

Wataru Ando and Norihiro Tokitoh

Department of Chemistry
University of Tsukuba
Tsukuba, Ibaraki 305
Japan

1.1. Introduction

Among the selenium-containing heterocycles, the 1,2,3-selenadiazole ring is generally a labile system, which shows versatile utility in organic synthesis [1]. Unlike the 1,2,3-thiadiazole ring system, the corresponding 1,2,3-selenadiazoles are easily pyrolyzed with loss of nitrogen and selenium to give high yields of alkynes [2]. However, 1,2,3-selenadiazoles fused to a ring that is smaller than eight-membered do not undergo ready acetylene formation and the denitrogenated intermediate can dimerize to 1,4-diselenin derivatives, which in turn can lose a selenium atom to afford selenophene derivatives [2a, 3]. In some cases, the selenoketenes and/or 1,3-diselenetane derivatives can be formed by thermal denitrogenation of 1,2,3-selenadiazoles followed by Wolff rearrangement of the resulting α-selenoketocarbene intermediates [4]. Photolysis of the 1,2,3-selenadiazole ring system has also been well investigated for the purpose of generation and matrix isolation of selenirenes and selenoketenes [5].

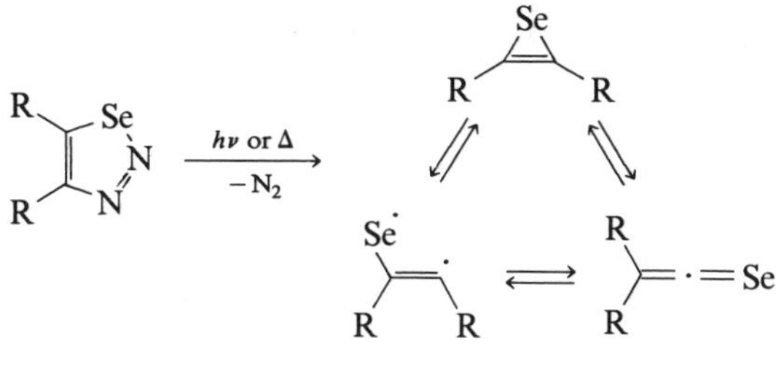

1.2. Results and Discussion

1.2.1. Photolysis of 1,2,3-Selenadiazole 1 [6, 7]

The sterically protected bicyclic 1,2,3-selenadiazole **1** was readily denitrogenated by irradiation with light of either $\lambda > 365$ nm or $\lambda = 254$ nm. However, the photochemically generated intermediates showed quite different reactivities. Regioselective cycloaddition products **2a** or **2b** of the zwitterionic intermediate **3** were obtained in good yields by photolysis of **1** with light of $\lambda > 365$ nm in the presence of excess amount of olefins such as acrylonitrile or methyl acrylate, whereas irradiation of **1** with light of $\lambda = 254$ nm resulted in polymerization of the olefin, suggesting formation of radical intermediate **4**. The high reactivity of diradical **4** toward alkyl C—H bonds was confirmed by exclusive formation of diselenide derivative **5** in the photolysis of **1** with light of $\lambda = 254$ nm using hexane as a solvent.

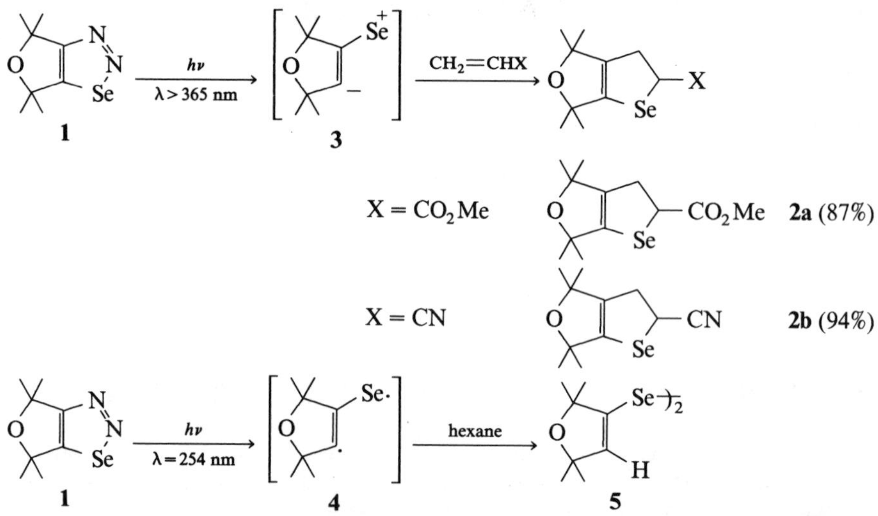

Photochemical reaction of **1** with thiophene and furan gave characteristic results in addition to the variation of the products with different light sources. When **1** was irradiated by a high-pressure mercury lamp in thiophene, three isomeric cycloaddition products, that is, [4 + 3] adduct **6** and [2 + 3] adducts **7** and **8**, were obtained. Among them, **6** was relatively labile and slowly isomerized into **7** at room temperature, probably due to ring strain. Photolysis of **1** in furan was also examined and was found to give a bicyclic aldehyde, **10**, the isolation of which suggests initial formation of the [4 + 3] adduct **11** or [2 + 3] adduct **11′** followed by the electrocyclic ring transformation via the conjugated selone **12**.

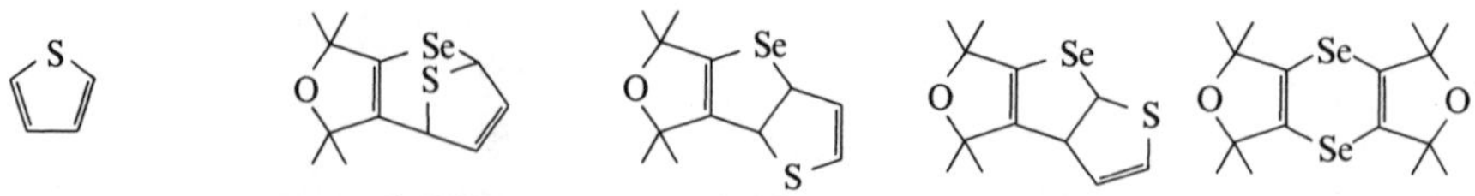

6 (11%)   7 (7%)   8 (2%)   9 (39%)

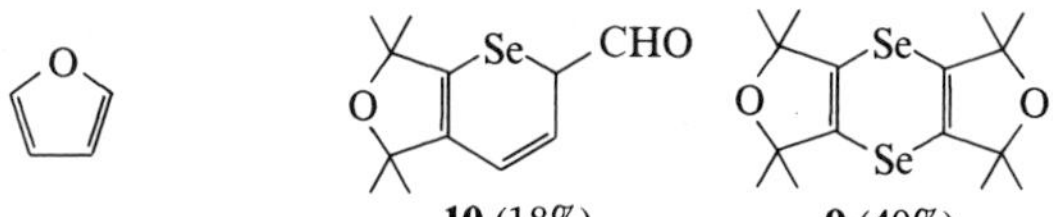

10 (18%)   **9** (40%)

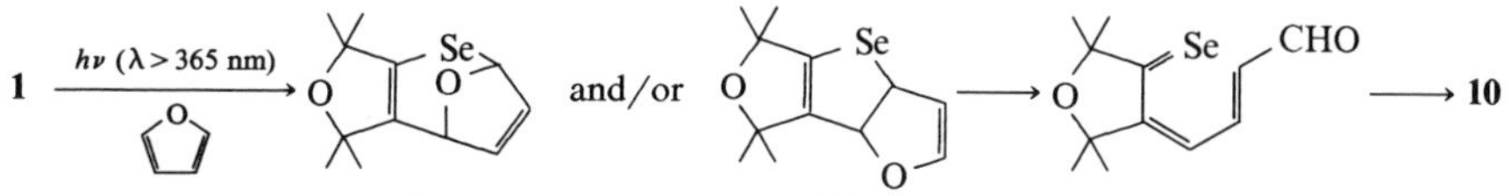

11   and/or   **11′**   →   **12**   → **10**

On the other hand, photolysis of **1** in furan with light of $\lambda = 254$ nm, using a low-pressure mercury lamp, resulted in trapping of the intermediary selenirene **13** to afford the selenirane **14** as a single stereoisomer along with the diselenide **15** and aldehyde **10**. The structure of the furan adduct **14**, the first example of a stable selenirane derivative, was confirmed by the molecular symmetry reflected in the NMR spectra, by the high resolution mass spectra, and by chemical proof. Adduct **14** underwent a facile reaction with tris(dimethylamino)phosphine to give quantitatively the corresponding deselenated product **16** with tris(dimethylamino)phosphine selenide. The selenirane **14** is stable in dry benzene at room temperature but decomposes slowly into polymeric products in the presence of a trace amount of acid.

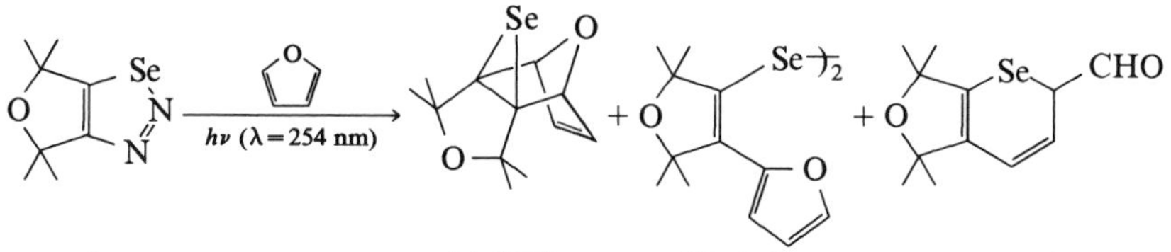

1   **14** (12%)   **15** (41%)   **10** (11%)

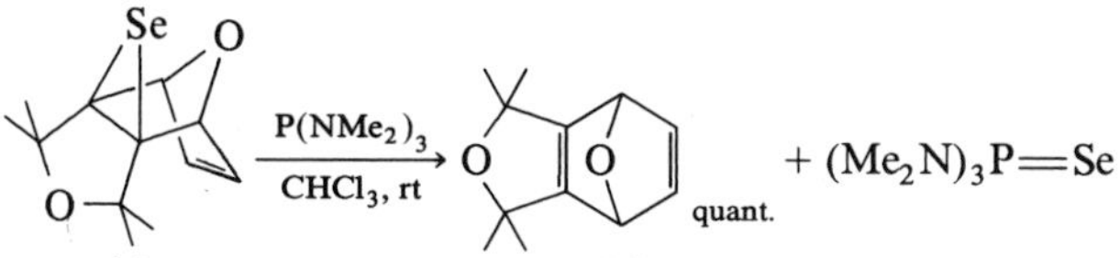

14   **16**   **17**

In the photochemical reaction of **1** with light of $\lambda = 254$ nm in 2,5-dimethylfuran, two types of hydrogen abstraction products (**5** and **18**) of the radical intermediate **4** were obtained rather than a cycloadduct of selenirene. Since only the 1,4-diselenin **9** was produced by the photolysis of **1** with light of $\lambda > 365$ nm in hexane or 2,5-dimethylfuran, the formation of **5** and **18** from **1** by irradiation with a low-pressure mercury lamp implies the coexistence of the selenirene and diradical intermediates **13** and **4**.

1 18 (14%) 5 (20%)

4 13

19 20 (11%) 23 (30%) 21 (12%) 24 (22%)

22

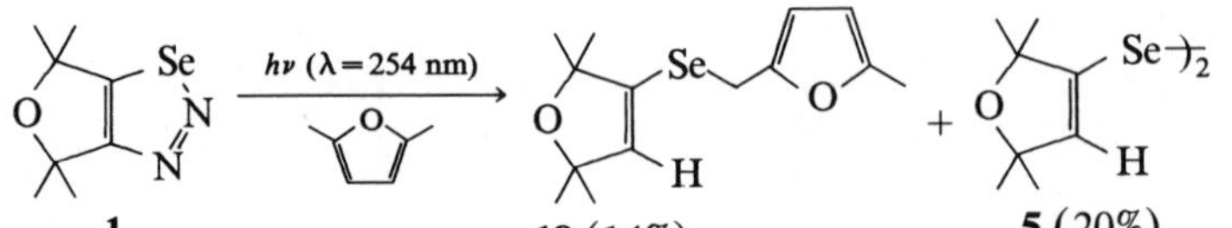

The analogous 1,2,3-thiadiazole **19**, which was unchanged by irradiation with light of $\lambda > 365$ nm, was readily photolyzed by the low pressure mercury lamp in either furan or 2,5-dimethylfuran to afford the Diels–Alder type cycloadducts **20** and **21** of the intermediary thiirene **22** along with the diradical adducts **23** and **24**, respectively.

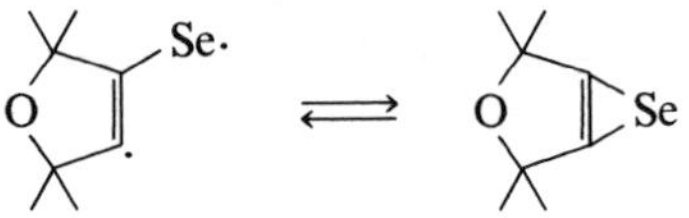

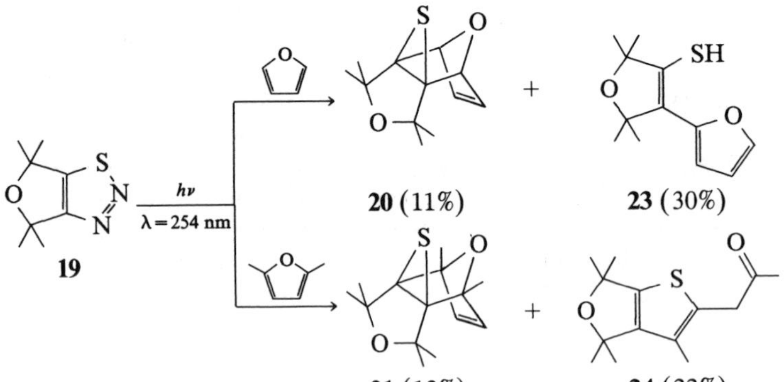

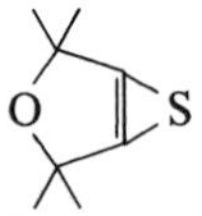

1.2.2. Photolysis of Other Fused 1,2,3-Selenadiazoles [8]

The photochemical reactions of some other types of fused 1,2,3-selenadiazoles, such as **25** and **26**, were also examined with a view of elucidating the effect of the ring size and steric protection by their fused ring components on the nature of the intermediates and products.

1.2.2a. Six-Membered Ring Fused System. Six-membered ring fused 1,2,3-selena-diazole **25** was photolyzed in furan with light of either $\lambda = 254$ nm or $\lambda > 365$ nm to give a complex mixture of 1,4-diselenin **27**, 1,3-diselenole **28**, selenophene **29**, and the cycloaddition product with furan **30** as shown below.

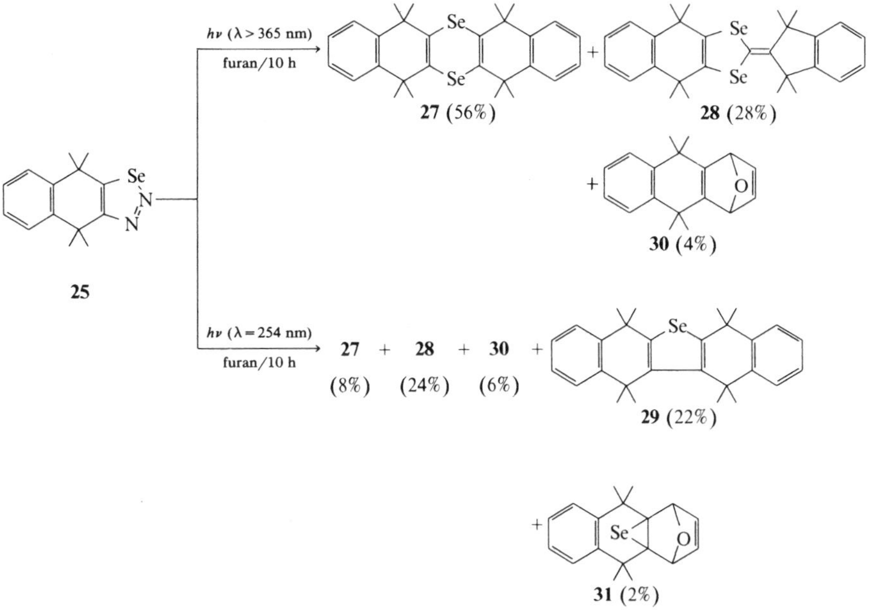

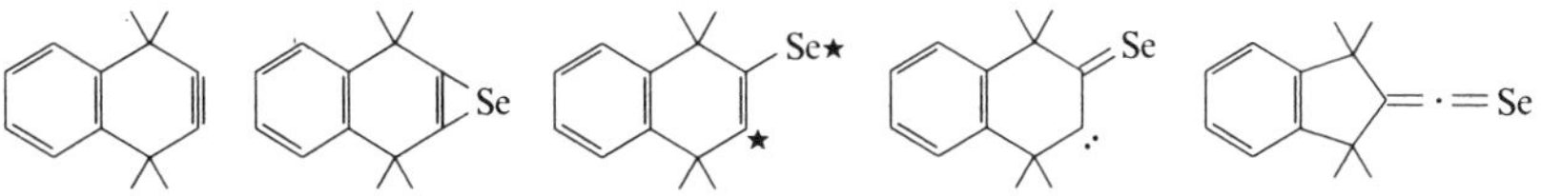

Since another furan adduct was isolated in 2% yield in the case of $\lambda = 254$ nm, with a molecular composition of $C_{18}H_{20}OSe$ as judged by high resolution mass spectroscopy, the formation **30** might be rationalized by the initial generation and subsequent deselenation of selenirane **31**. However, there is no denying the direct cycloaddition of intermediate unstable six-membered cycloalkyne **32** with furan.

1.2.2b. Seven-Membered Ring Fused System [8b]. When the selenadiazole fused with the dibenzocycloheptatriene skeleton **26** [9] was irradiated in furan, furan adduct **35** was formed as shown below along with 1,4-diselenin **33** and selenophene **34**, regardless of the light source, though in very low yields.

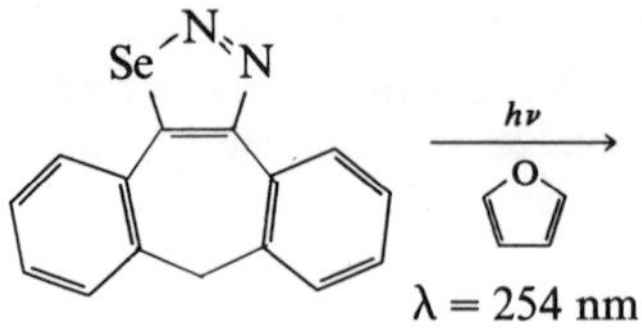

26

λ = 254 nm
λ > 365 nm

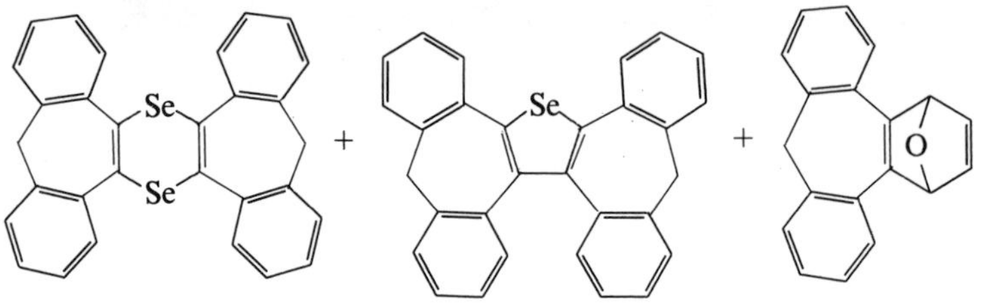

33	34	35
11%	21%	1.3%
7%	16%	1.2%

These results suggest that the mode of substitution and ring size of the fused part of 1,2,3-selenadiazoles have a considerable influence on the distribution and stability of the reaction intermediates and products in their photolysis.

1.2.2c. Formation of Unsymmetrically Substituted 1,4-Diselenin in the Photolysis of Selenadiazoles. Photolysis of a benzene solution of an equimolar mixture of two types of 1,2,3-selenadiazoles, **1** and **25**, with light of λ > 365 nm resulted in a cross-coupling reaction of the two different denitrogenated selenium-containing intermediates leading to the formation of a new type of unsymmetrically substituted 1,4-diselenin **36** in addition to the expected self-dimerization products such as **9**, **27**, and **28**.

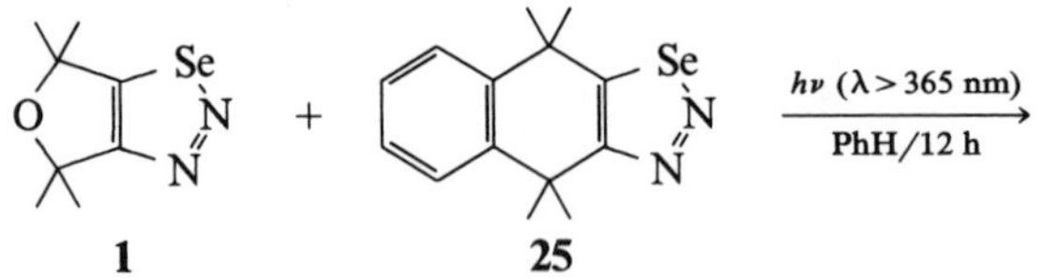

1 **25**

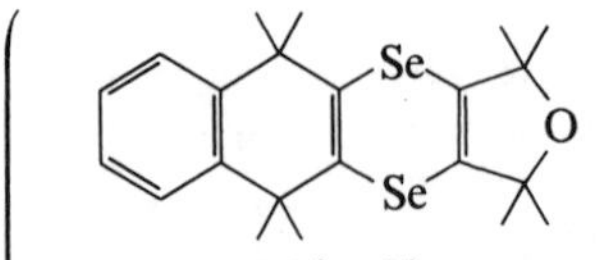

36 (27%)

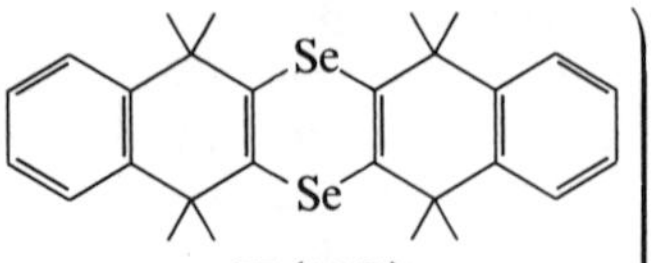

27 (17%)

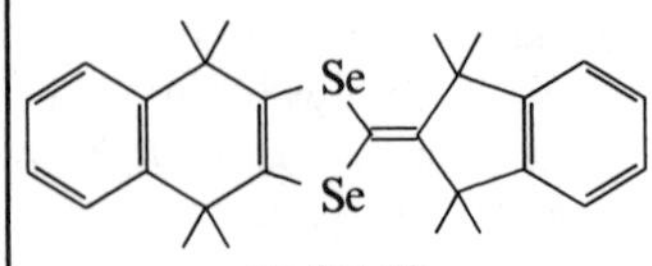

28 (21%)

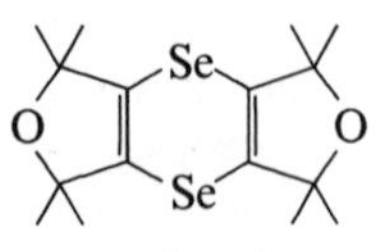

9 (21%)

The photochemical coupling reaction thus described is a useful and convenient method for synthesis of many kinds of diselenin derivatives, because only self-dimerization proceeds in thermolysis of the mixed selenadiazoles due to the wide difference between their decomposition temperatures.

1.3. Thermolysis of 1,2,3-Selenadiazoles

1.3.1. Five-Membered Ring Fused System [7]

1,2,3-Selenadiazole **1** was also found to be very labile thermally, and it readily loses nitrogen even at 80°C in benzene to afford the 1,4-diselenin **9** in 96% yield. In contrast, other 1,2,3-selenadiazoles and 1,2,3-thiadiazoles, for example six-membered ring fused system **25**, do not decompose at temperatures below 160 °C.

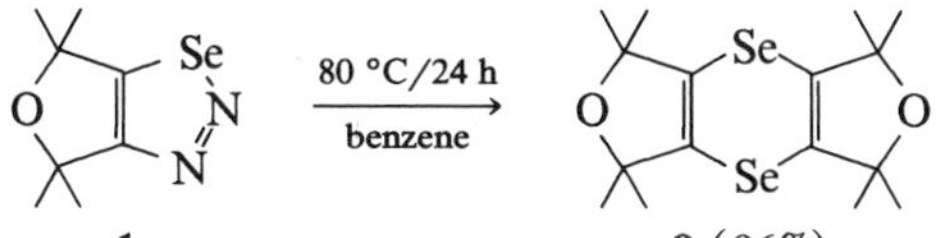

Thermolysis of **1** in the presence of several kinds of olefins, thiocarbonyl compounds, and thiiranes gave the corresponding dihydroselenophene derivatives **2a** and **2b**, 1,3-thiaselenoles **38–40**, and 1,4-thiaselenin derivatives **41** and **42** regiospecifically as shown in Table 1.1. Reaction of **1** with molten anthracene at 220 °C also produced a cycloadduct **37** along with the 1,4-diselenin **9**.

1.3.2. Six-Membered Ring Fused System 25 [8a]

Rather complicated results were obtained in the thermolysis of **25**. When **25** was heated in *o*-dichlorobenzene at 160 °C for 4 h, 1,3-diselenole **28**, and 1,3-diselenetane **43** were isolated along with 1,4-diselenin **27** and selenophene **29**. Formation of **28** and **43** suggests that the generation of the selenoketene **44** via Wolff rearrangement of selenoketocarbene intermediate is involved in thermolysis of this system.

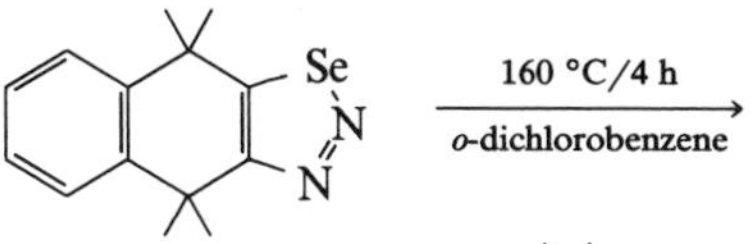

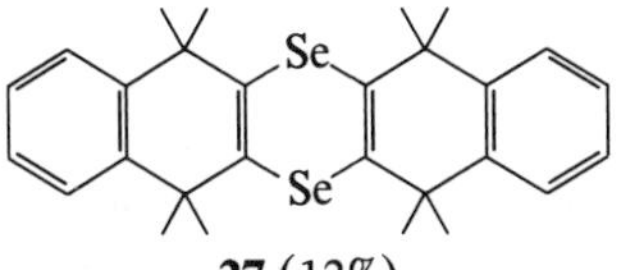

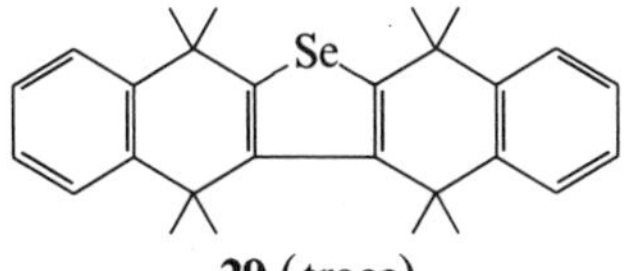

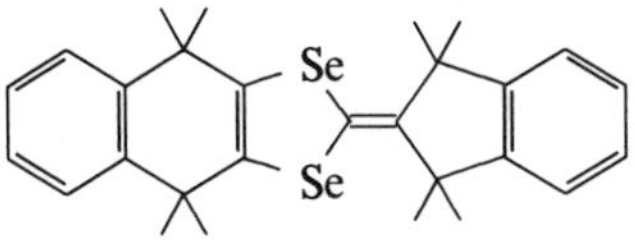

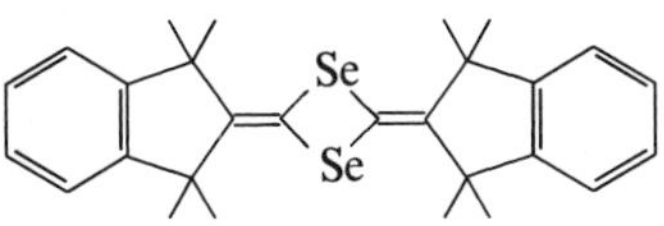

Table 1.1 ■ Thermolysis of 1,2,3-Selenadiazole 1 in the Presence of Trapping Reagents

Run	Trapping reagents	Reaction conditions	Products and yields
1	None (benzene)	80 °C/24 h	**9** (96%)
2	CH$_2$=CHCN	77 °C/48 h	**2a** (85%)
3	CH$_2$=CHCO$_2$Me	86 °C/12 h	**2b** (91%)
4	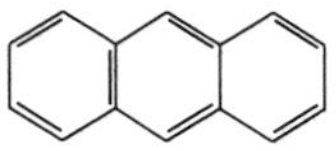	220 °C/5 min	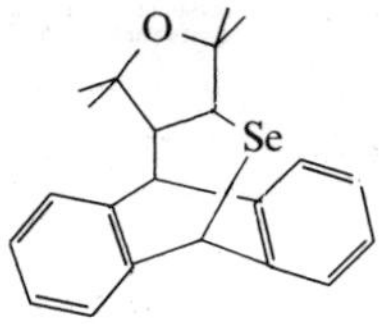 +**9** (30%) **37** (26%)
5	Ph$_2$C=S	80 °C/24 h	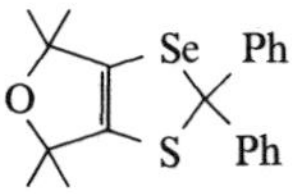 **38** (70%)
6	CS$_2$	90 °C/24 h	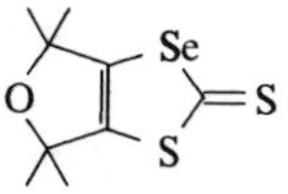 **39** (75%)
7	PhNCS	90 °C/12 h	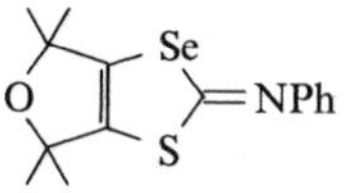 **40** (61%)
8	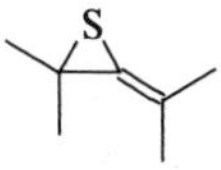	80 °C/24 h	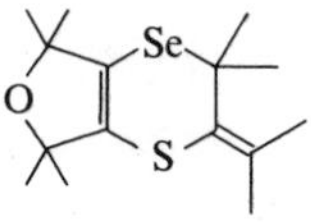 +**9** (65%) **41** (31%)
9	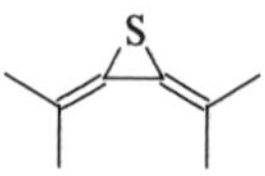	80 °C/24 h	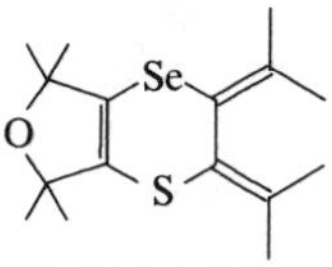 +**9** (73%) **42** (222%)

A trapping experiment of the flow pyrolyzed mixture of the benzene solution of **25** at 530 °C with an excess amount of diphenyldiazomethane solution (hexane) resulted in formation of the corresponding 2-alkylidene-1,3,4-selenadiazoline **45**, which is a 1,3-dipolar cycloadduct of intermediate selenoketene **44**.

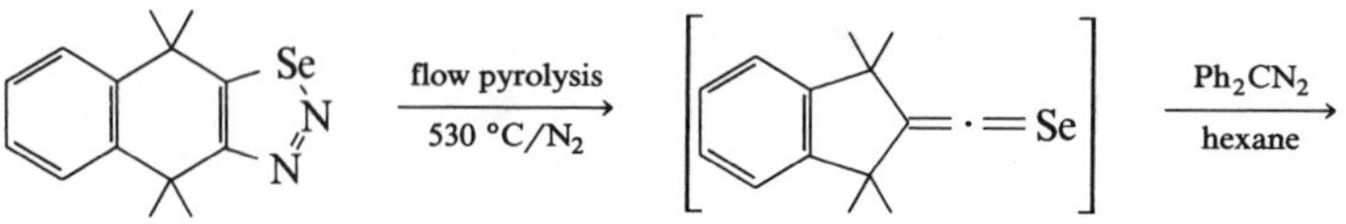

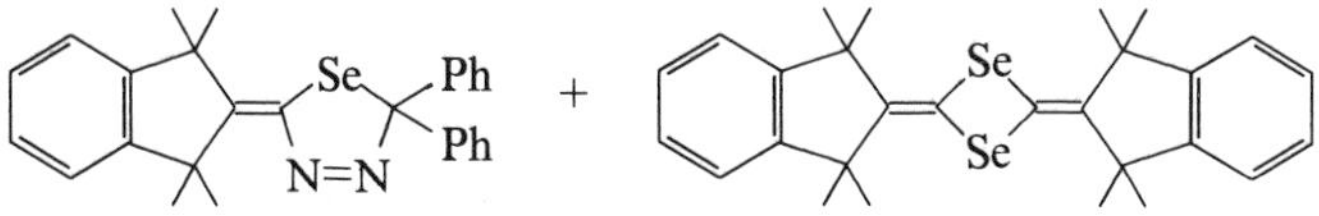

45 (20%)   **43** (54%)

Although we did not succeed in the isolation or spectroscopic detection of the selenoketene **44**, by flow pyrolysis of the analogously substituted 1,2,3-thiadiazole **46** we obtained the corresponding thioketene **47** quantitatively. Isolated thioketene **47** was found to dimerize slowly to the 1,3-dithietane derivative **48** in chloroform at room temperature after a few hours.

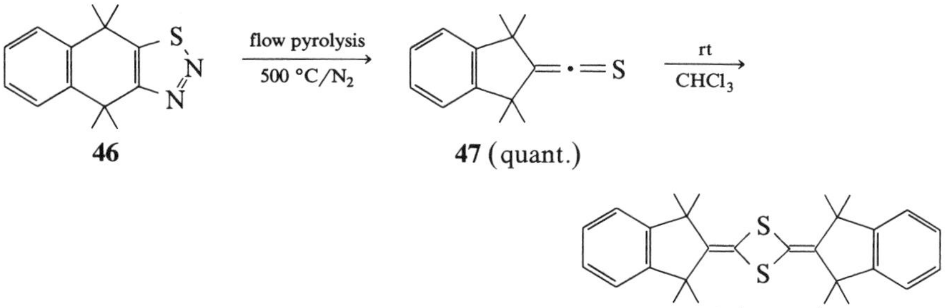

46   **47** (quant.)

48 (quant.)

1.4. Reactions of 1,2,3-Selenadiazoles with Nucleophiles

In contrast to the active research on the photolysis and thermolysis of 1,2,3-selenadiazole ring systems, little has been reported on their reactions with nucleophiles. Meier and co-workers [11] have already described the reactions of cycloalkeno-1,2,3-selenadiazoles with butyllithium to give the corresponding cycloalkyne and/or butyl selenide derivatives via nucleophilic attack of butyl anion on the selenium atom followed by the extrusion of nitrogen. Because of the notorious instability of cyclopentyne, the selenium atom was expected to be preserved in the reaction of the 1,2,3-selenadiazole **1** with nucleophiles.

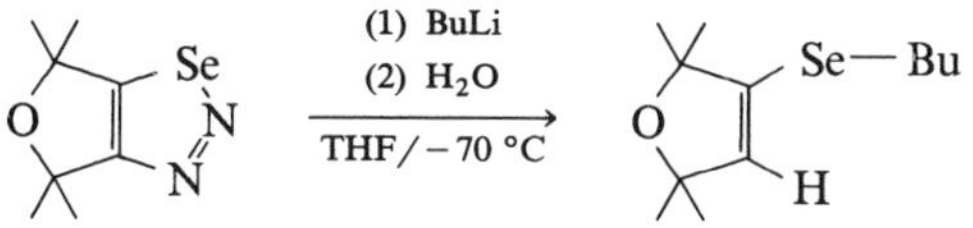

1   **49** (53%)

Treatment of a tetrahydrofuran solution of **1** with an equimolar amount of *n*-butyllithium at $-70\ ^\circ$C afforded the butyl selenide **49** in 53% yield. A similar butyl selenide **50** was obtained in 47% yield in the reaction of selenadiazole **25** with *n*-butyllithium. This type of reaction is also general in the case of rather soft nucleophiles, such as phosphites and disulfides, and it readily proceeded below the decomposition temperature of **1** to give the corresponding selenides as shown in Table 1.2.

These reactions can be rationalized by the initial attack of the nucleophiles on the selenium atom leading to the zwitterion **51** followed by the loss of nitrogen and the intramolecular nucleophilic attack of alkenyl anion **52**.

Table 1.2 ▪ Reactions of 1,2,3-Selenadiazole 1 with Nucleophiles

Entry	Nucleophiles	Reaction conditions	Products and yields
1	BuLi/H₂O	− 70 °C/30 min	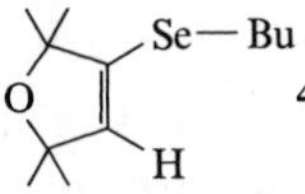 Se—Bu, 49 (53%), H
2	P(OMe)₃	50 °C/24 h	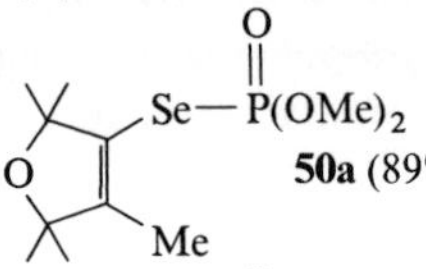 Se—P(OMe)₂ (O), 50a (89%), Me
3	P(OEt)₃	50 °C/36 h	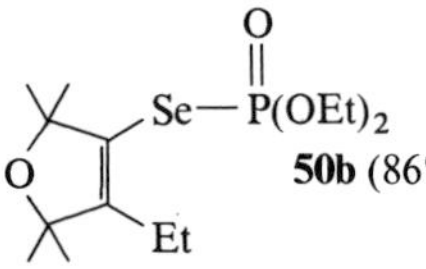 Se—P(OEt)₂ (O), 50b (86%), Et
4	PhSH	rt/12 h	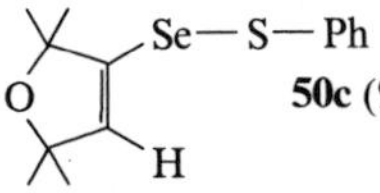 Se—S—Ph, 50c (98%), H
5	PhSSPh	60 °C/48 h	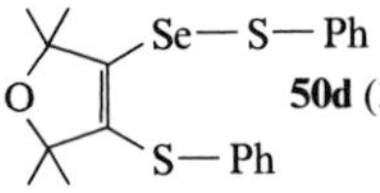 Se—S—Ph, 50d (32%), S—Ph
6	MeSSMe	60 °C/24 h	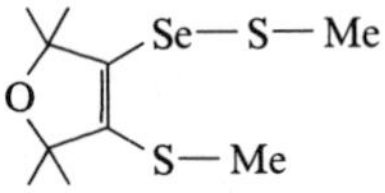 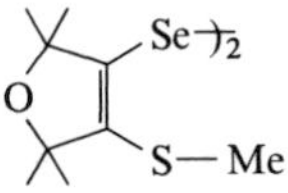 Se—S—Me, S—Me, 50e (25%); Se-)₂, S—Me, 50f (25%)

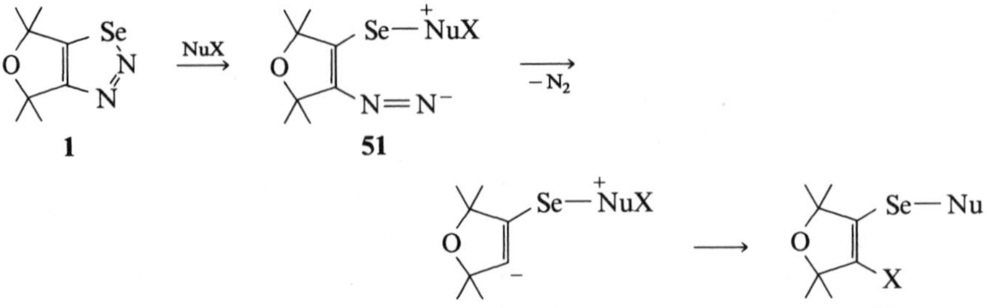

Furthermore, treatment of the selenadiazole **1** with diphenylcyclopropenethione, which is sufficiently nucleophilic to cleave the Se—N bond of 1,2,3-selenadiazole, also led to characteristic cyclization of the initially formed nitrogen-containing intermediate **54**, leading to a novel seven-membered heterocycle **55** in 90% yield [12].

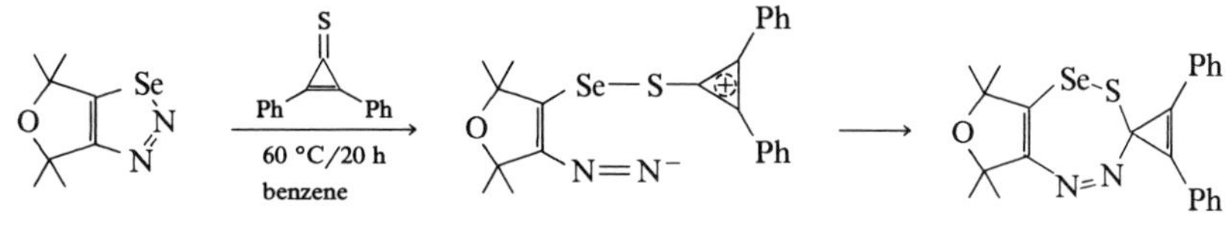

1   **54**   **55** (90%)

Reaction of the six-membered ring fused 1,2,3-selenadiazole **25** with triethyl phosphite proceeds in a different fashion. The spectral data on the product **57** confirmed the extrusion of the selenium atom, reflecting the high leaving ability of triethyl selenophosphate and the relative stability of cyclohexyne intermediate **32** compared to the corresponding cyclopentyne derivative.

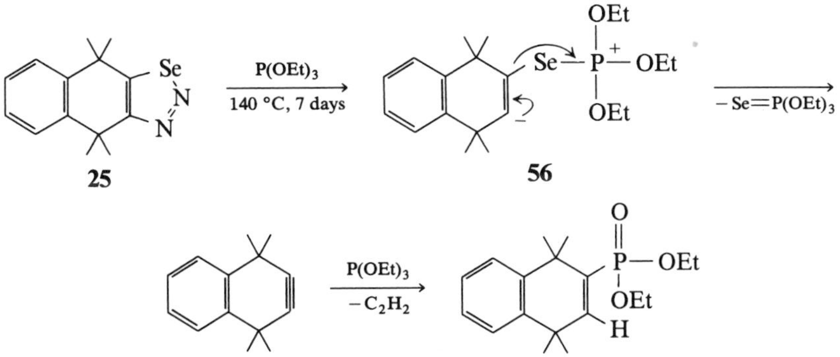

25   **56**

32   **57** quant.

1.5. Cyclopolychalcogenide Formation in the Reactions of 1,2,3-Selenadiazoles

1.5.1. Reactions of 1,2,3-Selenadiazole 1 with Elemental Sulfur [13]

The chemistry of cyclic polysulfides has attracted much attention not only for their unique physical and chemical properties but also for their biological activity [14]. Some 1,2,3,4,5-pentathiepins and 1,2,3-trithioles have been synthesized and characterized [15]. However, most examples of these compounds have been restricted to the aromatic and heteroaromatic ring fused systems, probably due to their much lower stability than the linear polysulfides. The intrinsic nature of these interesting heterocycles without the influence of the fused aromatic ring still remains to be studied. Furthermore, there have been no reports on their selenium analogues, such as 1,2,3,4,5-pentaselenepin and 1,2,3-triselenole ring systems.

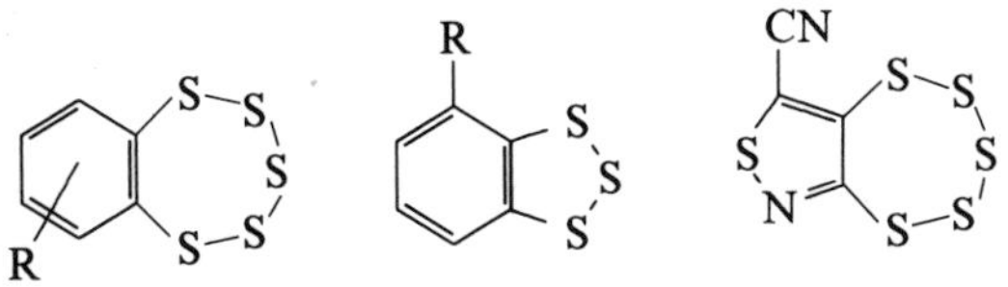

We present here a novel formation of stable cyclopolysulfides and cyclopoly-selenides by the thermal reactions of fused 1,2,3-selenadiazoles and 1,2,3-thiadiazoles with elemental sulfur and selenium.

1.5.1a. Reaction of Five-Membered Ring Fused System [13]. When 1,2,3-thiadia-zole **17** was treated with excess (equimolar amount as S_8) of molten sulfur at 120 °C for 15 h, two kinds of cyclic polysulfides, that is, 1,2,3,4,5-pentathiepin **58** and 1,2,3,4-tetra-thiin **59** [16], were isolated as stable crystalline products in 22% and 18% yields, respectively. Similarly, 1,2,3,4-tetrathia-5-selenepin (**60**, 39%), 1,2,3,4-tetrathiin (**59**, 19%), and 1,2,3,4,5-pentathiepin (**58**, 10%) were readily obtained in the reaction of 1,2,3-selenadiazole **1** with molten sulfur at 120 °C for 10 min.

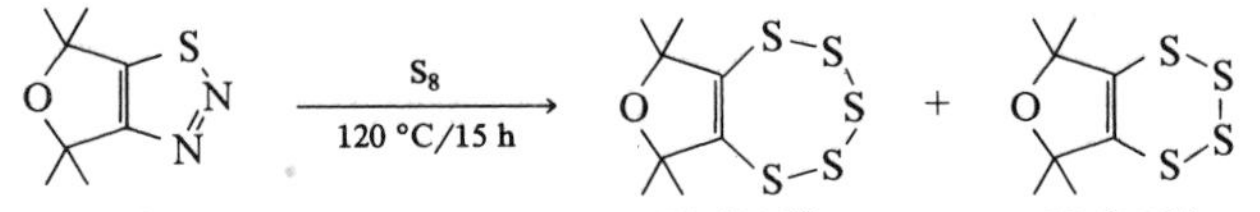

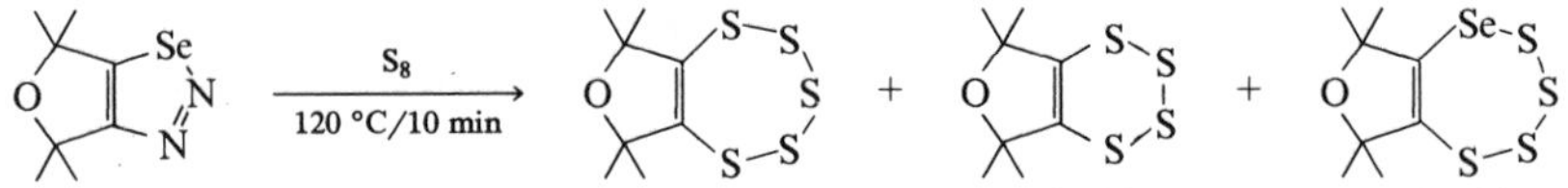

The ratio of the products in these reactions was affected by the reaction time: Prolonged heating of the mixture resulted in an increase of polymeric products.

1.5.1b. Reaction of 1,2,3-Selenadiazole 25 [17]. 1,2,3-Selenadiazole **25** was also sulfurized with molten sulfur at 140 °C for 2 days to give the 1,2,3-trithiole **61** and 1,2,3,4,5-pentathiepin **62** in a pure and stable crystalline form after silica gel column chromatography followed by HPLC purification. The remarkable stability of **61** here obtained is worthy of note in comparison with that of aromatic fused 1,2,3-trithioles.

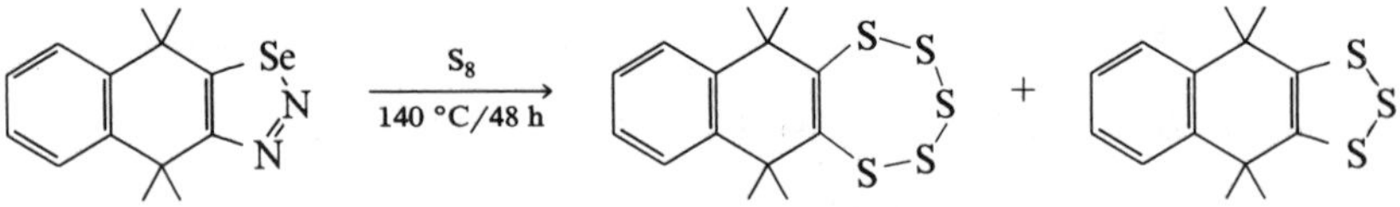

Thermolysis of the pentathiepin **62** at 140 °C in *o*-dichlorobenzene readily led to an equilibrium mixture of **61**, **62**, and S_8 (the **61/62** ratio was 3 : 1 as judged by NMR spectroscopy), which was alternatively generated by heating the *o*-dichlorobenzene solution of 1,2,3-trithiole **61** mixed with S_8 at the same temperature.

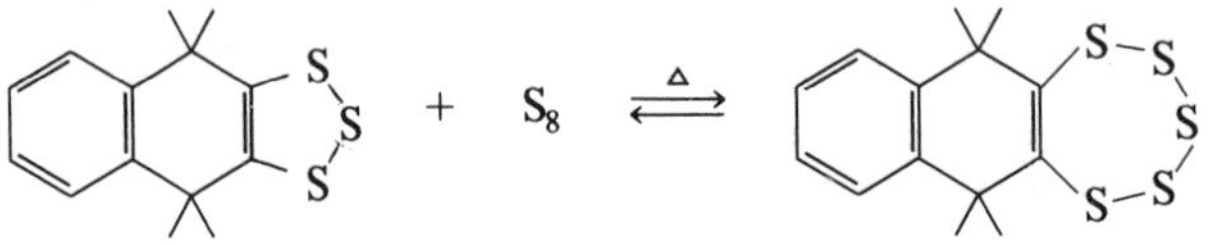

The equilibrium ratio certainly shows the relative stability of **61** and should be compared to the Chenard equilibrated cyclopolysulfides, the ratio of which has been reported as almost 1 : 1 [18].

Although the 1,2,3-selenadiazole **63** was quite inert to elemental selenium alone, even in a polar solvent such as N,N-dimethylformamide, below its decomposition temperature, the selenation of **25** was readily achieved by activated selenium in tributylamine used as a solvent. Chromatographic separation and recrystallization from benzene afforded the novel 1,2,3-triselenole **63** as brownish orange crystals (mp, 140–140.5 °C) in 25% yield.

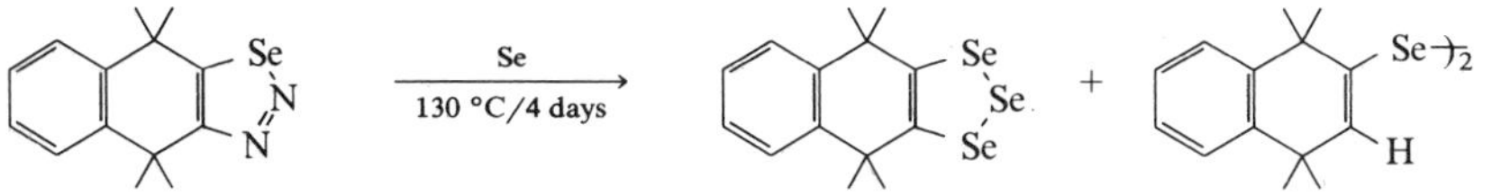

	63	**64**
(1) in Bu$_3$N/DMF	17%	15%
(2) in Bu$_3$N	25%	0%

1,2,3-Triselenole **63** was very stable even at its melting point and slowly thermolyzed in o-dichlorobenzene at 150 °C to give the 1,3-diselenole derivative **28** in 48% yield via the coupling of the initially formed deselenated intermediates **64a** and **64b** with the unstable selenoketene **44**, which was the rearranged product from **64b**.

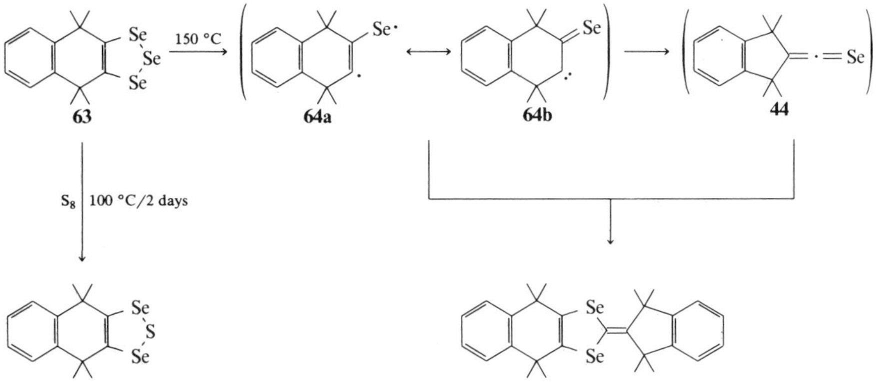

65 (70%)   **28** (48%)

Interestingly, the thermal reaction of the 1,2,3-triselenole **63** with elemental sulfur in o-dichlorobenzene at 100 °C for two days resulted in exclusive formation of the 1,2,5-thiadiselenole **65** as orange crystals (mp, 128–129 °C) in 70% yield.

The structures of new cyclopolychalcogenides such as **61**, **62**, **63**, and **65** thus obtained were determined by ^{1}H NMR, ^{13}C NMR, UV, and MS spectra as well as by elemental analysis. Compounds **63** and **65** are of great interest as the first examples of 1,2,3-triselenole and 1,2,5-thiadiselenole ring systems. The isolation of these new types of heterocycles in a stable crystalline form is one of the successful results of the effective steric protection of the fused tetramethyl-substituted dihydronaphthalene ring. However, the subtle steric requirement and ring strain of this fused system seem to disfavor the formation of the other expected cyclic polyselenides, such as 1,2,3,4-tetraselenin or 1,2,3,4,5-pentaselenepin.

1.5.1c. Sulfurization of Seven-Membered Ring Fused 1,2,3-Selenadiazole 26 [8b].

As another substrate, 8,8-dibenzo[3,4;6,7]cyclohepta[1,2-d][1,2,3]selenadiazole **26** was subjected to the thermal reaction with molten sulfur to afford the more complicated mixture of cyclic polysulfides. After exhaustive chromatographic separation and recrystallization, we could isolate three kinds of cyclopolysulfides, that is, 1,2,3-trithiole **66**, 1,2,3,6,7,8-hexathiecin **67**, and 1,2,3,4,5,6,7,8,9-nonathiacycloundecene **68**, in a stable crystalline form together with an inseparable mixture of two kinds of unidentified cyclopolysulfides.

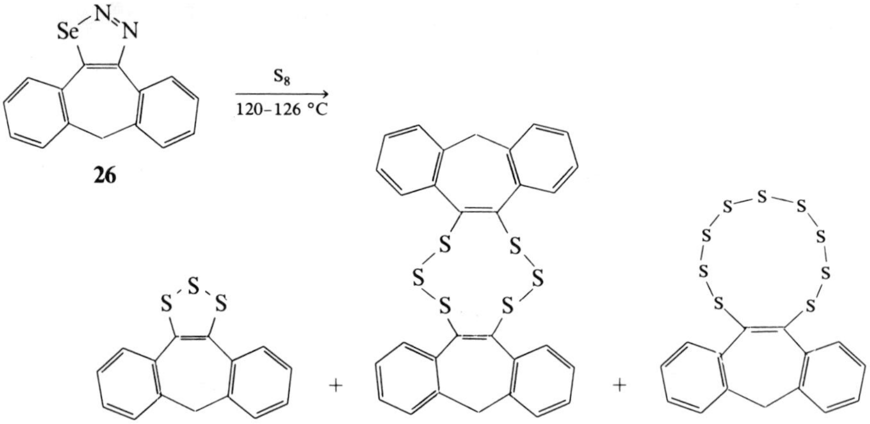

66	**67**	**68**
37%	3.3%	12%
orange needles	yellow crystals	yellow crystals
mp 215 °C	mp 255 °C	mp 155–157 °C
(decomp.)	(decomp.)	(decomp.)

Of these compounds 1,2,3,6,7,8-hexathiecin **67** was found to be rather labile, giving the desulfurized 1,2,3-trithiole **66** quantitatively by slight heating during the recrystallization. The formation and structure of **68** are of great interest inasmuch as **68** represents the first cyclic polysulfide containing a longer chain of sulfur atoms than that of elemental sulfur. In these cases the conjugation of the sulfur linkage with the aromatic ring might contribute to the stabilization of the cyclopolysulfide skeleton instead of the insufficient steric protection. In summary, we have established a facile and convenient synthetic route to a variety of cyclic polysulfides by the thermal reactions of 1,2,3-selenadiazoles with molten sulfur, which has the advantage of suppressing the hydrogen abstraction of the intermediary radical species from the reaction media.

1.6. A Novel Formation of 1,2-Diselone in the Reactions of 1,2,3-Selenadiazoles

Considerable interest has been focused on the formation of 1,2-dithione and its tautomer, 1,2-dithiete [19]. However, there have been no reports on their selenium analogues, that is 1,2-diselone and 1,2-diselenete [20]. It seems to us that the introduction of a selenium atom into the selenadiazole system would provide an ideal method to produce the 1,2-diselone. We now describe a novel formation of 1,2-diselone by the thermal reaction of 1,2,3-selenadiazole **1** with 4-*t*-butyl-1,2,3-selenadiazole **69** as a good precursor of selenium atom [21].

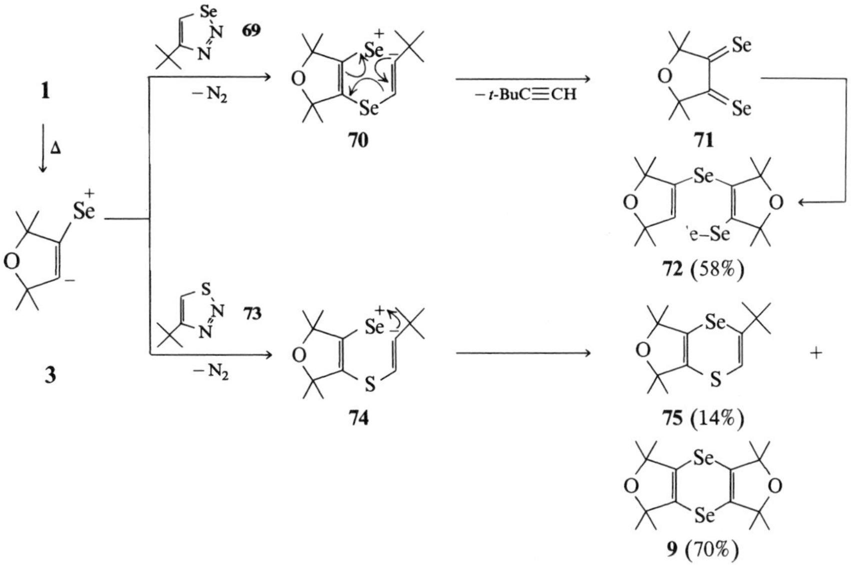

1,2,3-Selenadiazole **1** was heated with an equimolar amount of 4-*t*-butyl-1,2,3-selenadiazole **69** in benzene at 80 °C for 24 h to afford 1,2,5-triselenepin (**72**, 58%) together with 1,4-diselenin **9**, while in the absence of **69** only the latter product was obtained in a high yield. Considering the inertness of **1** to elemental selenium under these conditions, the formation of 1,2,5-triselenepin **72** might be rationalized by nucleophilic attack of the initially generated zwitterionic intermediate **3** on the selenium atom of **1** followed by the elimination of *t*-butylacetylene leading to the 1,2-diselone **71**. This nucleophilic nature of the reaction was strongly supported by the fact that the 1,4-thiaselenin derivative **75** was isolated in the reaction of **1** with 4-*t*-butyl-1,2,3-thiadiazole **73** as a result of more favorable cyclization than the elimination probably due to the rather rigid C—S bond.

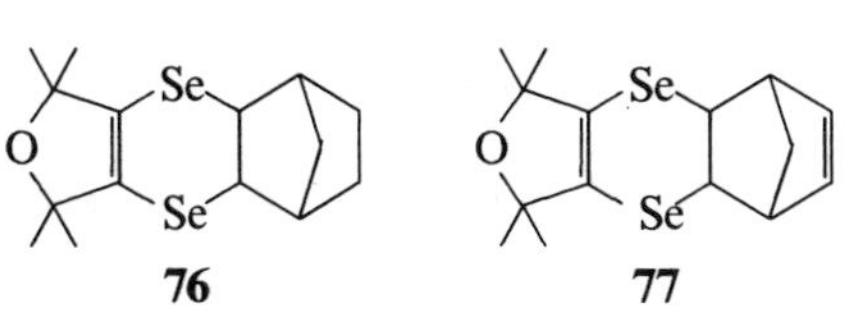

The intermediary 1,2-diselone **71** was trapped by the reaction in the presence of reactive olefin, such as norbornene or norbornadiene, to give the corresponding cycloaddition products of **71**, that is, the dihydro-1,4-diselenin derivatives **76** and **77** in 25% and 18% yields, respectively.

In the presence of acenaphthylene, not the expected cycloadduct **78** but a novel and interesting cyclic polyselenide, that is, the 1,2,3,4,5-pentaselenepin **79**, was obtained together with 1,4-diselenin **9**, 1,2,5-triselenepin **72**, and dihydroselenophene **80**.

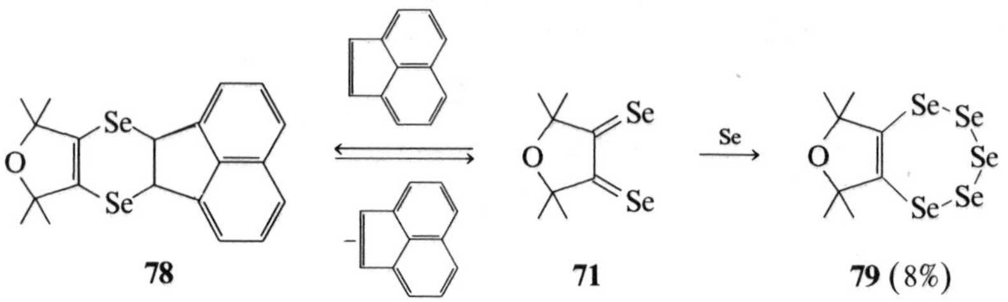

Since no pentaselenepin **79** was formed without acenaphthylene, this particular type of olefin should play an important role as a carrier of the 1,2-diselone unit. During the cycloaddition and retro reaction between **71** and **78**, the 1,2-diselone **71** reacts with the nascent selenium derived from 4-*t*-butyl-1,2,3-selenadiazole or 1,2-diselenetane **81**, the [2 + 4] adduct of 1,2-diselone and 1,2-diselenete. The 1,2,3,4,5-pentaselenepin **79**, which has been an unknown ring system so far as we know, was isolated as stable yellow-orange needles with a decomposition point at 130 °C.

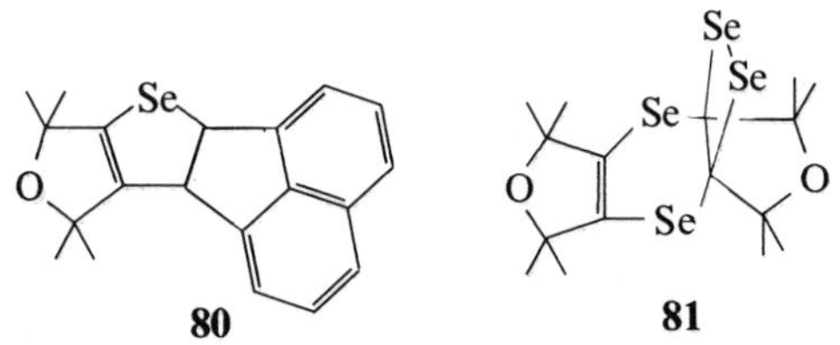

1.7. Facile Formation of 1,2-Dithione and Its Selenium Analogues by the Photolysis of Fused Cyclic Polychalcogenides

1.7.1. Photolysis of Five-Membered Ring Fused System [13, 22]

The successful and ready formation of a variety of stable cyclic polychalcogenides, as described in the previous sections, prompted us to examine their potential utility as precursors of 1,2-dithiones and their selenium analogues by photochemical dechalcogenation. Irradiation of the fused cyclopolysulfide **59** in benzene with light of $\lambda > 365$ nm led to the moderate yield of 1,4-dithiin **82**. When the photolysis was carried out in the presence of norbornene, dihydro-1,4-dithiin derivative **83a** was obtained in a high yield as a cycloadduct of the intermediary 1,2-dithione **85** with the host olefin.

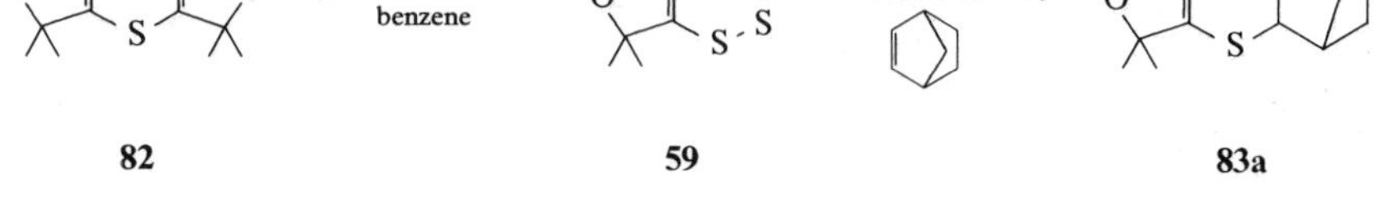

82   59   83a

Similarly, the photolysis of **58** in the presence of various olefins gave the corresponding adducts **83b–83e** as shown in Table 1.3. Since the tetrathiin **59** was found to be the primary product in the course of the photolysis of the pentathiepin **58** as judged by NMR spectroscopy, the formation of the cycloadducts from **58** is also rationalized by the loss of S_2 unit from once-generated tetrathiin **59** followed by cycloaddition reaction of the resulting 1,2-dithione **85** with external olefins.

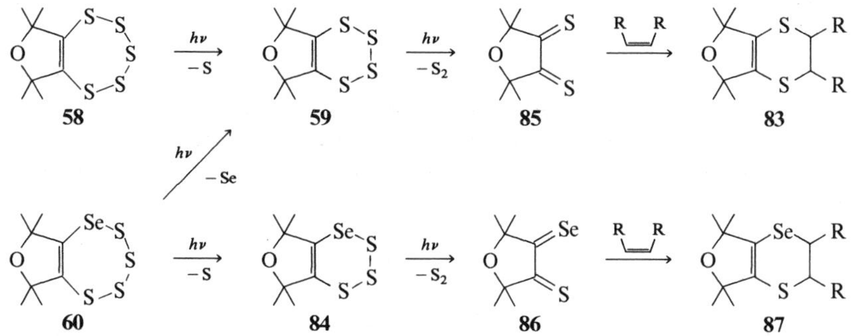

Irradiation of the 1,2,3,4-tetrathia-5-selenepin **60** in the presence of olefins (see Table 1.4) gave the dihydro-1,4-dithiins **83a**, **83c**, and **83f** as the major products by way

Table 1.3 ▪ Photolysis of 1,2,3,4,5-Pentathiepin 58 in the Presence of Olefins

Entry	Olefin	Products and yields
1	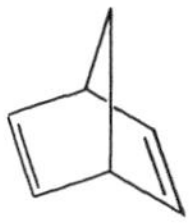	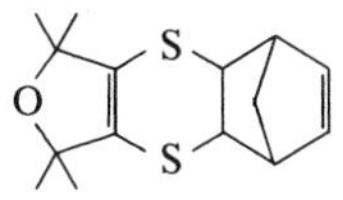 **83b** (70%)
2	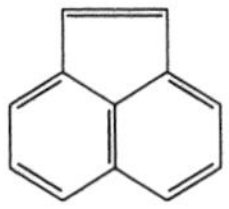	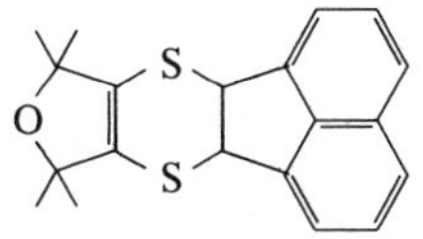 **83c** (27%) 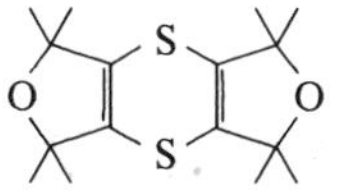 **82** (8%) 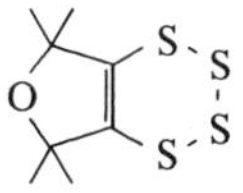 **59** (42%)
3	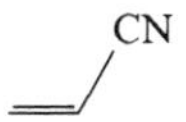 CN	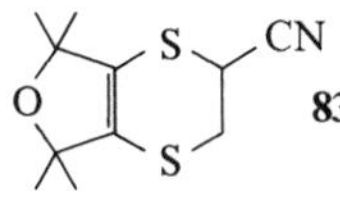 **83d** (85%)
4	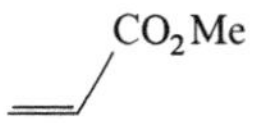 CO$_2$Me	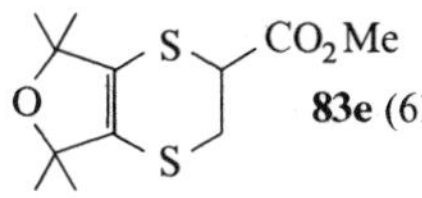 **83e** (61%)
5	None	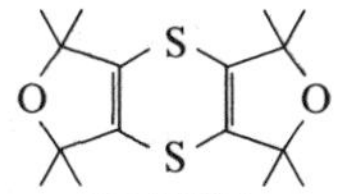 **82** (20%) 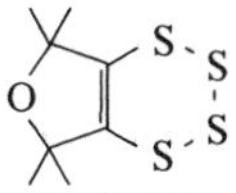 **59** (70%)

Table 1.4 ▪ Photolysis of 1,2,3,4-Tetrathia-5-selenepin 60 in the Presence of Olefins

Olefins	Products and Yields			
Norbornene	**83a** (77%)	**87a** (19%)		
Acenaphthylene	**83c** (45%)	**87c** (7%)	**82** (5%)	**59** (26%)
N-Phenylmaleimide	**83f** (58%)[a]	**87f** (15%)[a]		

[a] Structures **83f** and **87f** correspond, respectively, to **83** and **87** where R,R = — C(O)NPhC(O) — .

of 1,2-dithione **85** along with the dihydro-1,4-thiaselenin derivatives (**87a**, **87c**, and **87f**), which was almost certainly interpreted as the competitive formation of 1,2-selenoxothione **86** and the preference of deselenation to desulfurization in the first step in the photochemical reaction of **60**.

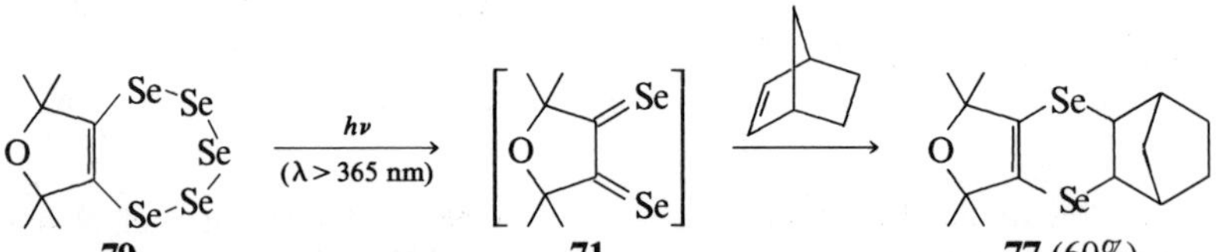

Photodeselenation of 1,2,3,4,5-pentaselenepin **79** was performed in the presence of norbornadiene to give the expected adduct **77** of 1,2-diselone **71** in 60% yield with precipitation of red amorphous selenium. In addition, the transfer of the 1,2-dithione and 1,2-selenoxothione units were achieved very easily by the photolysis of the cycloadducts **83c** and **87c** in the presence of norbornene to give in quantitative yields the rather stable cycloadducts **83a** and **87a**, respectively.

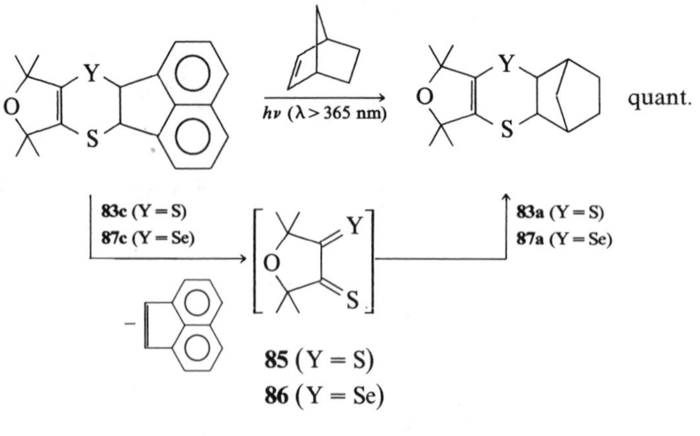

1.7.2. Photolysis of Six-Membered Ring Fused System

The photochemical dechalcogenation of 1,2,3-trithiole **61**, 1,2,3-triselenole **63**, and 1,2,5-thiadiselenole **65** were also examined [17]. With light of either λ > 365 nm or

$\lambda = 254$ nm, **61** and **63** were photolyzed very slowly in benzene at room temperature (about 20% of substrate was consumed after irradiation for ten days) to afford the 1,4-dithiin **88** and 1,4-diselenin **89**, respectively.

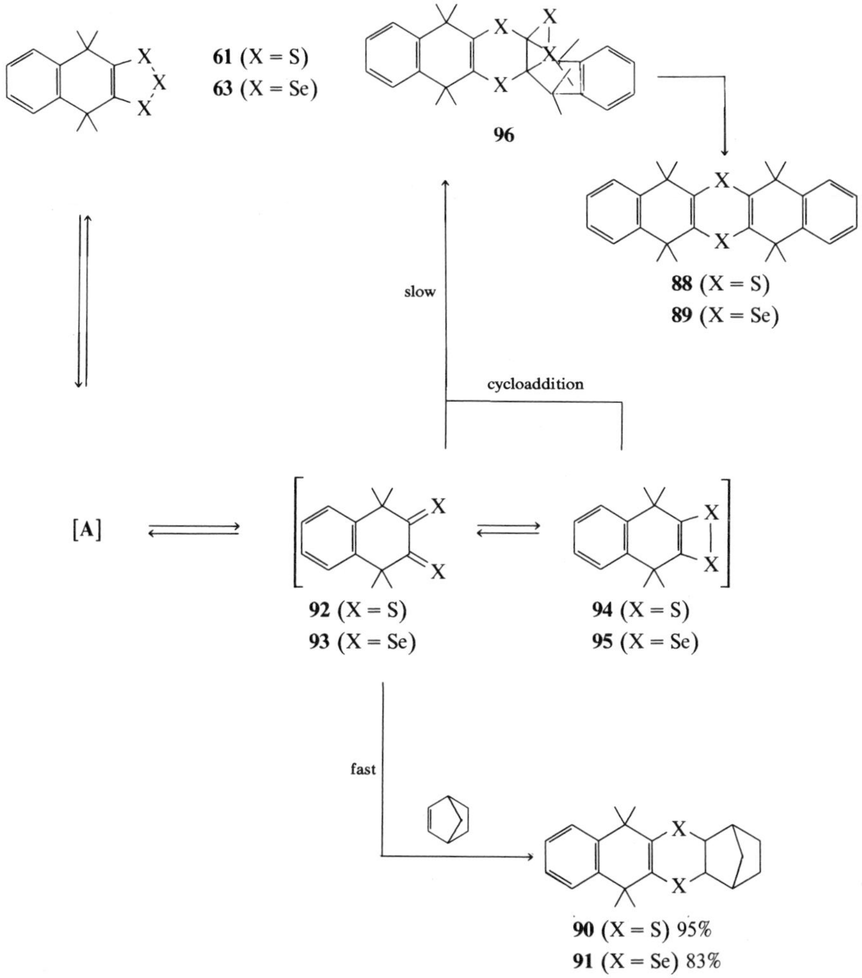

61 (X = S)
63 (X = Se)

96

88 (X = S)
89 (X = Se)

slow

cycloaddition

[A]

92 (X = S)
93 (X = Se)

94 (X = S)
95 (X = Se)

fast

90 (X = S) 95%
91 (X = Se) 83%

On the contrary, in the presence of reactive olefin such as norbornene the photolysis of **61** and **63** readily proceeded to give the corresponding cycloadducts **90** and **91** in 95% and 83% yields, respectively. These results suggest the possible formation of some initial intermediate [A] that can be reversely transformed into the starting cyclopoly-chalcogenides, prior to the 1,2-dithione **92** and 1,2-selenoxothione **93**. This assumption was also confirmed by the fact that the 1,4-thiaselenin derivative **97** was isolated together with the 1,4-diselenin derivative **91** in the photolysis of 1,2,5-thiadiselenole **65** in the presence of norbornene. The formation of these two types of adducts, **91** and **97**, implies the competition of deselenation with desulfurization following the initial ring-opening step of the photolysis of **65**.

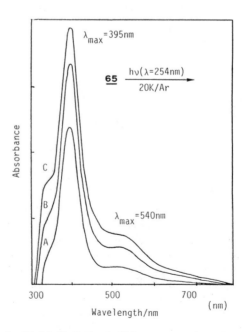

1.7.3. Spectroscopic Studies on the Photolysis of Cyclopolychalocogenides Using Low-Temperature Matrices [23,24]

To elucidate the mechanism of the photochemical dechalocogenation of the series 1,2,3-trithiole **61**, 1,2,3-triselenole **63**, and 1,2,5-thiadiselenole **65**, the photolysis of **65** with light of $\lambda = 254$ nm was performed in an argon matrix at 20 K [25] and was monitored by absorption spectroscopy to show two characteristic absorption maxima at 395 and 540 nm (Figure 1.1) [26].

Figure 1.1 ■ Photolysis of 1,2,5-thiadiselenole **65** in argon matrix at 20 K; difference electronic spectra after irradiation with light of $\lambda = 254$ nm (A, 2 min; B, 5 min; C, 18 min).

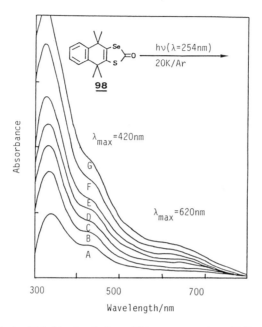

Figure 1.2 ▪ Photolysis of 1,3-thiaselenole-2-one **98** in argon matrix at 20 K; difference electronic spectra after irradiation with light of λ = 254 nm (A, 3 min; B, 8 min; C, 18 min; D, 28 min; E, 38 min; F, 48 min; G, 70 min).

On the other hand, 1,3-thiadiselenole-2-one **98** was photodecarbonylated under similar conditions with the appearance of shoulder peaks at 420 and 620 nm in the electronic spectra (Figure 1.2), which are certainly attributable to the 1,2-thiaselenete **99** and 1,2-selenoxothione **100** intermediates, respectively.

Furthermore, when photochemical retrocycloaddition of dihydro-1,4-diselenin derivative **101** was carried out with light of λ = 254 nm in a 3-methylpentane glass matrix at 77 K, only a broad absorption maximum at 465 nm, which is assignable to 1,2-diselenete intermediate **95**, was observed.

The dissimilarity of the electronic spectra obtained by the photolysis of **65**, **98**, and **101** in low-temperature matrices as well as the results of a comparison of the wavelength values of each absorption maximum of **65**, **98**, and **101** with those of known organosulfur and organoselenium compounds [27], suggest that the two absorption maxima observed at 395 and 540 nm in the case of **65** might be attributable to the initial intermediates, such as the thioselenoxide **103** and/or the spirothiaselenirane **104**, respectively.

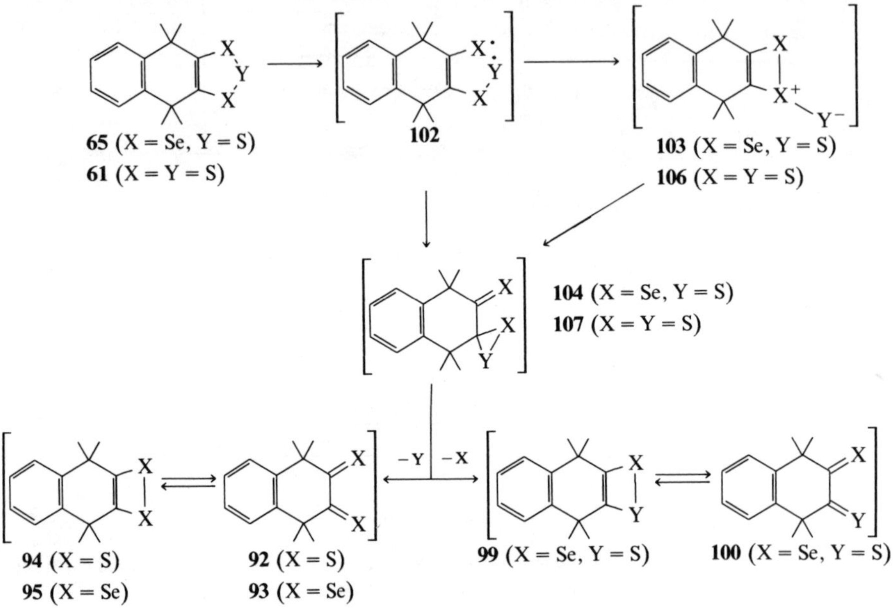
X
Y
X
65 (X = Se, Y = S)
61 (X = Y = S)
102
103 (X = Se, Y = S)
106 (X = Y = S)
104 (X = Se, Y = S)
107 (X = Y = S)
−Y
−X
94 (X = S)
95 (X = Se)
92 (X = S)
93 (X = Se)
99 (X = Se, Y = S)
100 (X = Se, Y = S)

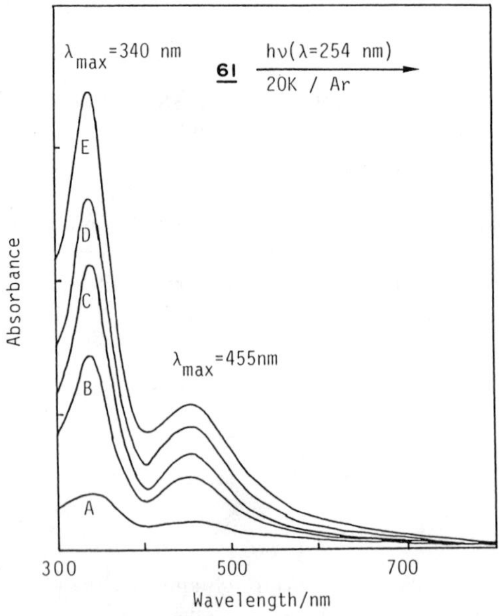
λmax=340 nm
61
hν(λ=254 nm)
20K / Ar
E
D
C
B
A
λmax=455nm
Absorbance
300
500
700
Wavelength/nm

Figure 1.3 ■ Photolysis of 1,2,3-trithiole **61** in argon matrix at 20 K; difference electronic spectra after irradiation with light of λ = 254 nm (A, 0.5 min; B, 1.5 min; C, 2.5 min; D, 3.5 min; E, 7.5 min).

The competitive photochemical dechalocogenation from **65** can be rationalized by assuming the intermediacy of the spirothiaselenirane **104** and its instability leading to a facile extrusion of a chalcogen atom as shown in the above scheme.

The photolysis of 1,2,3-trithiole **61** performed in an argon matrix at 20 K with light of λ = 254 nm showed two absorption maxima at 340 and 455 nm (Figure 1.3). On the other hand, photodecarbonylation of 1,3-dithiole-2-one **105** in a 3-methylpentane glass matrix gave a quite different spectrum, having two absorption maxima at 370 and 580 nm (Figure 1.4), which correspond to the 1,2-dithiete and 1,2-dithione intermediates **94** and **92**.

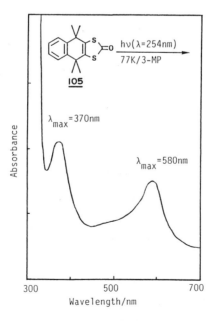

These results are also interpreted with the possible formation of the thiosulfoxide **106** and/or spirodithiirane **107**, as in the case of 1,2,5-thiadiselenole **65**, leading to the dechalcogenative formation of 1,2-dithiete and 1,2-dithione intermediates **94** and **92**.

Figure 1.4 ■ Photolysis of 1,3-dithiole-2-one **105** in 3-MP glass matrix at 77 K; difference electronic spectrum after irradiation with light of λ = 254 nm for 2 min.

106   107

1.8. Summary

The chemistry of 1,2,3-selenadiazoles and 1,2,3-thiadiazoles has been widely explored by use of sterically protected fused systems as substrates, resulting in facile and characteristic formation of many kinds of interesting organoselenium and organosulfur compounds. Of particular note are the isolation of the selenirane derivative **14**, the first example of a stable carbon–carbon–selenium three-membered ring system [28], by the cycloaddition of selenirene **13** and the ready formation of unusually stable cyclopoly-chalcogenides via the thermal reactions of 1,2,3-chalcogenadiazoles with elemental sulfur and selenium. In these systems the reaction intermediates might be invested with enough stability and capacity to produce a variety of novel and characteristic chalco-gen-containing compounds not only by the steric protection of the bulky substituents, but also by the ring strain caused by their fused structure leading to a suppression of the deselenative cycloalkyne formation.

The photolysis of the cyclopolychalocogenides, which are readily derived from the fused 1,2,3-selenadiazole and 1,2,3-thiadiazole systems, was found to be a useful method for the formation of a novel 1,2-dithione and its selenium analogues. Spectro-scopic studies on the photolysis of cyclopolychalcogenides in low-temperature matrices using electronic spectra implicated the initial formation of hitherto unknown interme-diates, for example, the thioselenoxide **103** and spirothiaselenirane **104** in the case of 1,2,5-thiadiselenole **65**.

The new results that we report here on fused 1,2,3-chalcogenadiazoles should prove useful for the synthesis and investigation of organochalcogen compounds.

References

1. (a) Lalezari, I.; Shafiee, A.; Yaplani, M. *Angew. Chem., Int. Ed. Engl.*, **1970**, *9*, 464. (b) Lalezari, I.; Shafiee, A.; Yaplani, M. *J. Org. Chem.*, **1971**, *36*, 2836. (c) Golgolab, H.; Larezari, I. *J. Hetrocycl. Chem.*, **1975**, *12*, 801. (d) Shafiee, A.; Larezari, I.; Mirrashed, M.; Nercesian, D. *J. Heterocycl. Chem.*, **1977**, *14*, 567; (e) Pennanen, S. I. *J. Heterocycl. Chem.*, **1977**, *14*, 745. (f) Larezari, I.; Shafiee, A.; Khorrami, J.; Soltani, A. *J. Pharm. Sci.*, **1978**, *67*, 1336. (g) Larezari, I.; Shafiee, A.; Golgolab, H. *J. Heterocycl. Chem.*, **1973**, *10*, 655. (h) Larezari, I.; Shafiee, A.; Rabest, F.; Yaplani, M. *J. Heterocycl. Chem.*, **1973**, *10*, 953. (i) Gilchrist, T. L.; Mente, P. G.; Rees, C. W. *J. Chem. Soc., Perkin Trans. 1*, **1972**, 2165, and references cited therein.
2. (a) Larezari, I.; Shafiee, A.; Rabest, F.; Yaplani, M. *J. Heterocycl. Chem.*, **1972**, *9*, 141. (b) Meier, H.; Menzel, I. *J. Chem. Soc., Chem. Commun.*, **1971**, 1059.
3. (a) Meier, H.; Layer, M.; Combrink, W.; Schniepp, S. *Chem. Ber.*, **1976**, *109*, 1650. (b) Golgolab, H.; Larezari, I. *J. Heterocycl. Chem.*, **1975**, *12*, 801. (c) Meier, H.; Voigt, E. *Tetrahedron*, **1972**, *28*, 1987.

4. Holm, A.; Berg, C.; Bjerre, C.; Bak, B.; Svanholt H. *J. Chem. Soc., Chem. Commun.*, **1979**, 99.
5. Krantz, A.; Laureni, J. *J. Am. Chem. Soc.*, **1977**, *99*, 4842.
6. Ando, W.; Kumamoto, Y.; Tokitoh, N. *Tetrahedron Lett.*, **1986**, *27*, 6107.
7. Ando, W.; Kumamoto, Y.; Tokitoh, N. *J. Phys. Org. Chem.*, **1988**, *1*, 317.
8. (a) Ishizuka, H. Thesis, University of Tsukuba, **1989**. (b) Okano, Y. Thesis, University of Tsukuba, **1989**.
9. Lorch, M.; Meier, H. *Chem. Ber.*, **1982**, *114*, 2382.
10. To excess of molten anthracene was added, in portions, the 1,2,3-selenadiazole **1** as solids.
11. (a) Peterson, H.; Kolshorn, H.; Meier, H. *Angew. Chem., Int. Ed. Engl.*, **1978**, *17*, 461. (b) Meier, H. *Synthesis*, **1972**, 235.
12. Ando, W.; Kumamoto, Y.; Ishizuka, H.; Tokitoh, N. *Tetrahedron Lett.*, **1987**, *28*, 4707.
13. Ando, W.; Kumamoto, Y.; Tokitoh, N. *Tetrahedron Lett.*, **1987**, *28*, 4833.
14. (a) Morita, K.; Kobayashi, S. *Chem. Pharm. Bull.*, **1967**, *15*, 988. (b) Rohman, R.; Safe, S.; Taylor, A. *J. Chem. Soc. C.*, **1969**, 1665. (c) Still, I. W. J.; Kutney, G. W. *Tetrahedron Lett.*, **1981**, *21*, 1939. (d) Harpp, D. N.; Smith, R. A. *J. Am. Chem. Soc.*, **1982**, *104*, 6045. (e) Whitesides, G. M.; Houk, J.; Patterson, A. K. *J. Org. Chem.*, **1983**. *48*, 112. Also see references cited in these reports.
15. (a) Chenard, B. L.; Miller, T. J. *J. Org. Chem.*, **1984**, *49*, 1221. (b) Feher, F.; Langer, M. *Tetrahedron Lett.*, **1971**, 2125. (c) Sato, R.; Saito, S.; Chiba, H.; Goto, T.; Saito, M. *Chem. Lett.*, **1986**, 349, and references cited therein.
16. Only one example has been reported for the isolation of 1,2,3,4-tetrathiin: Krespan, C. G.; Brasen, W. R. *J. Org. Chem.*, **1962**, *27*, 3995.
17. Tokitoh, N.; Ishizuka, H.; Ando, W. *Chem. Lett.*, **1988**, 657.
18. Chenard, B. L.; Harlow, R. L.; Johnson, A. L.; Vladuchick, S. A. *J. Am. Chem. Soc.*, **1985**, *107*, 3871.
19. (a) Dittmer, D. C.; Kuhlmann, G. E. *J. Org. Chem.*, **1970**, *35*, 4224. (b) Burros, B. C.; De'Aath, N. J.; Denney, D. B.; Denney, D. Z.; Kipnis, I. J. *J. Am. Chem. Soc.*, **1978**, *100*, 7300. (c) Orahovatz, A.; Levinson, M. I.; Carroll, P. J.; Lakshmikantham, M. V.; Cava, M. P. *J. Org. Chem.*, **1985**, *50*, 1550.
20. (a) Davison, A.; Shawl, E. T. *Inorg. Chem.*, **1970**, *9*, 1820. (b) Diel, F.; Schweig, A. *Angew. Chem., Int. Ed. Engl.*, **1987**, *26*, 343.
21. Ando, W.; Kumamoto, Y.; Tokitoh, N. *Tetrahedron Lett.*, **1987**, *28*, 5699.
22. Kumamoto, Y. Ph.D. thesis, University of Tsukuba, **1988**.
23. Ando, W.; Ishizuka, H.; Kumamoto, Y.; Tokitoh, N.; Yabe, A. *Nippon Kagaku Kai Shi*, **1989**, *8*, 1447.
24. Tokitoh, N.; Ishizuka, H.; Yabe, A.; Ando, W. *Tetrahedron Lett.*, **1989**, *30*, 2955.
25. The cryostat system used is the same one reported by Chapman and co-workers: MacMahon, R. J.; Chapman, O. L.; Hayes, R. A.; Hess, T. C.; Krimmer, H. P. *J. Am. Chem. Soc.*, **1985**, *107*, 7597.

Organosilicon Chemistry: A Synergistic Relation between Theory and Experiment

Yitzhak Apeloig

Department of Chemistry
Technion—Israel Institute of Technology
Haifa 32000
Israel

2.1. Introduction

This chapter is an attempt to demonstrate how important and useful theory can be as a research tool in the field of organosilicon chemistry. I will naturally concentrate on the work of my own group, focusing on silenes $R_2Si = CR'_2$, on cationic reactive intermediates, and on some intriguing novel structures of the general formula $SiR_2R'_2$ that have no precedent among isoelectronic organic molecules.

The application of theory, particularly quantum-chemical methods, to silicon chemistry is a relatively recent endeavor. Because silicon is a member of the second row of the periodic table, calculations were considered to be a difficult problem to handle in the early applications of quantum-chemical methods. This was true both for ab-initio methods, for which the relatively large number of electrons posed severe limitations on the calculations, and for semiempirical methods (and even purely empirical methods, such as force-field calculations) where the scarcity of experimental data prevented the development of reliable parametrization. Consequently, the theoretical study of silicon compounds lagged by 10–15 years behind the analogous theoretical developments for the chemistry of carbon compounds. Recently, the situation has changed dramatically due to the technological breakthroughs in computer hardware and the development of new theoretical methods and related computer programs and software. In the last decade and particularly in the last five years it has become possible to carry out ab-initio calculations (using appropriate levels of theory) for silicon compounds of sizable complexity, which can serve as models for compounds that are of interest to chemists working in this field.

Since the time that the necessary computational methods became available theory played an extremely important (and in some cases even crucial) role in the development of silicon chemistry [1], especially in laying the foundations for the understanding of compounds that have multiple bonding to silicon. The fact that stable compounds with multiple bonds to silicon were unknown experimentally until ten years ago placed theory in a special, possibly unique, position in the development of this field [1]. Thus, in many cases the theoretical predictions preceded the experimental studies and in other cases theory identified and pointed out erroneous interpretations of experimental data. In general, the experimental and theoretical aspects of the field have grown synergistically. In this respect the history of organosilicon chemistry and of organic chemistry are quite different. Organic chemistry was a mature and well-developed field when reliable computational methods for tackling small organic compounds became available. On the other hand, many areas in the field of organosilicon chemistry, such as the study of molecules with multiply bonded silicon or of reactive intermediates, were experimentally in their infancy at the point when computational methods in chemistry came of age and could be applied reliably to a variety of problems. Furthermore, given the scarcity of quantitative experimental fundamental data (such as thermochemistry, geometry, etc.) for many organosilicon compounds, quantum-mechanical calculations have been almost the sole source of primary information employed to derive patterns of understanding. These circumstances placed theory in a special position in the development of organosilicon chemistry.

2.2. Geometries of Multiply Bonded Silicon Compounds

2.2.1. Silenes

Before considering organosilicon reactive intermediates, about which little is known experimentally, it is appropriate to demonstrate the power and reliability of current computational methods by presenting a structural problem from the field of multiply bonded silicon compounds where experimental data are available. The problem was first pointed out by Schaefer [2], who found that calculations at various levels of theory, ranging from relatively simple to quite sophisticated ones, predict that the singlet silaethylene is planar with a C=Si bond length ranging around 1.69–1.72 Å. These results were in conflict with a report from a daring electron-diffraction experiment, which concluded that the C=Si bond length in the transient unstable $(CH_3)_2Si=CH_2$ is 1.83 ± 0.04 Å [3]. In an attempt to resolve the discrepancy, Schaefer studied the effect of methyl substitution but found that it was very small and thus it could not be responsible for the discrepancy [2]. Schaefer concluded that the C=Si bond distance in $H_2Si=CH_2$ and in $(CH_3)_2Si=CH_2$ is about 1.70 Å and that the experimental–theoretical discrepancy is so large that the unavoidable conclusion is that the interpretation of the electron-diffraction data was erroneous [2]. At about the same time Brook and his co-workers succeeded in isolating the first stable silene (1) and in measuring its X-ray structure [4]. The C=Si bond length in 1 was determined to be 1.764 Å, an intermediate value between the calculated and experimental C=Si bond lengths in $(CH_3)_2Si=CH_2$. However, as 1 is heavily substituted and also twisted by 16° about the Si=C bond, it was not clear how to extrapolate from its structure to the structure of $(CH_3)_2Si=CH_2$ and the structures of other simple silenes. Therefore, Schaefer did not consider the relatively long $C=Si$ bond in **1** to be in conflict with the theoretical results [2].

1

We became interested in this problem at about the same time and have attempted to understand the reasons for the longer $C=Si$ bond length in **1** relative to that calculated for $(CH_3)_2Si=CH_2$, by studying systematically the effect of various substituents on the $C=Si$ bond distance [5]. The results are given in Table 2.1.

The effect of the substituents on the $C=Si$ bond length are best understood in terms of their effect on the polarization and thus ionicity of this bond [5]. Substituents that increase the polarization of the $C=Si$ bond, such as OH substitution at silicon, shorten this bond and vice versa. In particular, it is important to note that silyl substitution at the carbon end and silyloxy substitution at the silicon end of the $C=Si$ bond both lengthen this bond significantly. Furthermore, calculations for disubstituted silenes showed that the effect of substituents on the $C=Si$ distance is additive [5]. This finding is important because it allows us to predict the double bond length in polysubstituted silenes whose complexity precludes direct calculations at a reliable level of theory. On the basis of this additivity and the calculated substituent effects in Table 2.1, it was possible to calculate the $C=Si$ bond length in $(H_3Si)_2Si=CCH_3(OSiH_3)$, which serves as an electronic model compound for the stable silene **1**, as follows [5].

Table 2.1 ■ 3-21G Optimized $C=Si$ Bond Lengths (in Å) in the Substituted Silenes $H_2C=SiHR$ and $RCH=SiH_2$

| R | $H_2C=SiHR$ | | $RCH=SiH_2$ | |
	Bond Length	Δr^a	Bond Length	Δr^a
H	1.718	0	1.718	0
CH_3	1.716	−0.002	1.725	0.007
SiH_3	1.725	0.007	1.721	0.003
OH	1.705	−0.013	1.746	0.028
$OSiH_3$	1.705	−0.013	1.749	0.031
F	1.698	−0.020	1.730	0.012
CN	1.711	−0.007	1.727	0.009
NH_2	1.702	−0.016	1.731	0.013

^{a}Change in the $C=Si$ bond length relative to $H_2C=SiH_2$

Predicted $r(C = Si)$ in $(H_3Si)_2Si = CCH_3(OSiH_3)$ = {1.718 [$r(C = Si)$ in $H_2C = SiH_2$] + 2(1.725 − 1.718) + (1.725 − 1.718) + (1.749 − 1.718) [i.e., changes in $r(C = Si)$ due to the effect of two H_3Si, one CH_3, and one $OSiH_3$ substituents, respectively]} = 1.770 Å. Corrections for the addition of polarization functions and correlation effects are expected to shorten this bond by approximately 0.013 Å to about 1.757 Å [5]. This calculated $C = Si$ bond length is very close to the experimental value of 1.764 Å determined for **1** [4]. The excellent experimental–theoretical agreement also suggests that elongation of the $C = Si$ bond by steric repulsions between the bulky substituents or as a result of the observed 16° twisting about the $C = Si$ bond are of small importance.

The preceding calculations showed that the geometry of **1** is consistent with a $C = Si$ bond length in simple silenes of about 1.70 Å, but not with the value of 1.83 Å deduced from the electron-diffraction measurements [3]. The theoretical–experimental dispute was finally resolved in 1985 by Wiberg et al. [6a,6b] in favor of theory. These authors solved the crystal structure of **2** and determined that $r(C = Si) = 1.702 \pm 0.005$ Å [6a,6b]. The calculated $C = Si$ bond distance in $(CH_3)_2Si = C(SiH_3)_2$, an electronic model for **2**, is 1.707 Å (see calculation below), in excellent agreement with experiment.

Predicted $r(C = Si)$ in $(CH_3)_2Si = C(SiH_3)_2$ = {1.718 [$r(C = Si)$ in $H_2C = SiH_2$] + 2(1.721 − 1.718) [change due to two H_3Si substituents] + 2(1.716 − 1.718) [change due to two CH_3 substituents]} = 1.720 Å − 0.013 [corrections for the deficiencies of the 3-21G basis set] = 1.707 Å.

2

The last word on this issue was put forward a few months ago by Gutowsky et al. [6c], who reported that the microwave rotational transitions for $(CH_3)_2Si = CH_2$ agree with the calculated $Si = C$ double bond length of 1.692 Å [2,5] rather than with the electron-diffraction value of 1.83 Å [3].

2.2.2. Complexes

Wiberg and co-workers have recently observed that silenes and silanimines form stable adducts (e.g., **3** and **4**) amenable to X-ray diffraction analysis, with various π-donors, such as tetrahydrofuran (THF) and NMe_3 [6d–6f]. We have studied computationally the corresponding model adducts with water (**5** and **6**). Both **5** and **6** were characterized to be minima on the potential energy surface by their Hessian matrices, and their optimized geometries are shown in Figure 2.1.

3

$t\text{-}Bu_2Si = NSi(t\text{-}Bu)_3 \cdot THF$

4

$H_2C = SiH_2 \cdot H_2O$ $HN = SiH_2 \cdot H_2O$ $H_2Si = SiH_2 \cdot H_2O$

5   **6**   **7**

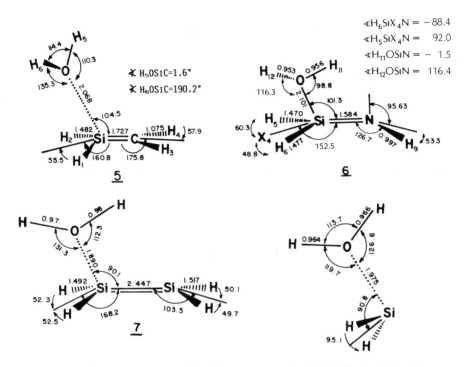

Figure 2.1 ■ Optimized geometries of the silene–water complex **5** (3-21G), the silanimine–water complex **6** (6-31G*), the disilene–water complex **7** (6-31G*), and a silylene–water complex (3-21G).

The computational results are generally in good agreement with the experimental findings. The water nucleophile bonds to the silicon end of the double bond in both **5** and **6**. Attempts to locate a complex with the water molecule bonded to the carbon or to the nitrogen atoms were unsuccessful. The calculations reproduce also the intimate changes that occur upon complexation in the geometry of the organosilicon compound. Thus, both theory and experiment reveal significant pyramidalization at the silicon atom and a lengthening of the central multiple bond upon complexation. The changes observed experimentally are somewhat larger than the calculated ones. For example, in the parent systems **5** and **6**, the C=Si and Si=N bonds are lengthened upon complexation with water by 0.01 Å, compared with lengthening by 0.045 and 0.020 Å in **3** and **4**, respectively. The experimental–theoretical differences result to a large extent from the simplicity of the models used in the calculations. For example, the silyl substituents in **3** and **4** stabilize the negative charge that develops upon complexation at the carbon (i.e., in **3**) or the nitrogen (i.e., in **4**) atoms promoting stronger complexation and larger geometrical changes in the experimental systems than in **5** and **6**. The dative O→Si bond distances are also significantly longer in **5** and **6** than in **3** or **4**, suggesting weaker complexation in the parent molecules. The dative O→Si bond distance is expected to be especially sensitive to the addition of polarization functions and to the effect of electron correlation. Indeed, geometry optimizations for **5** with the small polarized 3-21G* basis set give a silicon–oxygen distance of 1.988 Å compared with 2.068 Å at 3-21G. However, even the shorter 3-21G* distance is longer than the experimental silicon–oxygen distance in **3** of 1.878 Å.

Both complexes are calculated to be moderately bound, that is, **5** by 9.0 kcal mol^{-1} (3-21G//3-21G) and **6** by 10.9 kcal mol^{-1} (6-31G*//6-31G*). These relatively low complexation energies are in accord with Wiberg's suggestion that **3** and **4** react via the "free" silaethene or silanimine, which are formed by primary dissociation of the THF adducts [6d].

The disilene–water complex **7** is especially intriguing. Upon complexation disilene undergoes remarkable geometrical changes, as shown in Figure 2.1. Both silicon atoms undergo very strong bending [7] and the silicon–silicon bond lengthens dramatically to 2.44 Å—a distance longer than the silicon–silicon distance in disilane (2.383 Å at 3-21G//3-21G; exp. 2.340 Å in $Me_3SiSiMe_3$)! Thus, complexation with water breaks not only the π component of the Si — Si bond, as expected on the basis of simple valence-bond Lewis structures, but also strongly affects the character of the σ(Si — Si) bond. The disilene–water complex **7** can also be viewed as a complex between a silylene and a silylene–water complex, as shown schematically in structure **8**. According to this description, the left-hand side in **8** describes a complex formed by interaction between the empty 3p orbital of a silylene (Si^1) and an electron lone pair on the water oxygen. This part of the complex has indeed a geometry that is similar to that of an isolated silylene–water complex (see Figure 2.1). In **8** a second silylene (Si^2) is bonded to the $H_2Si — OH_2$ complex via interaction between its empty 3p orbital and the lone-pair electrons on Si^1. These bonding interactions are quite strong. At 3-21G//3-21G, **8** is more stable than [$H_2Si = SiH_2 + H_2O$] by 23.7 kcal mol^{-1} and than [$H_2Si — OH_2 + H_2\ddot{S}i$] by 66.4 kcal mol^{-1}. We are continuing our studies of these interesting complexes and extending our calculations to other π-donors, such as NH_3 and F^- [6d–6f].

8

These examples show that calculations can produce very reliable geometries for organosilicon compounds. This is important because for many types of organosilicon compounds such experimental data is very limited or not available.

The calculation of accurate energies (e.g., relative thermodynamic stabilities) is in general a more difficult theoretical task than the calculation of molecular geometries. However, many examples that clearly demonstrate the reliability of ab-initio calculations (when performed at an appropriate level of theory) for energetic comparisons of organosilicon compounds are already available [1].

2.3. Silicenium Ions

Although silicenium ions R_3Si^+ have been known and studied in the gas phase for more than 20 years [8], and more recently it was claimed that several such species have been generated in solution [9–11], very little is yet known experimentally about their properties, such as their geometries or the effect of substituents on their thermodynamic stabilities. This is therefore a field where theory can be of enormous importance, providing reliable information for a wide variety of derivatives.

2.3.1. Substituent Effects on Stability

2.3.1a. General. We have recently calculated, at a uniform, relatively high level of theory (MP3/6-31G*//6-31G*) the effect of substituents from the first and second rows of the periodic table on the stability of silicenium ions [12a]. The effect of the substituents R on the stabilities of silicenium ions is evaluated via isodesmic eq. (2.1) and for the analogous carbenium ions, via eq. (2.2). Equation (2.3) compares the stabilities of silicenium ions with the analogous carbenium ions in a hydride transfer process. The results are presented in Table 2.2.

$$H_2SiR^+ + SiH_4 \longrightarrow SiH_3^+ + SiH_3R \qquad (2.1)$$

$$H_2CR^+ + CH_4 \longrightarrow CH_3^+ + CH_3R \qquad (2.2)$$

$$H_2SiR^+ + CH_3R \longrightarrow CH_2R^+ + H_3SiR \qquad (2.3)$$

The use of isodesmic equations such as eqs. (2.1)–(2.3), where the same types of bonds appear on both sides of the equation, reduces significantly the errors associated with the deficiencies of the theoretical method [13]. Indeed, the MP3/6-31G*//6-31G* results that include the effect of polarization functions as well as of electron correlation are in many cases similar, even quantitatively, to our previous split-valence 3-21G calculations [12b,12c]. This observation is important because it shows that reliable calculations can be carried out also for larger cations, for example, $C_6H_5Si(CH_3)_2^+$, where the use of basis sets larger than 3-21G is not practical.

The results in Table 2.2 lead to two important conclusions:

1. Silicenium ions and carbenium ions respond qualitatively in an analogous manner to the effects of substituents. Substituents that strongly stabilize carbenium ions (e.g., $R = NH_2$) also strongly stabilize silicenium ions, and vice versa.
2. All the substituents examined, in particular π-donors, are less effective in stabilizing silicenium ions than carbenium ions. For example, an amino group, the substituent most effective in stabilizing both cations, stabilizes the methyl cation by 97.8 kcal mol^{-1} and the silyl cation by only 36.8 kcal mol^{-1}. Also, for $R = OH$ and SH, the stabilization in $RSiH_2^+$ is only about 30% of that in RCH_2^+. The reduced effect for Si$^+$ results mainly from a poorer orbital overlap for silicon than for carbon, which is also reflected in the reluctance of silicon to participate in multiple bonding [1]. In line with this picture, σ-donor substituents, such as Li, BeH, and SiH$_3$, are only slightly less effective in stabilizing Si$^+$ compared with C$^+$ (Table 2.2).

The observation that π-conjugation is much less effective for silicenium ions than for carbenium ions carries over also to stabilization via hyperconjugation. While a

Table 2.2 ■ Calculated Energies (kcal mol^{-1}, MP3/6-31G*//6-31G*) for Eqs. (2.1)–(2.3)^a

Substituent YH$_n$ ($n = 0$–3)^b	Equation^c (2.1)	Equation^c (2.2)	Equation^c (2.3)
H	0.0	0.0	54.9^d
Li	58.1	77.3	35.7
BeH	14.4	16.2	53.1
BH$_2$, planar^e	8.3	2.7	60.4
BH$_2$, perp.^{f}	13.9	25.8	43.0
CH$_3$	15.1	34.1	35.9^g
NH$_2$, planar^e	36.8	97.8	−6.1
NH$_2$, perp.^{f}	11.9	15.9	50.9
OH, planar^e	17.9	62.7	10.1
F	−2.3	21.5	31.1
SiH$_3$	12.6	17.7	49.8
PH$_2$ planar^e	13.5^h	60.0	8.4
SH, planar^e	18.4	60.9	12.4
SH, perp.^{f}	0.4	8.1	47.3
Cl	2.0	26.6	30.3

^{a}The neutral monosubstituted silanes and methanes are taken in their most stable conformations.
^{b}The notation used in the footnotes is as follows: H$_n$YXH$_2{}^+$, where X = Si, C; Y = Be, B, C, N, O, F, Si, P, S, Cl.
^{c}A positive value indicates that the reaction is endothermic as written.
^{d}Experimental value = 53 kcal mol^{-1} [15].
^{e}The dihedral angle HYXH is 0° or 180°.
^{f}The dihedral angle HYXH is 90° or −90°.
^{g}Experimental value = 24.8 kcal mol^{-1} [15].
^{h}Stabilization is higher (16.0 kcal mol^{-1}) in a conformation where the phosphorus is allowed to pyramidalize.

β-silyl group stabilizes a carbenium ion very effectively [14], the analogous stabilization of a silicenium ion is small. Thus, at MP2/6-31G*//3-21G$^{(*)}$ eq. (2.4) is exothermic by 33.1 kcal mol^{-1} compared with only 10.3 kcal mol^{-1} for eq. 5 (MP3/6-31G*//6-31G*). Furthermore, in H$_3$SiCH$_2$SiH$_2{}^+$ the barrier to rotation around the C — Si$^+$ bond, which measures the contribution of hyperconjugation to the total stabilization, is only 6.5 kcal mol^{-1} (MP2/6-31G*//6-31G*) compared with 31.4 kcal mol^{-1} in H$_3$SiCH$_2$CH$_2{}^+$ (MP2/6-31G*//3-21G$^{(*)}$). These computational results suggest that the solvolyses of compounds such as R$_3$SiC(SiMe$_3$)$_2$SiMe$_2$X, which were studied extensively by Eaborn's group [10], are not enhanced significantly by hyperconjugative stabilization by the β-(C — Si) bonds.

$$H_3SiCH_2CH_3 + CH_3CH_2{}^+ \longrightarrow H_3SiCH_2CH_2{}^+ + CH_3CH_3 \qquad (2.4)$$

$$H_3SiCH_2SiH_3 + CH_3SiH_2{}^+ \longrightarrow H_3SiCH_2SiH_2{}^+ + CH_3SiH_3 \qquad (2.5)$$

Are these computational results reliable? Unfortunately, very little experimental data is available for comparison [8]. However, a recent ion cyclotron resonance (ICR) study of methyl-substituted silicenium ions [15] is in very good agreement with our calculations. Thus, at MP3/6-31G*//6-31G* (6-31G*//6-31G* values in parenthesis) we calculate that H_3Si^+ is stabilized by the first, second, and third methyl substitutions by 15.1 (13.8), 12.9 (11.9), and 10.6 (10.0) kcal mol^{-1} [12d], respectively, to be compared with the experimental values of 15.5, 15.8, and 9.6 kcal mol^{-1} [15], respectively. The theoretical–experimental [15] agreement is good also for eq. (2.3). Thus, the $R_nSiH_{3-n}^+$ cations are calculated (6-31G*//6-31G*) to be more stable than the $R_nCH_{3-n}^+$ cations (experimental values [15] in parentheses) by 27.4 (24.8) kcal mol^{-1} for $n = 1$, 17.4 (21.8) kcal mol^{-1} for $n = 2$, and 12.9 (12.9) kcal mol^{-1} for $n = 3$.

2.3.1b. Cationic Rearrangements. The fact that alkyl-substituted tertiary silicenium ions are more stable than the isomeric tertiary carbenium ions was utilized by our group in designing an experiment that led to the first demonstration of a solvolytically generated silicenium ion [11]. The calculations showed that the tertiary silicenium ion $Me_3CSiMe_2^+$ is more stable than the tertiary carbenium ion $Me_3SiCMe_2^+$ by 9.7 kcal mol^{-1} (3-21G//3-21G). As the barriers for methyl migration were also calculated to be very small it was expected that $Me_3SiCMe_2^+$ should rearrange easily to $Me_3CSiMe_2^+$. Following this prediction, the analogous adamantyl derivative **9** was synthesized and reacted in a variety of solvents [11]. Product studies indeed suggested that the initially generated carbenium ion **10** rearranged to the silicenium ion **11** [eq. 2.6].

$$(2.6)$$

Related studies in the gas phase have also revealed similar rearrangements, interconverting α- and β-silyl–substituted carbocations and silicenium ions [16a]. Recently, we have performed in collaboration with Professor Schwarz' group in Berlin additional gas-phase experiments [16b] that beautifully demonstrate the predictive power of theory and the importance of stereoelectronic factors in the operation of the β-silicon hyperconjugative effect. According to the calculations, $Me_3SiCMe_2^+$ (**12**) is less stable than both $Me_3CSiMe_2^+$ (**13**) and $Me_3SiCMe_2CH_2^+$ (**14**), both of which have comparable stabilities. The calculations predict that a very small barrier of less than 1 kcal mol^{-1} separates **12** from **13**. In contrast, a significant barrier of about 15 kcal mol^{-1} exists for the rearrangement of **13** to **14**. This barrier results from the fact that a 1,2-hydrogen shift in **13** produces the eclipsed conformation **14E** (i.e., the dihedral angle $HC^+CSi = -30°$), which lies about 30 kcal mol^{-1} higher in energy than conformation **14P** (i.e., $\sphericalangle HC^+CSi = 90°$). In **14P**, but not in **14E**, the empty $2p(C^+)$ orbital and the C — Si bond are properly aligned to interact [14] and for fragmentation to [$MeCH{=}CH_2 + Me_3Si^+$]. Cation **13** cannot fragment directly to produce Me_3Si^+, because the accompanying fragment, the carbene $Me_2C{:}$, is very high in energy.

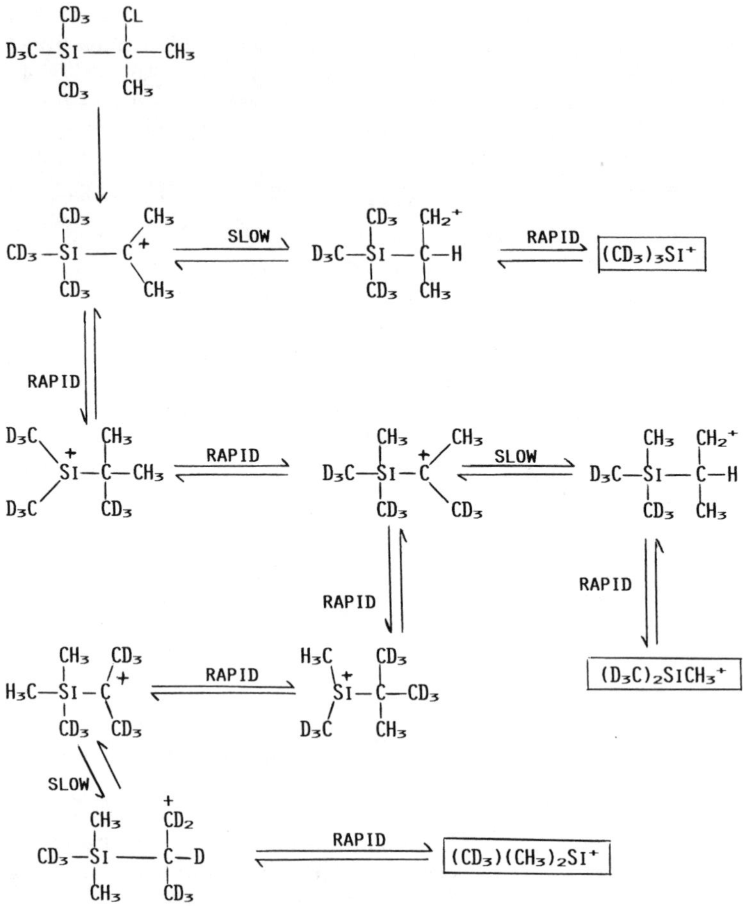

Scheme 2.1 ■ A proposed sequence of rearrangements leading to a complete scrambling of the CD_3 label in the gas-phase dissociative ionization of $(CD_3)_3SiC(CH_3)_2Cl$.

In conclusion, the calculations predict that, in the gas phase, cation **12** should undergo methyl rearrangement (**12 → 13**) much faster than hydride shift to **14**, although both reactions are exothermic to a similar extent. To test this prediction, we have synthesized the deuterated chloride $(CD_3)_3SiC(CH_3)_2Cl$. Dissociative ionization in the gas phase of this chloride gave the cations $(CD_3)_3Si^+$, $(CD_3)_2(CH_3)Si^+$, and $(CD_3)(CH_3)_2Si^+$ in the ratios of 14:67:19, respectively. These results indicate that nearly full scrambling of the CD_3 groups (i.e., 10:60:30, respectively) occurs prior to the hydride shift that produces the β-silyl–substituted cation **14**, as shown in Scheme 2.1 and as predicted computationally.

2.3.1c. SH and OH Substitution. Two substituents, OH and SH, deserve special attention in relation to the recent report by Lambert and co-workers on the observation of a relatively long-lived $(i\text{-}PrS)_3Si^+$ cation [9]. According to a private communication from Professor Lambert, experiments similar to those that produced $(i\text{-}PrS)_3Si^+$ did

did not lead to the analogous $(i\text{-PrO})_3\text{Si}^+$, but it was not clear if this failure is due to thermodynamic or kinetic factors. We decided to answer this question computationally.

Lambert has intuitively attributed the successful generation of $(i\text{-PrS})_3\text{Si}^+$ to its high thermodynamic stability, resulting from effective conjugation between the empty $3p$-orbital on Si and the $3p$-lone pair of sulfur and to the high polarizability of sulfur [9]. Such conjugation could be expected qualitatively to be less effective for oxygen. Our initial calculations did not give a clear answer regarding the thermodynamic stability of H_2SiOH^+ versus that of H_2SiSH^+, because at the STO-3G calculational level, SH and OH stabilize silicenium ions to a similar extent, whereas OH was substantially more stabilizing than SH at the 3-21G level [12c]. Calculations using more reliable methods (MP3/6-31G*//6-31G*) show that OH and SH stabilize $\text{SiH}_3{}^+$ *almost to the same degree*, that is, by 17.9 and 18.4 kcal mol^{-1}, respectively [Table 2.2, eq. (2.1)]. This conclusion remains practically unchanged with R = MeS versus MeO and with R = i-PrS versus i-PrO. Thus, at least for the parent cations, the qualitative conclusion that SH is more stabilizing than OH is not supported by the calculations. Similarly, it was found that OH and SH stabilize $\text{CH}_3{}^+$ to the same degree [17].

It is interesting to note, as observed for carbenium ions [17], that there is no correlation between the stability of the silicenium ion and the ability of the substituents to disperse the positive charge. Figure 2.2 presents the calculated 6-31G* electron transfers in the π- and σ-frameworks, between the SH (or OH) and the H_2Si^+ units. Relative to H_2Si^+, both sulfur and oxygen are π-donors and σ-acceptors. However, sulfur is a better π-donor and a poorer σ-acceptor than oxygen, so that overall, the SH substituent disperses the positive charge better than OH (OH actually withdraws electrons from the H_2Si^+ unit). In both cations the Si—R (R = SH, OH) bond is shorter than in the corresponding neutral molecules, but the change is larger for R = SH (Figure 2.2), indicating stronger interactions. Although SH is superior to OH in dispersing the positive charge, conjugatively as well as inductively, the thermodynamic stabilities of H_2SiSH^+ and H_2SiOH^+ are nearly identical.

In order to model more closely the experimentally studied cations [9], the corresponding disubstituted and trisubstituted silicenium ions were also studied. The results,

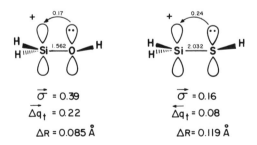

Figure 2.2 ■ 6-31G* σ- and π-charge transfers from R to H_2Si^+ in H_2SiR^+ (R = OH, SH). π-Electron transfers were calculated from the gross orbital populations of the $3p(\text{Si}^+)$ orbital. σ-Electron transfers were calculated using the equation $\sigma = 1 - \pi_t - q(\text{SiH}_2)$, where π_t is the π-electron transfer to the H_2Si group, $q(\text{SiH}_2)$ is the charge on SiH_2, Δq_t is the total charge transfer between R and the SiH_2 fragment, and ΔR is the difference in the Si—R bond length in H_2SiR^+ versus H_3SiR.

which are based on MP2/6-31G*//3-21G calculations (recent MP3/6-31G*//6-31G* calculations give similar results: see Table 2.3) of appropriate isodesmic equations (i.e., eqs. (2.7) and (2.8) for the disubstituted cations and neutrals, respectively, and eqs. (2.9) and (2.10) for the trisubstituted systems) are shown schematically in Figure 2.3. To assist the comparison, H_3SiSH and H_3SiOH were arbitrarily placed in Figure 2.3 at the same energy level.

$$H_2SiR_2 + H_2SiR^+ \longrightarrow HSiR_2^+ + H_3SiR \qquad (2.7)$$

$$H_2SiR_2 + SiH_4 \longrightarrow 2H_3SiR \qquad (2.8)$$

$$HSiR_3 + HSiR_2^+ \longrightarrow R_3Si^+ + H_2SiR_2 \qquad (2.9)$$

$$HSiR_3 + H_3SiR \longrightarrow 2H_2SiR_2 \qquad (2.10)$$

A careful analysis of the data in Figure 2.3 provides important insights into the differences between the sulfur and oxygen trisubstituted silicenium ions. As stated previously, the relative dissociation energies of H_3SiOH and H_3SiSH to the corresponding cations and a hydride ion (E_I^O and E_I^S, respectively, in Figure 2.3) are nearly equal [eq. (2.1) in Table 2.2, compare R = OH and R = SH]. However, this is not the case for the disubstituted molecules, and particularly for the trisubstituted molecules. Hydride transfer from the sulfur-substituted silane to the oxygen-substituted silicenium ion becomes progressively *more exothermic* on going from the monosubstituted to the disubstituted and to the trisubstituted systems (Figure 2.3). Thus, reaction (2.11) is exothermic by only 2.6 kcal mol^{-1}, whereas reaction (2.12) is exothermic by 8.6 kcal

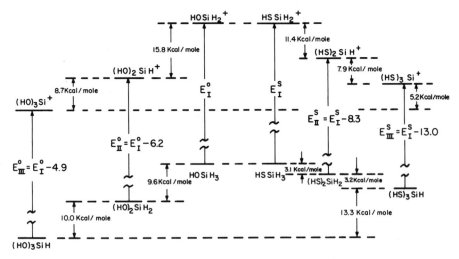

Figure 2.3 ■ Schematic description of the stabilization of a silicenium ion and of the corresponding silanes by mono-, di-, and tri-OH substitutions and SH substitutions, based on MP2/6-31G*//3-21G calculations of appropriate isodesmic equations (e.g., eqs. (2.7) and (2.8) for the disubstituted systems). E^O and E^S are the dissociation energies of the neutral oxygen- and sulfur-substituted neutrals, respectively, to the corresponding cations and hydride ion. Subscripts I, II, and III denote monosubstitution, disubstitution, and trisubstitution, respectively. H_3SiOH and H_3SiSH are placed artificially at the same energy level.

mol^{-1}. The conclusion is that $(HS)_3Si^+$ is significantly more stable than $(HO)_3Si^+$ and therefore thermodynamic factors might indeed be responsible for the failure to generate $(i\text{-}PrO)_3Si^+$ by a hydride exchange reaction.

$$HSi(OH)_2^+ + H_2Si(SH)_2 \longrightarrow HSi(SH)_2^+ + H_2Si(OH)_2 \qquad (2.11)$$

$$(HO)_3Si^+ + (HS)_3SiH \longrightarrow (HS)_3Si^+ + (HO)_3SiH \qquad (2.12)$$

Further analysis of the computational results adds important insights. The relatively high exothermicity of eq. (2.12) *does not* result from the fact that three SH substituents stabilize a silicenium ion center much more effectively than three OH substituents. Rather, it is mainly a reflection of the stronger stabilizing anomeric interactions in $(HO)_3SiH$ compared with that in $(HS)_3SiH$ [18]. Thus, reaction (2.8) is more endothermic for R = OH (9.6 kcal mol^{-1}) than for R = SH (3.1 kcal mol^{-1}) by 6.5 kcal mol^{-1} and the same applies for eq. (2.10). Part of the ground-state stabilization in the neutral precursors (which adopt a gauche conformation maximizing the anomeric effect [18]) is lost upon ionization to the cations, which adopt an all-planar structure that maximizes the interactions between the empty $3p(Si^+)$ orbital and the lone pairs on S or O. The loss of these anomeric interactions is energetically more important for oxygen substitution than for sulfur substitution and, consequently, eq. (2.11) is more exothermic than eq. (2.12).

These calculations also provide insights into the recent study of the kinetics of hydride transfer from various organosilanes to carbenium ions [eq. (2.13)]. The experimental results are shown in Figure 2.4 [19].

$$X_n Me_{3-n} SiH + Ph_3C^+ SbF_6^- \longrightarrow X_n Me_{3-n} Si^+ SbF_6^- + Ph_3CH \qquad (2.13)$$

The reactivity order observed for reaction (2.13) as a function of X is very intriguing. For example, it was found that the reaction of $(EtO)_3SiH$ with $Ph_3C^+SbF_6^-$

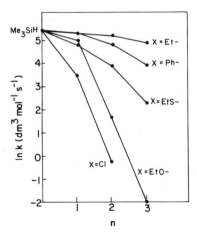

Figure 2.4 ▪ Structure–reactivity dependence in the reaction of $Ph_3C^+SbF_6^-$ with silyl hydrides of the general formula $X_n Me_{3-n} SiH$, X = Et⁻, Ph⁻, EtS⁻, EtO⁻, and Cl⁻. Rate constants in methylene chloride at 25 °C. Reproduced with permission from [19].

Table 2.3 ■ Calculated Energies (MP3/6-31G*//6-31G*) for the Equation
$$X_n SiH_{3-n}^+ + SiH_4 \longrightarrow X_n SiH_{4-n} + H_3 Si^+$$

X = CH₃[a]		X = SH			X = OH			X = Cl	
n	Energy	Energy	Δ(CH₃)[b]	n	Energy	Δ(CH₃)[b]	Δ(SH)[c]	Energy	Δ(CH₃)[b]
1	15.5	18.4	−3.3	1	17.9	−2.4	0.5	−2.0	17.5
2	31.3 (15.8)[d]	27.1 (8.7)[d]	4.2	2	26.5 (8.6)[d]	4.8	0.6	−1.1 (0.9)[d]	32.4
3	40.9 (9.6)[d]	31.2 (4.1)[d]	9.6	3	27.8 (1.3)[d]	13.1	3.4	3.2 (4.3)[d]	37.7

[a]Experimental values from [15]. Calculated values are very similar (see text).
[b]Stabilization by X relative to methyl.
[c]Stabilization by OH relative to SH.
[d]Additional stabilization provided by the second and third substituents.

is about seven orders of magnitude *slower* than that of Et₃SiH [19]! Assuming that the Hammond postulate [20] is valid (i.e., that a faster rate reflects hydride transfer to produce a more stable silicenium ion), this implies that $(EtO)_3Si^+$ is much less stable than Et_3Si^+. On similar grounds it can be concluded that also $(EtS)_3Si^+$ is less stable than Et_3Si^+. The opposite order of stability is surprising from analogy with the corresponding carbenium ions, where $(EtO)_3C^+$ and $(EtS)_3C^+$ are both substantially more stable than Et_3C^+ [20]. The apparently peculiar kinetic results for eq. (2.13) are easily understood with the aid of the calculations. According to the calculations, two or three methyl substituents indeed stabilize a silicenium ion more effectively than two or three SH or OH substituents [Table 2.3, see $\Delta(CH_3)$ values]. The thermodynamic differences between the series are particularly large for the trisubstituted cations, where (relative to hydride transfer) $(HO)_3Si^+$ is less stable than $(HS)_3Si^+$ and $(CH_3)_3Si^+$ by 3.4 and 13.1 kcal mol⁻¹, respectively, and Cl_3Si^+ is the least stable cation in the series (Table 2.3). These results are in conflict, as previously stated, with the qualitative conclusion, which attributed special stability to the $(i\text{-}PrS)_3Si^+$ cation [9]. In full harmony with these calculations, the experimental order of reactivity in reaction (2.13) (for both R_3SiH and R_2SiH_2 silanes) is Et > EtS > EtO > Cl [19]. As expected, only a fraction of the difference in the thermodynamic stabilities of the cations (Table 2.3) is reflected in the corresponding transition states and thus in the rates of hydride transfer reactions (Figure 2.4). We are currently studying the transition states for hydride transfer reactions between carbenium ions and silicenium ions and hope to shed more light on the intimate details (e.g., synchronous hydride transfer versus single-electron transfer [19]) of this important process—the only reaction known that leads to observable silicenium ions.

2.4. Disilenes—Highly Fluxional Molecules

One of the major advances in silicon chemistry in the last decade has been the preparation of stable disilenes [21]. However, the study of these compounds, and in particular their dynamic behavior, is still in its infancy [21]. This is therefore a field of research where theory can serve as a valuable guide for analyzing the existing data and for planning new experiments.

One of the most intriguing properties of disilenes revealed by theory is the fact that the classical structure **15** lies in a relatively shallow energy minimum, especially in comparison with the isoelectronic ethylene. $H_2Si=SiH_2$ is calculated to be only several kcal mol^{-1} lower in energy than H_3SiHSi: and the barrier that separates these species is also relatively low, about 17 kcal mol^{-1} above **15** [1]. More surprisingly, Kohler and Lischka [22] found that in addition to **15** and H_3SiHSi: the hydrogen bridged structures **16** and **17** are also local minima on the lowest singlet Si_2H_4 surface, and they lie only about 15 kcal mol^{-1} higher in energy than **15**. In the analogous C_2H_4 system similar hydrogen-bridged structures lie (at a similar level of theory) 194 kcal mol^{-1} higher in energy than $H_2C=CH_2$.

Two recent experimental studies [23,24] attracted our attention to this unique feature of disilenes. West and co-workers discovered the intriguing rearrangement shown in eq. (2.14) and suggested that it proceeds via a doubly bridged structure **18**, similar to **16** [23].

$$(2.14)$$

Jutzi et al. [24] reported the sequence of reactions shown in Scheme 2.2 and suggested that they proceed via the silylene **19** which couples to form the disilene **20** which then dimerizes to produce the isolable **21**. The intermediate was assigned as **20** on the basis of its ^{29}Si NMR spectrum [24] (see below).

In light of the calculated small energy difference between **15** and **16** (or **17**) and the fact that a structure analogous to **16**, with all hydrogens replaced by fluorines, was found to be the lowest energy structure of Si_2F_4 [25], we wondered if it is possible that the intermediate observed by Jutzi et al. [24] had a bridged structure **22** rather than the classical disilene structure **20**. This possibility was not considered in the experimental paper [24].

In an attempt to clarify this point we have undertaken a systematic study of the potential energy surface of $H_2Si_2R_2$ systems, where R = H, BH$_2$, CH$_3$, NH$_2$, OH, F, and aryl [26]. Only the $H_2Si_2F_2$ system, which is relevant to Jutzi's experiments [24]

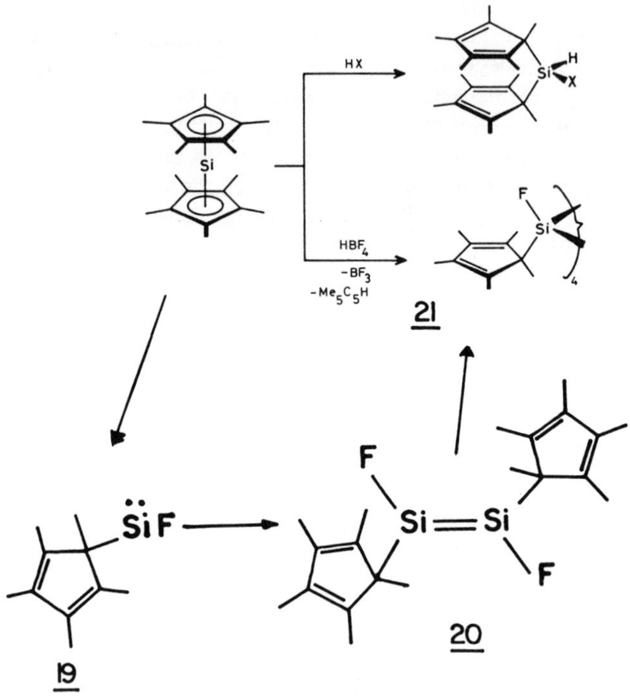

Scheme 2.2 ■ Proposed reaction scheme for the production of **21**, from [24].

will be considered here. The optimized 6-31G* geometries for the species of interest are shown in Figure 2.5.

The parent disilene **15** has a trans-bent structure, but the planar structure is only slightly (by above 1–2 kcal/mol^{-1}) higher in energy [1]. Bending is more pronounced in 1,2-difluorodisilene **23** (43.5° in cis-**23**, Figure 2.5). In the trans hydrogen-bridged Si$_2$H$_4$ **16** the Si----Si distance is 2.609 Å, much longer than in **15** or even in the singly bonded Si$_2$H$_6$ (2.342 Å at 3-21G$^{(*)}$). The Si----Si distance is slightly longer (2.679 Å) in the hydrogen-bridged Si$_2$H$_2$F$_2$ molecule **24** and much longer (2.907 Å) in the fluorine-bridged structure **25**.

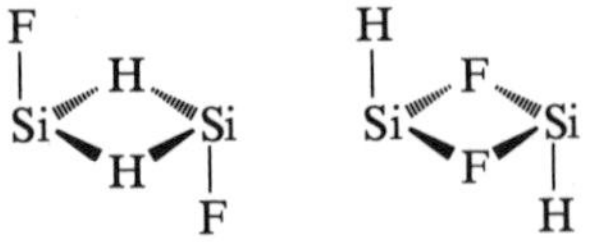

The bonding in the unusual bridged structures is of interest. Mulliken population analysis and examination of the orbitals of **16**, **24**, and **25** show that there is essentially no bonding between the silicon atoms, in agreement with the relatively long Si---- Si distance. The bridged and the classical disilenes can be described as being formed via

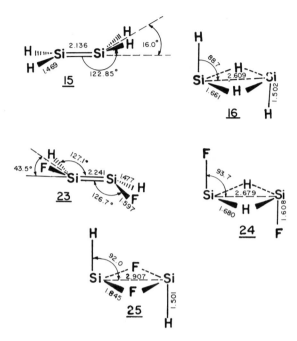

Figure 2.5 ■ 6-31G* optimized geometries of disilene, cis-1,2-difluorodisilene and the corresponding bridged structures.

two different approaches of two silylenes, as shown in Scheme 2.3. A skewed C_{2h} approach, in which the empty $3p$ orbital of one silylene interacts with the Si — H bond of the second silylene, leads to the bridged structure; a direct C_s approach, in which the empty $3p$ orbital of one silylene interacts with the lone pair of the second silylene, leads to the classical structure.

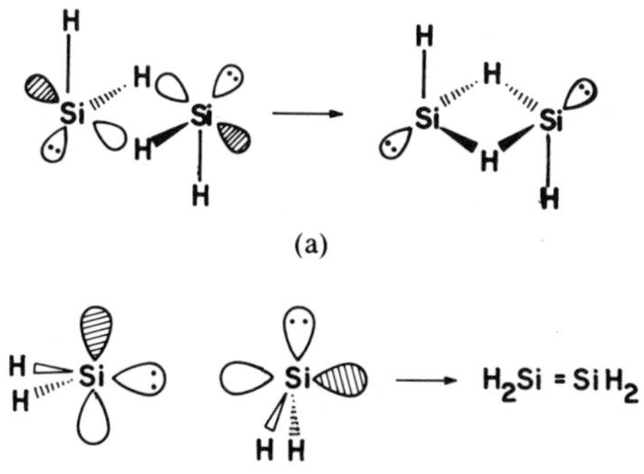

(a)

(b)

Scheme 2.3 ■ Schematic representations of two possible dimerizations of two silylenes: (a) leading to the bridged disilene (C_{2h} approach); (b) leading to the classical disilene (C_s approach).

Table 2.4 ■ Relative Energies (kcal mol^{-1}) of Bridged and Classical Si$_2$H$_4$ and Si$_2$H$_2$F$_2$ Molecules

Molecule	\multicolumn{4}{c}{Relative Energy}			
	3-21G//3-21G	6-31G*//6-31G*	MP3/6-31G*//6-31G*	MNDO
15	0.0	0.0	0.0	0.0
16	26.0	23.5	25.8	55.4
2 × H$_2$Si^a	44.0	48.4	62.4	53.7
23	0.0	0.0	0.0	0.0
24	2.8	4.8	6.1	14.4
25	−29.9	5.0	9.5	−16.3
2 × HFSi^b	15.0	20.9	32.8	15.6

^{a}The energy of two isolated H$_2$Si: molecules.
^{b}The energy of two isolated HFSi: molecules.

The major point of interest in the context of Jutzi's experiment is the relative energy of the bridged and classical structures of the difluoro-system. The results of the calculations for the model molecules **15**, **16**, and **23–25** are given in Table 2.4. For the bridged species I include only the trans structures, as the cis isomers are very close in energy.

For the unsubstituted disilene the energy difference between the classical and the bridged structures is relatively insensitive to the theoretical level used [except for the semiempirical modified intermediate neglect of differential overlap (MNDO) method, which gives a higher separation than the ab-initio methods]. At MP3/6-31G*//6-31G* the bridged **16** is by 25.8 kcal mol^{-1} less stable than the classical **15**, but it is bonded by 34.2 kcal mol^{-1} relative to two isolated silylenes. Fluorine substitution reduces considerably the energy differences between the classical and the bridged structures, when either the fluorines or the hydrogens are bridging. At MP3/6-31G*//6-31G*, **24** and **25** lie only 6.1 and 9.5 kcal mol^{-1}, respectively, above the classical disilene **23**. Both are also strongly bound relative to two FHSi: silylenes. Examination of Table 2.4 reveals some of the computational pitfalls associated with these unusual structures. Thus, at 3-21G//3-21G the fluorine-bridged **25** is calculated to be 29.9 kcal mol^{-1} more stable than **23**. Similarly, MNDO finds **25** to be substantially more stable than either **23** or **24**. Calculations for (CH$_3$)$_2$Si$_2$F$_2$ (**26** and **27**), using both 3-21G and MNDO shows that methyl substitution has only a small effect on the energy difference between the disilene and the fluorine-bridged structure.

26   **27**

The theoretical conclusion that can be reached at this point is that for difluoro dialkyl–substituted disilenes the classical structure is the lowest in energy but that the fluorine bridged structure is merely about 6–10 kcal mol^{-1} higher in energy. This rather low energy difference suggests the possibility that *dimerization of the silylene* **19** *leads to the fluorine-bridged structure* **22**, which at low temperatures does not rearrange to the more stable disilene **20**. The NMR data, in particular the J_{SiF} spin–spin coupling constant, are consistent with the suggestion that the intermediate species observed by Jutzi et al. [24] is the fluorine-bridged **22**. In order to account for the observation that the ^{29}Si absorption at -34.7 ppm is a triplet, while a doublet of doublets is expected for structure **20**, Jutzi et al. had to assume that $^1J_{SiF} = {}^2J_{SiF} = 341$ Hz [24]. This assumption is unlikely on the following grounds: The few known $^2J_{SiF}$ spin–spin coupling constants are significantly lower [27] than the value of 341 Hz suggested for $^2J_{SiF}$ in **20** [24]. On the other hand, this value is consistent with many known $^1J_{SiF}$ spin–spin coupling constants [27]. Also, for C–F spin–spin coupling constants, $^1J_{CF}$ are always much larger than $^2J_{CF}$ (also in 1,2-difluoro–substituted alkenes) by a factor of 5–8 [28]. A fast equilibrium between **20** and **22** is also unlikely because it requires an observed coupling constant which is the average of $^1J_{SiF}$ and $^2J_{SiF}$ which is expected [27] to be much smaller than 341 Hz. On the other hand, the observed splitting is fully consistent with the fluorine-bridged structure **22**, which is expected to exhibit a triplet ^{29}Si signal. Analysis of the position of the ^{29}Si NMR chemical shift is less straightforward and it can be associated with either structure **20** or **22**.

The possibility that the species observed in Jutzi's experiment is bridged was not considered in the short paper that described the experimental observations [24]. A more complete theoretical analysis, including calculations of the transition states for the rearrangement of the bridged structures to the classical isomers, is required before more definite conclusions can be reached regarding the structure of the intermediate observed in Jutzi's experiment. However, the calculations point clearly to the fact that disilenes, especially with substituents that have lone-pair electrons, are expected to be highly fluxional molecules that can adopt very unusual structures, including ones that do not obey the classical van't Hoff rules of valency. We hope that this discussion will prompt further experimental efforts in this field.

Acknowledgment. I wish to thank the excellent collaborators who participated in both the experimental and the theoretical aspects of our research on organosilicon compounds. Their names appear in the references. Our research in this field was supported by three consecutive grants from the United States–Israel Binational Science Foundation (BSF) and by grants from the Vice-President of Research at the Technion, to whom we are grateful. The numerous discussions that we had with Professor R. West and his group have contributed significantly to our organosilicon research.

References

1. For a comprehensive review on the contributions of theory to organosilicon chemistry see the chapter by Apeloig, Y., in *The Chemistry of Organic Silicon Compounds*, Patai, S., Rappoport, Z., eds., Wiley, Chichester, 1989, pp. 57–225.
2. Schaefer, H. F. III. *Acc. Chem. Res.*, **1982**, *15*, 283.
3. Mohaffy, P. G.; Gutowski, R.; Montgomery, L. K. *J. Am. Chem. Soc.*, **1980**, *102*, 2854.
4. Brook, A. G.; Nyburg, S. C.; Abdesaken, F.; Gutenkunst, B.; Gutenkunst, G.; Kallury, R. K. M. R.; Poon, U. C.; Chang, Y. M.; Wong-Ng, W. *J. Am. Chem. Soc.*, **1982**, *104*, 5667.
5. (a) Apeloig, Y.; Karni, M. *J. Am. Chem. Soc.*, **1984**, *106*, 6676. (b) Apeloig, Y; Karni, M. *J. Chem. Soc., Chem. Commun.*, **1984**, 768.

6. (a) Wiberg, N.; Wagner, G.; Müller, G. *Angew. Chem., Int. Ed. Engl.*, **1985**, *24*, 229. (b) Wiberg, N.; Wagner, G.; Riede, J.; Müller, G. *Organometallics*, **1987**, *6*, 32. (c) Gutowsky, H. S.; Chen, J.; Hajduk, P. J.; Keen, J. D.; Emilsson, T. *J. Am. Chem. Soc.*, **1989**, *111*, 1901. (d) Wiberg, N.; Wagner, G.; Reber, G.; Riede, J.; Müller, G. *Organometallics*, **1987**, *6*, 35. (e) Wiberg, N.; Wagner, G.; Müller, G.; Riede, J. *J. Organomet. Chem.*, **1984**, *271*, 381. (f) Wiberg, N.; Schurz, K.; Reber, G.; Müller, G. *J. Chem. Soc., Chem. Commun.*, **1986**, 591.

7. Both cis and trans bending occur. The trans-bent structure has a similar geometry to the cis-bent structure shown in Figure 2.1, but it is 2.8 kcal mol^{-1} higher in energy at 3-21G//3-21G.

8. For a review see Schwarz, H., in *The Chemistry of Organic Silicon Compounds*, Patai, S., Rappoport, Z., eds., Wiley, Chichester, 1989, pp. 445–510.

9. Lambert, J. B.; Schulz, N. J., Jr.; McConnell, J. A.; Schilf, W. *J. Am. Chem. Soc.*, **1988**, *110*, 2201, and references therein.

10. (a) Eaborn, C. in *Organosilicon and Bioorganosilicon Chemistry: Structure, Bonding, Reactivity and Synthetic Applications*, Ellis Horwood, Chichester, 1985, and references therein. (b) Eaborn, C.; Jones, K. L.; Lickiss, P. D. *J. Chem. Soc., Chem. Commun.*, **1989**, 595.

11. (a) Apeloig, Y.; Stanger, A. *J. Am. Chem. Soc.*, **1987**, *109*, 272. (b) Apeloig, Y. *Studies in Organic Chemistry*, **1986**, *31*, 33.

12. (a) Marin-Aharoni, A., unpublished results. For previous studies of silicenium ions at lower levels of theory see the following: (b) Apeloig, Y.; Schleyer, P. v. R. *Tetrahedron Lett.*, **1977**, 4647. (c) Apeloig, Y.; Godleski, S. A.; Heacock, D. J.; McKelvey, J. M. *Tetrahedron Lett.*, **1981**, *22*, 3297. (d) Braude, V., unpublished results.

13. Hehre, W. J.; Radom, L.; Schleyer, P. v. R.; Pople, J. A. *Ab Initio Molecular Orbital Theory*, Wiley-Interscience, New York, 1986.

14. (a) Ibrahim, M. R.; Jorgensen, W. L. *J. Am. Chem. Soc.*, **1989**, *111*, 819. (b) Wierschke, S. G.; Chandraeskhar, J.; Jorgensen, W. L. *J. Am. Chem. Soc.*, **1985**, *107*, 1496. (c) Apeloig, Y.; Arad, D. *J. Am. Chem. Soc.*, **1985**, *107*, 5285.

15. Shin, S. K.; Beauchamp, J. L. *J. Am. Chem. Soc.*, **1989**, *111*, 990.

16. (a) Apeloig, Y.; Karni, M.; Stanger, A.; Schwarz, H.; Drewello, T.; Czekay, G. *J. Chem. Soc., Chem. Commun.*, **1987**, 989. (b) Drewello, T.; Burgers, P. C.; Zummack, W.; Apeloig, Y.; Schwarz, H. *Organometallics*, **1990**, *9*, in press.

17. Apeloig, Y.; Karni M. *J. Chem. Soc., Perkin Trans. II*, **1988**, 625.

18. Apeloig, Y.; Stanger, A. *J. Organomet. Chem.*, **1988**, *346*, 305.

19. Chojnowski, J.; Fortuniak, W.; Stanczyk, W. *J. Am. Chem. Soc.*, **1987**, *109*, 7776.

20. See, for example, Lowry, T. H.; Richardson, K. S. *Mechanism and Theory in Organic Chemistry*, 3d ed., Harper & Row, New York, 1987, pp. 212–214.

21. For recent reviews see the following: (a) Chapter by Raabe, G.; Michl, J. in [1] pp. 1015–1142. (b) Raabe, G.; Michl, J. *Chem. Rev.*, **1985**, *85*, 419. (c) West, R. *Angew. Chem., Int. Ed.*, **1987**, *26*, 1201.

22. Kohler, H.-J. *Z. Chem.*, **1984**, *24*, 155. (b) Kohler, H.-J.; Lischka, H. *Chem. Phys. Lett.*, **1984**, *112*, 33.

23. Yokelson, H. B.; Maxka, J.; Siegel, D. H.; West, R. *J. Am. Chem. Soc.*, **1986**, *108*, 4329.

24. Jutzi, P.; Holtman, U.; Bogge, H.; Muller, A. *J. Chem. Soc., Chem Commun.*, **1988**, 305.

25. (a) Nagase, S.; Kudo, T., Ito, K., in *Applied Quantum Chemistry*, Smith, V. H., Jr., Schaefer, H. F., III, Morokuama, K., eds., Reidel, Dordrecht, 1986, pp. 249–267. (b) Trinquier, G.; Malrieu, J.-P. *J. Am. Chem. Soc.*, **1987**, *109*, 5303.

26. (a) Maxka, J.; Apeloig, Y., manuscript in preparation. (b) Maxka, J.; Apeloig, Y. *J. Chem. Soc., Chem. Commun.*, **1990**, in press.

27. Marsmann, H. in *NMR Basic Principles and Progress*, Diehl, P., Fluck, E., Kosfeld, R., eds., Vol. 17, Springer-Verlag, Berlin and New York, 1981, p. 122.

28. Wehrli, F. W.; Marchand, A. P.; Wehrli, S. *Interpretation of Carbon-13 NMR Spectra*, 2d ed., Wiley, New York, 1988, p. 92.

Electron-Rich Bonding at Low-Coordination Main Group Element Centers

Anthony J. Arduengo, III, and David A. Dixon

E. I. du Pont de Nemours & Co.
Central Research and Development Department
Box 80328
Wilmington, Delaware 19880-0328

3.1. Introduction

Our recent studies of the chemistry of the main group elements have centered on creating unusual bonding arrangements for these elements to examine extremes in behavior. Specifically, we have developed ligands [1–5] that are capable of stabilizing main group elements in electron-rich, low-coordinate environments.

For a given valence electron count at a main group element, chemists usually expect certain "standard" geometries and coordination numbers [6]. The well-known eight-electron carbon center is usually associated with a four-coordinate idealized tetrahedral geometry, although three-coordinate pyramidal anions and three-coordinate π-bonded systems are also common. A 10-electron carbon center is most commonly associated with the SN_2 reaction transition state [a five-coordinate trigonal bipyramidal (TBP) geometry]. For phosphorus a valence electron count of 10-e suggests a five-coordinate TBP geometry such as that found for PCl_5. Relatively unusual or uncommon compounds such as π-bonded heavy main group element centers [7] or compounds showing alternative geometries (e.g., square pyramidal versus trigonal bipyramidal structures at a five-coordinate center [8]) pose interesting synthetic and theoretical challenges. These more unusual bonding arrangements frequently attract the attention of many chemists studying the main group elements. To categorize the relationship between electron count and coordination number at main group centers, Martin and Arduengo [9] developed the "N-X-L" system of classifying bonding arrangements. This system is useful in cataloging compounds of the main group elements and helps to identify structural and electronic correlations among unusual bonding arrangements. The 10-electron phosphorus molecules can thus be listed as a set of 10-P-L bond schemes where L, the coordination number of phosphorus, is varied while the electron count and atomic center are held constant. The 10-P-5 centers are found in the well known phosphoranes such as PCl_5. The 10-P-4 centers are represented by the recently studied phosphoranides [10]. Logically, one might ask the question, what of 10-P-3 systems?

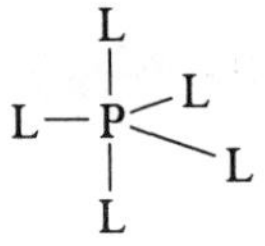

10-P-5	10-P-4	10-P-3
Phosphorane	Phosphoranide	Phosphorandiide

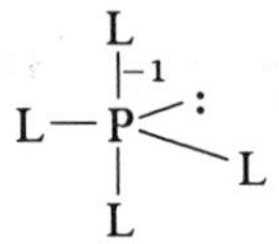

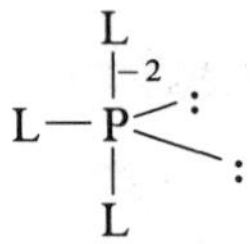

3.2. Tricoordinate Hypervalent Pnictogens

3.2.1. Related Molecules

The connection between 10-P-3 centers and the isoelectronic 10-S-3 centers is clear from their N-X-L designations. Although 10-P-3 centers were unknown prior to 1984 [1,11], 10-S-3 centers have long been known as components of the trithiapentalenes [12] (1) and more recently in sulfuranides [13] such as 2. The 10-P-3 center can also be viewed as the two-electron reduction product of an 8-P-3 center. The addition of two electrons to an eight-electron tricoordinate phosphorus center could result in the formation of a second lone pair of electrons at phosphorus. This reduced species should adopt a *pseudo*-trigonal bipyramidal (Ψ-TBP) geometry at phosphorus with both lone pairs in the equatorial plane to give a T-shaped structure.

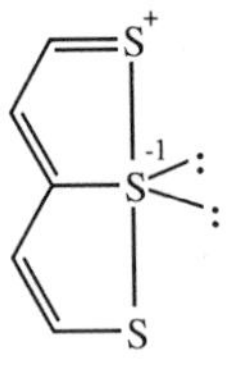

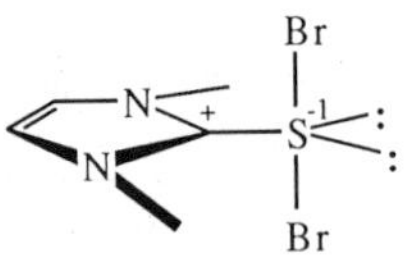

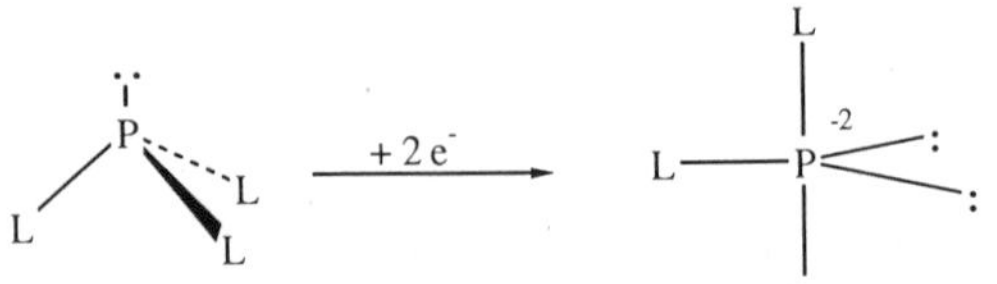

3.2.2. Ligand Design

The stability of 10-S-3 bonding systems and the concept of two-electron reduction of 8-P-3 systems led to the choice of the tridentate diketoamine ligand system 3 as a means to stabilize the 10-P-3 bonding array. In addition to being a powerful reducing ligand, **3** also stabilizes the 10-P-3 center with three additional features: The first is a five-membered ring linkage of apical and equatorial sites of the Ψ-TBP geometry [14]. Second, electronegative elements are placed in apical positions. Third, there is charge compensation for the *formally* dianionic phosphorus center. This latter charge compensation is a natural consequence of the hypothetical intramolecular oxidation–reduction reaction that transforms an 8-P-3 center into a 10-P-3 center. It is important, as we have previously stressed, to realize that the atomic charges assigned in the 10-P-3 ADPO (see [15] for a definition) structure are *formal*, not actual, charges [4]. The significance of formal charges that arise from the application of conventional valence bond rules for structure depiction has been discussed in detail elsewhere [16]. Indeed, we have shown with preliminary ab-initio calculations that the actual charge at phosphorus in 10-P-3 ADPO is positive, as would be expected from the presence of the hypervalent (linear three-center, four-electron) bond [4]. Nonetheless, the formal charges in the traditional valence bond depiction of 10-P-3 ADPO do provide insight into the structure and chemistry of this unusual molecule.

3	**10-P-3 ADPO**

The successful design and synthesis of a molecule that contains a 10-P-3 center has led to the synthesis of arsenic [2] and antimony [3] analogs. Additionally, unusual bonding environments that take advantage of the unique structural and electronic properties of ligand **3** have been synthesized at bismuth [17], tin [18], germanium [19], silicon [19], lead [19], gallium [19], indium [19], and aluminum [19] centers. The chemical reactivity and physical properties of 10-P-3 ADPO, 10-As-3 ADAsO, and 10-Sb-3 ADSbO have been reported [4] and have contributed significantly to the development of an understanding of these novel compounds. Theoretical studies [4] of the 10-Pn-3 ADPnO (Pn = pnictogen*) molecules have allowed the formulation of more accurate descriptions of the bonding in these molecules. The postulate of an edge inversion mechanism [20] was the result of detailed theoretical examination of the ADPnO molecules and was subsequently demonstrated experimentally [21].

The study of uncommon or unusual bonding arrangements generally results in the acquisition of substantial new information on molecules and their bonding and reactivity. In this regard, the study of the ADPnO molecules is no exception: Detailed study of the ADPnO molecules is continuing to yield interesting new science.

---

*The term "pnictogen" is used to refer to the main group 5 (group 15) elements (N, P, As, Sb, and Bi). The term is derived from the Greek word *pniktos* (suffocate).

3.3. Valence Bond Descriptions of the ADPnO Structures

The validity of various extremes in bonding descriptions is worth consideration. Some of the descriptions that may be considered are the classical valence bond description of a 10-P-3 phosphorandiide structure **4a**, a localized π-bonding structure **4b**, distorted phosphine structure **4c** and the internally solvated pnictinidine (6-Pn-1) **4d**. Although some of these models have been previously discussed [1,4], they warrant review in light of recent theoretical and experimental advances on related systems (vide infra).

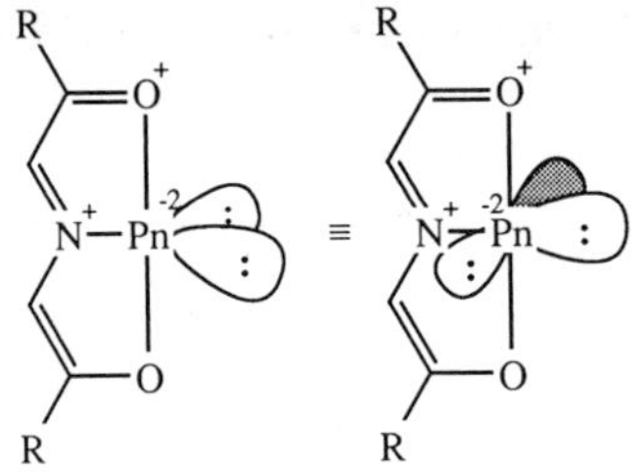

4a 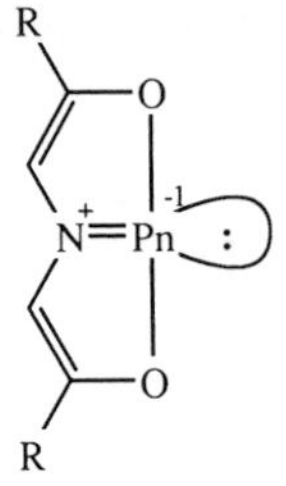   **4b**

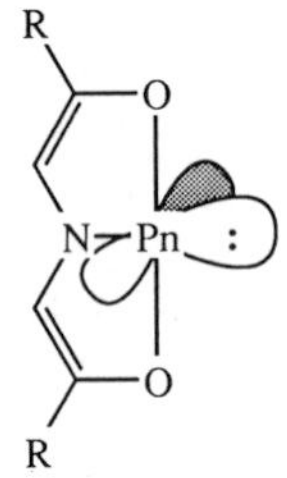

4c 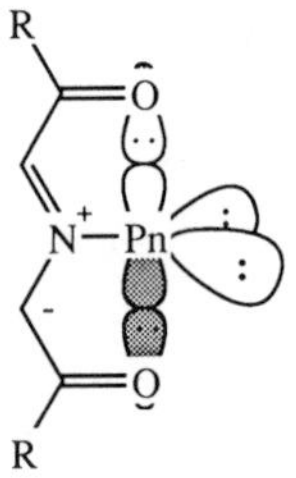   **4d**

3.3.1. Representation of Structural and Spectroscopic Properties

To a large extent all four of the hypothetical structures **4a–4d** are actually the same. They merely focus on different aspects of the 10-Pn-3 ADPnO bonding arrangement. The structures labeled as **4a** are equivalent representations of the pnictandiide moiety, differing only in the description of the lone pairs as equivalent or as a σ and a π set [4]. Structures **4b** and **4c** differ from **4a** only in the location (or delocalization) of electron density. Structure **4c** suggests that the electron density from the ligand backbone, which was to have effected the two-electron reduction of the pnictogen center, did not do so.

The failure of the ligand backbone to transfer electron density to the pnictogen center leaves the planar structure **4c** with internal inconsistencies. Because structure **4c** is planar, symmetry considerations prevent participation of the out-of-plane pnictogen p orbital in the σ-bonding system about the pnictogen. The coaxial Pn — N bond and pnictogen in-plane lone pair require the use of an in-plane pnictogen p orbital and the pnictogen s orbital to accommodate these valence electrons. This means that the O—P—O bonding array is a linear three-center four-electron (3c, 4e) hypervalent bond as would be expected. This electron-rich (orbitally deficient) hypervalent bond is incompatible with the presence of a vacant p orbital at the pnictogen center. Reassignment of the in-plane s- and p-orbital functions at the pnictogen could eliminate the need for the presence of a hypervalent bond but would require the rupture of the Pn—N bond because it would become a two-center, four-electron array.

The unstable arrangement in structure **4c** would be eliminated by a folding of the two five-membered rings about the Pn—N bond to allow mixing of the hypervalent O—P—O system with the formerly out-of-plane pnictogen orbital. This out-of-plane folding points to the identity of structure **4c** as the transition state for the edge inversion process [20–25] at the pnictogen center (vide infra) and identifies 10-P-3 ADPnO molecules as stable models for this transition state. Thus, the electronic assignments suggested by structure **4c** are inconsistent with the ground state planar tricoordinate nature of the pnictogen centers in the 10-Pn-3 ADPnO molecules. The electronic representation in structure **4c** could be reconciled with the observed geometry if virtual orbital participation could be invoked. The participation of higher order virtual orbitals (d orbitals) could stabilize the orbitally deficient O—Pn—O hypervalent bond and eliminate the driving force to fold the structure for σ–π mixing. However, it is not reasonable to invoke the participation of virtual orbitals before all the valence orbitals have been utilized. Additionally, it would be expected that the pnictogen trifluorides (PnF_3) would show planar T-shaped ground states as suggested by **4c** because the electronegative fluorines would be well suited to the hypervalent bonding arrangement and should enhance virtual orbital participation by their inductive effect.

Because the bonding scheme depicted in **4c** is not likely to be very important, it is reasonable to assume that there is some electron density in the out-of-plane p orbital in **4c**. This brings structure **4c** back toward structure **4a** and reduces the question to the extent of delocalization of the second (out-of-plane) pnictogen lone pair. In other words, are the ADPnO structures best represented by **4a** or by a combination of **4a** + **4b** or perhaps **4b** + **4c**? On the basis of preliminary ab-initio calculations [4], we have previously suggested that the out-of-plane pnictogen lone pair is indeed delocalized and this view is further supported by more extensive calculations (vide infra). This π-overlap–based delocalization would be expected to be largest for phosphorus and smallest at antimony so that 10-Sb-3 ADSbO is the best physical model of the extreme structure **4a**. Calculated atomic charges in 10-P-3 ADPO and 10-As-3 ADAsO also support these trends in π-delocalization, which actually oppose the σ effects but still manage to increase electron density at the heavier pnictogen center (vide infra). Because the presence of out-of-plane electron density at the pnictogen center is critical to obtaining the planar structure, the phosphorus system actually loses some stability because of the π-delocalization.

These trends in degree of delocalization are consistent with the observed NMR spectroscopic trends in the 10-Pn-3 ADPnO series. Charge separation increases from ADPO to ADAsO to ADSbO. Structures **4a** and **4b** are better models than **4c** for the ^{15}N NMR shifts of the ADPnO series because the nitrogen clearly has large iminium character ($\delta \sim -100$) [4]. The ^{13}C, ^{1}H, and ^{17}O NMR shifts are also better represented by structure **4a** than **4b** or **4c**. These NMR shifts indicate carbonyl-like and iminium-like bonding arrangements, not vinyl ethers or enamines [4]. The lengths of the Pn—N bonds in the ADPnO series are comparable to common equatorial Pn—N single bonds at 10-Pn-5 centers and do not suggest multiple bond character between these centers, as structure 4b would require [4]. However, extreme care must be taken when comparing bond lengths between two atomic centers in which hybridization changes occur. These changes in hybridization can cause changes in bond lengths solely due to effects on the σ bonds.

A demonstration of this hybridization or σ effect has been observed for three structures of PHF_2 in which no π interactions can influence the P—H bond length [20]. Pyramidal PHF_2 (5a) is calculated to have a P—H bond length (R_{P-H}) of 1.409 Å. The trigonal planar structure 5b, which is the transition state of vertex inversion (vide infra), gives $R_{P-H} = 1.352$ Å. This is consistent with the increase in s-orbital character from phosphorus in this P—H bond at the sp^2 hybridized transition state structure relative to the sp^3 hybridized ground state structure. The T-shaped structure, 5c, which is the transition state for edge inversion (vide infra), has $R_{P-H} = 1.397$ Å. While shorter than the P—H bond in 5a, the length of this bond in 5c relative to 5b demonstrates the demand of the σ lone pair at phosphorus for s-orbital character. Lengths of bonds in 10-Pn-3 and 10-Pn-5 systems will also be subject to such σ effects in addition to any π effects.

5a   5b   5c

One final point to be made regarding the validity of structures 4b or 4c concerns the replacement of the pnictogen center by a main group IV center. If the pnictogen center were transmuted to a main group IV center, a former pnictogen lone pair would be replaced by the additional substituent at the main group IV center. If structures 4b or 4c were the best representations of these bonding systems, a square planar structure would have to result because the only lone pair is in the plane. However, substitution of the pnictogen center by a Sn—Cl unit results in the Ψ-TBP structure that has been observed for ADSnO · Cl [18].

ADPnO   ADSnO·Cl

3.3.2. The Character of the hypervalent Bond in ADPnO Systems

Structure 4d brings the O—Pn—O hypervalent bond to prominence. This structure represents a singlet pnictinidine (6-Pn-1) center that is internally solvated by the carbonyl substitutents of an azomethine ylide. The interaction of carbonyl oxygen lone

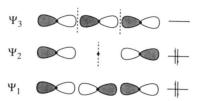

Figure 3.1 ■ Simple valence orbital representation of the three-center, four-electron hypervalent bond.

pairs of electrons with the vacant p orbital of the pnictinidine is actually the same interaction that is found in the lowest lying molecular orbital (Ψ_1) of a linear three-center, four-electron bond (Figure 3.1). The significant degree of interaction between the pnictogen and oxygens is evidence that this is more than just mere solvation. The oxygen–pnictogen distances in the 10-Pn-3 ADPnO series are slightly longer (< 6%) than corresponding (apical) distances in representative molecules that contain 10-Pn-5 centers [4]. Nonetheless, these distances indicate significant interaction between the pnictogen and oxygen atoms. The relationship between ^{15}N–^{31}P one-bond coupling constant and phosphorus coordination number, which we have previously observed for ADPO-derived compounds [4], would require a coupling constant of over 100 Hz if the phosphorus in ADPO were truly phosphinidine (6-P-1) in nature. The ^{15}N–^{31}P coupling constant in 10-P-3 ADPO molecules (generally about 80 Hz) is consistent with the tricoordinate environment of the phosphorus. The large upfield shifts (> 200 ppm) in the nuclear magnetic resonances for the ^{17}O centers of the ADPnO series relative to the starting ligand also suggest a high degree of interaction between the pnictogen and oxygen centers.

3.3.3. Structural Implication of Observed Chemistry

The chemistry that has been observed for the ADPnO series suggests that the best model for these compounds should provide a description that places two lone pairs of electrons at the pnictogen centers. Various Lewis acid adducts and oxidative addition products have been synthesized from these compounds. For the arsenic and antimony compounds, one lone pair is utilized in these reactions and the second remains at the pnictogen center and is stereochemically active. The phosphorus system does not follow this same trend but rather shows a dramatic change in the ligand backbone after reaction with oxidizing agents or Lewis acids. This change in the ligand backbone is manifested in upfield shifts in the ^{1}H ($\Delta\delta$, 1.5–2.0 ppm), ^{13}C ($\Delta\delta$, 10–28 ppm), ^{15}N ($\Delta\delta$, 140–170 ppm), and ^{17}O ($\Delta\delta$, 150 ppm) resonances, consistent with return of electron density from phosphorus to the ligand. The reason that phosphorus centers fail to maintain the additional electron density originally donated from the ligand backbone may be a combination of steric and electronic factors. The smaller size of phosphorus relative to arsenic and antimony should cause a preference for tetrahedral and TBP geometries over their Ψ-TBP and Ψ-octahedral alternatives. The electronic factors that influence this behavior are apparent from the calculations on the ADPO molecules and are discussed below.

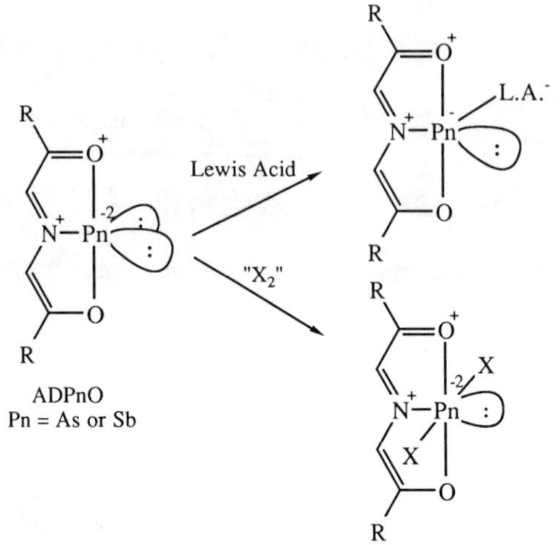

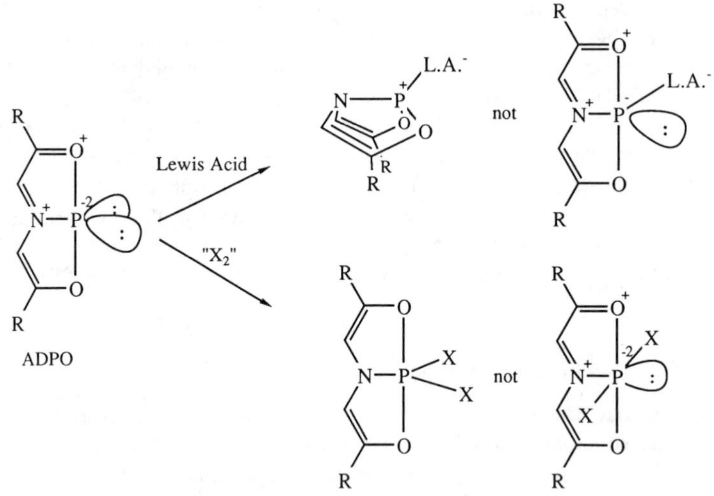

Either by maintaining the second lone pair at the pnictogen center or by dramatically changing the ligand backbone, the chemistry of the ADPnO compounds supports representation **4a** as the best single model. Although structure **4a** is the best overall representation of structure, spectroscopic properties, and chemistry for the ADPnO series, it is not without its shortcomings. The formal charges in **4a** must be tempered by the realization that the equal sharing of electrons between bonded centers, which is implicit in simple valence bond representations, is not accurate when the bonded centers differ in electronegativity. Hence, even simple σ bonds will be polarized. In the case of the ADPnO molecules the σ-bonded neighbors are all more electronegative than the central pnictogen. This will serve to polarize electron density away from the central pnictogen. The hypervalent bond will extensively polarize electron density away from the central pnictogen and toward the oxygens. This polarization in the hypervalent bond is largely the result of Ψ_2 (Figure 3.1), which places electron density only on the oxygens at the expense of the central pnictogen. Finally, some degree of π interaction may occur between the pnictogen center and its ligand. This is the basis for the delocalization of the pnictogen π–lone pair described above which is most pronounced for phosphorus. Such π interaction further decreases the electron density at the central phosphorus in 10-P-3 ADPO relative to its heavier analogs.

3.4. Calculated Geometries and Energies

The geometries of the di-*t*-butyl derivatives of ADPnO for Pn = P, As have been determined by X-ray crystallography, as has the structure of the di-adamantyl derivative of ADAsO. The X-ray geometries are compared with the calculated (see Appendix for details) geometries for the unsubstituted derivatives of ADPO and ADAsO in Table 3.1. The total energies are given in Table 3.2. There is excellent agreement between the calculated and experimental geometries. We first note that both molecules are planar experimentally and the second-derivative calculations show that the calculated structures are indeed minima with all real frequencies (all positive directions of curvature). The agreement between the calculated and symmetrized experimental structures (C_{2v}) is excellent for ADPO, with all bond distances agreeing to better than 0.02 Å and most angles to within 1°. The only differences in the geometry larger than 1° are for the angles at the nitrogen, and these agree to better than 2°. For ADAsO the agreement between the calculated and experimental structures is almost as good, with the largest differences found in the bond distances of the atoms bonded to the arsenic. The As—O bonds are calculated to be 0.018 Å too short and the As—N bond is calculated to be about 0.03 Å too short. The bond angles are all within 1° except for the angles at nitrogen, where the differences are about 2°.

Following the above valence bond discussions, it is initially somewhat surprising that the ADPnO molecules are planar if one considers them as 8-Pn-3 systems, which should be folded. In order to better understand the differences between the 8-P-3 and 10-P-3 structures for ADPO, we optimized ADPO in a folded geometry having only C_s symmetry. This was accomplished by folding the two five-membered rings about the P—N bond. The optimized geometry is given in Table 3.1. The C_s structure is lower in energy than the planar (C_{2v}) structure at the SCF level by 9.9 kcal mol^{-1}. However, inclusion of correlation makes the planar structure lower in energy by 13.9 kcal mol, a correlation energy correction of 23.8 kcal mol^{-1}. It is clear from this result that some form of a correlation energy correction, at least for the phosphorus system, is required to make the planar structure the most stable one in agreement with the experimental structure. We have previously observed sizable correlation energy corrections in calculations of edge inversion barriers (vide infra) [20]. Although the correlation energy correction to the vertex inversion barrier in PH$_3$ is small, 1.3 kcal mol^{-1} out of 35.0 kcal mol^{-1} at the MP-2 level with a DZP basis set, the correlation correction to the edge inversion barrier in PF$_3$ is substantially larger, 14.6 kcal mol^{-1} out of 53.8 kcal mol^{-1}. The correlation correction would be expected to be even larger for ADPO (as found) because there are more centers in the five-membered rings over which to delocalize.

For ADAsO, both the planar and folded structures are again minima on the SCF potential energy surface. However, the planar structure is now more stable than the

Table 3.1 ■ Calculated Geometries for ADPO, Planar (C_{2v}) and Folded (C_s), and ADAsO[c]

	ADPO			ADAsO			
Property	Expt.[a] (C_{2v})	Calc. (C_{2v})	Calc. (C_s)	Expt.[a] (C_{2v})	Expt.[b] (C_{2v})	Calc. (C_{2v})	Calc. (C_s)
	Bond distances (Å)						
Pn—O	1.814	1.804	1.652	1.976	1.976	1.958	1.787
Pn—N	1.703	1.688	1.763	1.839	1.842	1.813	1.866
N—C	1.378	1.394	1.438	1.372	1.376	1.383	1.442
O—C	1.330	1.316	1.387	1.308	1.313	1.300	1.380
C—C	1.340	1.348	1.328	1.363	1.358	1.360	1.330
C-3—H-3			1.065	1.066		1.064	1.067
C-4—H-4			1.069	1.065		1.070	1.067
	Bond angles (degrees)						
O—Pn—N	83.8	83.9	92.3	80.2	80.3	80.5	89.4
O—Pn—O	167.7	167.7	108.0	160.3	160.6	161.1	106.3
Pn—N—C	117.4	118.0	107.3	117.4	117.0	118.0	106.8
Pn—O—C	114.8	115.4	112.4	114.2	114.2	114.0	111.0
C—N—C	125.2	123.9	112.9	125.2	126.1	124.1	112.4
N—C—C	111.3	109.6	113.1	113.2	113.7	111.7	115.2
O—C—C	112.6	113.1	114.4	115.1	114.8	115.8	116.9
N—C—H			121.5	120.1		121.1	119.5
C-3—C-4—H			128.8	126.8		127.2	125.2
O—C—H			119.9	116.0		119.6	115.7
C-4—C-3—H			127.0	129.5		124.6	127.4

[a] Di-t-butyl.
[b] Di-adamantyl.
[c] Numbering system, when necessary is the same as for 10-P-3 ADPO.

Table 3.2 ▪ Total Energies in a.u.

Molecule	E(SCF)	E(MP-2)	$i\nu$ (cm^{-1})^a
ADPO (planar)	−698.625375	−699.621743	0
ADPO (folded)	−698.641178	−699.599598	0
H$_2$ · ADPO (planar)	−699.789512	−700.781380	96
H$_2$ · ADPO (folded)	−699.834875	−700.795805	0
H$_4$ · ADPO (quasi-planar)	−700.955412	−701.943011	197
H$_4$ · ADPO (folded)	−701.023883	−701.987443	0
ADAsO (planar)	−2591.547626	−2592.545033	0
ADAsO (folded)	−2591.546644	−2592.508356	0
ADSO$^+$(C_{2v})	−755.116302	−756.146236	394
ADSO$^+$(C_s)	−755.126897	−756.123050	0
8 (C_{2v})	−738.837848	−739.806933	299
8 (C_s)	−738.844042	−739.795180	0

$^a i\nu$ is imaginary frequency.

folded one by 0.6 kcal mol^{-1}. Inclusion of correlation makes the planar structure more stable (23.0 kcal mol^{-1}). Thus, the arsenic system (ADAsO) shows an even greater preference for planar over folded structures than was found for the phosphorus system (ADPO). The correlation correction between the planar and folded forms of ADAsO is 22.4 kcal mol^{-1}, slightly less than the value for ADPO.

The result for ADAsO points out an important feature of the true potential energy surface: Even though both arsenic structures are minima at the SCF level (separated by an energy barrier), it is unlikely that such a barrier exists at the correlated level. Hence, the folded minimum would disappear. The large difference in energy between the planar and folded forms is consistent with this conclusion. A similar result is expected for ADPO, namely, only a single minimum on the potential energy surface, although the lower energy difference between the planar and folded structures suggests a more flexible motion for out-of-plane bending. This correlation effect on the potential energy surface of ADPO may be the reason why folded 8-P-3 ADPO has not been observed even when ADPO is released from an 8-P-4 complex under low-temperature conditions [26].

The geometry of folded ADPO (C_s) is significantly different from the planar C_{2v} geometry. The P—N bond increases in the C_s structure as compared to the C_{2v} structure by 0.075 Å and the P—O bonds decrease by an even larger 0.152 Å. The P—O bonds are now shorter than the P—N bonds, as would be expected for a normally bonded 8-P-3 structure based on covalent radii. The C—O bonds in the folded structure increase by 0.071 Å and are now like a normal C—O single bond. The C—N bonds also increase significantly by 0.04 Å. The C=C bonds shorten slightly in the folded structure, increasing their double-bond character. The folded structure clearly shows less delocalization in the backbone as compared to the planar structure. The bond angles change significantly in the folded structure. Both the phosphorus and the nitrogen are pyramidal. The O—P—N bond angles are about 5° less than the F—P—F bond angle in PF$_3$ but the O—P—O bond angle is about 11° larger than the F—P—F angle. The P—N—C bond angles are decreased somewhat from a tetrahedral value, whereas the C—N—C angle is somewhat opened-up from the tetrahedral value. Because of the presence of the five-membered ring, the O—C—C and N—C—C angles are somewhat smaller than those typically observed at C=C double bonds but the rings are less strained in the folded structure than in the planar structure.

The folded structure for ADAsO follows the same patterns as found for folded ADPO. The As—O bond distances decrease by 0.171 Å and the As—N distance increases by 0.053 Å. The remaining distance and angle changes follow the results obtained for ADPO.

3.5. Edge Inversion Processes at Main Group Centers

3.5.1. Types of Inversion Processes

As mentioned previously, the electronic arrangement depicted in structure **4c** would not be expected to be a stable ground state structure but rather a model of a transition state for a previously unrecognized inversion process at a pnictogen center, which we have called *edge inversion* [20]. The edge inversion process is illustrated in Figure 3.2 along with the more conventional vertex inversion process. The edge inversion process proceeds by inversion of the edges of a tetrahedron through its center to form a square planar transition state. The traditional vertex inversion process occurs by inverting a vertex and its opposite face through the center of a tetrahedron to form a trigonal planar transition state.

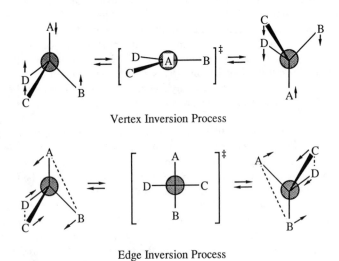

Vertex Inversion Process

Edge Inversion Process

Figure 3.2 ■ Illustrations of edge and vertex inversion processes. The gray spheres at the centers of the structures represent the central atoms.

3.5.2. Transition State Requirements

The overall outcome of the two inversion processes is the same but the reaction courses give rise to important differences. The vertex inversion process requires that one of the substituents (e.g., the group A in Figure 3.2) pass through the nucleus at the center of the tetrahedron. If this substituent is a lone pair of electrons, this is not a problem. If the group A were actually an atom, vertex inversion is not likely to occur. Edge inversion, on the other hand, is well suited to the inversion of both pyramidal three-coordinate centers like PF_3 and tetrahedral four-coordinate centers like SiF_4. The edge inversion process has been predicted [22, 24, 25] to occur at tetrahedral four-coordinate centers and has been demonstrated experimentally at a germanium center [27]. The discussion here will be limited to 8-Pn-3 pnictogen centers relevant to ADPnO chemistry.

For 8-Pn-3 centers the edge and vertex processes are preferred under opposing conditions. Vertex inversion is preferred when the substituents at the pnictogen center are electropositive, π-acceptors, and large (or the pnictogen center is small). Edge inversion is preferred when the substituents are electronegative, π-donors, and small (or the pnictogen center is large). The reasons for these preferences are apparent from examination of the two possible transition states. The transition state for vertex inversion will undergo a relief of steric compression as the geometry changes from pyramidal to trigonal planar. Thus, large substituents and small pnictogen centers favor this path. The edge-inversion transition state will cause more steric compression between the substituents as the geometry changes from pyramidal to T-shaped (pseudo-square planar) and smaller substituents or a larger pnictogen will thereby favor this pathway. The σ bonds at a pyramidal center undergoing vertex inversion will experience increased s-orbital contributions from the central pnictogen as the transition state is approached. This change from approximate sp^3 to sp^2 hybridization at the pnictogen center will be favored by more electropositive substituents. Conversely, as a pnictogen center undergoing edge inversion approaches its transition state, the in-plane σ bond become orbitally deficient (electron rich) as the formally vacant out-of-plane p orbital develops at the pnictogen center (see structure **4c**). This orbitally deficient bonding arrangement is stabilized by electronegative substituents. Finally, the π-acceptor/donor preferences for the substituents are the result of the need to stabilize an out-of-plane p-type lone pair of electrons at the vertex-inversion transition state and a vacant p orbital at the edge-inversion transition state. Among the pnictogen family, nitrogen would not be expected to invert by an edge inversion process due to its high electronegativity and small size. All of the remaining members of the family can be expected to invert by either an edge or vertex process depending on the nature of the substituents.

3.5.3. Demonstration of the Edge Inversion Process

The relationship between the edge inversion process and the ADPnO compounds has led to other discoveries concerning edge inversion processes. The degree of stabilization necessary to observe edge inversion processes has been studied for the pnictogen halides [20, 23, 28]. The fluorides have been predicted to prefer the edge inversion for P, As, and Sb, with decreasing barriers as one moves down the family. Saturated analogs of the ADPnO compounds are known to possess the folded 8-P-3 ground state. It has recently been shown that the barrier to edge inversion at the saturated ADPO analog **6** can be measured experimentally and is in excellent agreement with theory (vide infra).

6 $H_4 \cdot ADPO$ $H_2 \cdot ADPO$

In order to measure the edge inversion barrier for the first time, it was necessary to design a structure that would have an inversion barrier height appropriate for experimental measurement. For convenience we wanted to use NMR techniques (line shape analysis or magnetization transfer) to measure the inversion barrier. Reasoning that unsaturation in the five-membered rings had stabilized the planar form of ADPO, we performed calculations on the saturated compound, $H_4 \cdot ADPO$, for the planar and folded geometries. The planar structure was initially optimized in C_{2v} symmetry and was found to have three imaginary frequencies, one at $188i$ cm^{-1}, corresponding to the inversion motion, and two smaller ones at $84i$ and $34i$ cm^{-1}. These latter two frequencies correspond to torsions about the C—C bonds. We distorted the molecule along the $84i$ cm^{-1} direction of negative curvature and optimized the geometry as a transition state. This finally gave a structure (Figure 3.3) [29] that deviates slightly from

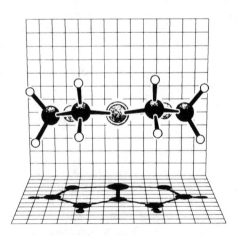

Figure 3.3 ■ KANVAS [29] drawing of quasi-planar $H_4 \cdot ADPO$.

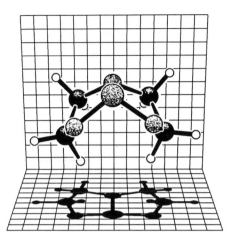

Figure 3.4 ■ KANVAS [29] drawing of ground-state $H_4 \cdot$ ADPO.

C_{2v} symmetry because of C—C torsions. We label this distorted structure as *quasi-planar* ('C_{2v}'). The folded structure was optimized in C_s symmetry and has the conformation shown in Figure 3.4 [29] for $H_4 \cdot$ ADPO. The quasi-planar structure is a transition state characterized by one negative direction of curvature (an imaginary frequency of $197i$ cm^{-1}). The folded structure is a true minimum and is 43.0 kcal mol^{-1} lower in energy than the transition state at the SCF level and is 27.9 kcal mol^{-1} lower at the MP-2 level. This suggested that the inversion barrier would be of the right size for an NMR measurement. The correlation correction to the barrier height is 15.1 kcal mol^{-1}, comparable to the value found for the edge inversion barrier in PF$_3$ but smaller than the stabilization found for ADPO. This is consistent with the lack of π-delocalization in the five-membered rings.

The geometries of the two $H_4 \cdot$ ADPO structures are given in Table 3.3 The ground state geometry (C_s) of $H_4 \cdot$ ADPO is clearly different from the geometry of 10-P-3 ADPO. The P—O bond distance in the quasi-planar form of $H_4 \cdot$ ADPO is significantly shorter than the bond in ADPO, indicating stronger interactions in the 3c,4e hypervalent bond. In fact the P—O bond in quasi-planar $H_4 \cdot$ ADPO is more than halfway to the bond distance in folded 8-P-3 ADPO (C_s). The P—N bond distance in quasi-planar $H_4 \cdot$ ADPO is shorter than the comparable quantity in planar 10-P-3 ADPO (C_{2v}). The C—O bonds in quasi-planar $H_4 \cdot$ ADPO are essentially single bonds and show none of the carbonyl character found in 10-P-3 ADPO. The C—N bonds are also longer in quasi-planar $H_4 \cdot$ ADPO as compared to 10-P-3 ADPO. One consequence of these results is that there are much smaller differences in the bond distances of the folded (C_s) and quasi-planar ('C_{2v}') structures of $H_4 \cdot$ ADPO as compared to ADPO, consistent with the presence of an inversion barrier in $H_4 \cdot$ ADPO. The differences in the bond angles of the folded and quasi-planar structures in $H_4 \cdot$ ADPO follow those in ADPO. In the folded structure the O—P—N bond angles are smaller than the O—P—O bond angle. The C—N—C bond angle in folded $H_4 \cdot$ ADPO is even larger than the same angle in folded ADPO, as are the P—N—C

Table 3.3 ▪ Calculated Geometries for Planar and Pyramidal $H_4 \cdot$ ADPO

Property	'C_{2v}' quasi-planar	C_s folded	Property	'C_{2v}' quasi-planar	C_s folded
	Bond dist. (Å)			Bond dist. (Å)	
P—O	1.716	1.634	C-4—H-4a	1.085	1.080
P—N	1.664	1.749	C-4—H-4b	1.083	1.081
N—C	1.458	1.476	C-3—H-3a	1.081	1.082
O—C	1.412	1.432	C-3—H-3b	1.082	1.078
C—C	1.544	1.539			
	Bond angles (deg.)			Bond angles (deg.)	
O—P—N	86.5	93.6	C-3—C-4—H-4a	112.1	112.5
O—P—O	172.4	106.0	C-3—C-4—H-4b	112.1	110.5
P—N—C	121.2	108.7	H-4a—C-4—H-4b	107.7	107.8
P—O—C	120.7	115.4	O—C-3—H-3a	109.8	109.8
C—N—C	117.1	115.9	O—C-3—H-3b	109.4	107.9
N—C—C	121.2	106.8	C-4—C-3—H-3a	111.9	112.1
O—C—C	106.0	104.5	C-4—C-3—H-3b	111.5	113.3
N—C-4—H-4a	109.8	110.6	H-3a—C-3—H-4b	108.3	109.1
N—C-4—H-4b	110.2	109.1			

angles. The bond angles at oxygen remain large but the angles at the carbons decrease. Since the carbons are nominally tetrahedral, strain can be relieved in the five-membered rings by decreasing these angles a few degrees.

The inversion barrier for 6 was determined by magnetization transfer experiments [21]. The diastereotopic protons on the carbons adjacent to nitrogen appear as an ABX pattern in the ^{1}H NMR spectrum. Selective inversion of the downfield group and subsequent analysis of the recovery profile led to an activation enthalpy ($\Delta H^{\ddagger}$) of 23.4 kcal mol^{-1}.

When comparing experimentally determined and calculated inversion barriers, an important consideration is the conformation of the two five-membered rings in the ground state. The five-membered rings in $H_4 \cdot$ ADPO (C_s) are both folded inwards (as shown in Figure 3.3) to adopt a bis-endo conformation. However, the structure chosen for experimental measurements (6) has adamantyl groups at the carbons bonded to the oxygens and clearly there will be a large steric interaction between the adamantyls in this conformation. Therefore, the molecule probably adopts a conformation with one

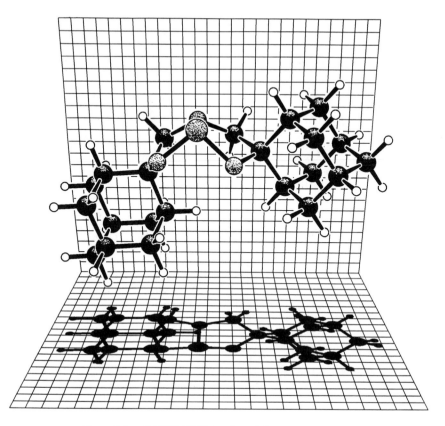

Figure 3.5 ■ KANVAS [29] drawing of the ground state of **6**.

ring flipped up. In order to test this hypothesis, we optimized structure **6** containing two adamantyl substituents at the STO-3G level. As shown in Figure 3.5 [29], **6** does indeed adopt the proposed conformation with one ring bent down and one ring up. The experimental value for the edge inversion barrier is 23.4 kcal mol^{-1} as compared to the calculated value of 27.9 kcal mol^{-1} for the unsubstituted compound. Part of this difference of 4.5 kcal mol^{-1} is due to the fact that the adamantyl-substituted molecule is in a less stable conformation (higher energy ground state), which will lower the barrier to inversion. Of course, adamantyl and hydrogen are not the same substituent so there may be 1–2 kcal mol^{-1} of an electronic substituent effect on the inversion barrier.

Complete saturation of the ligand backbone in the ADPnO molecules is clearly sufficient to change the preference for ground-state structure to the more classical 8-Pn-3 folded arrangement and thus allows the verification of the edge inversion process. The question of ground-state structure also has been investigated for compounds derived from ligands similar to **3** in which only a single ring is saturated. In order to obtain more understanding about how the saturation of the ring affects the structure of the molecule and the inversion barrier, we optimized the structure of the

Table 3.4 ▪ Calculated Geometries for Planar and Folded $H_2 \cdot ADPO$

Property	C_s planar	C_1 folded	Property	C_s planar	C_1 folded
	Bond dist. (Å)			Bond dist. (Å)	
P—O-2	1.852	1.665	C-6—C-7	1.546	1.544
P—O-8	1.693	1.622	C-4—H-4	1.066	1.067
P—N	1.664	1.756	C-3—H-3	1.068	1.066
N—C-4	1.396	1.433	C-6—H-6a	1.081	1.079
N—C-5	1.465	1.482	C-6—H-6b	1.081	1.079
O-2—C-3	1.317	1.385	C-7—H-7a	1.080	1.079
O-8—C-7	1.413	1.437	C-7—H-7b	1.080	1.078
C-3—C-4	1.349	1.332			
	Bond angles (deg.)			Bond angles (deg.)	
O-2—P—N	82.7	92.2	C-3—C-4—H-4	129.0	126.5
O-8—P—N	88.1	93.8	O-2—C-3—H-3	120.0	116.6
O—P—O	170.8	107.4	C-4—C-3—H-3	127.4	129.4
P—N—C-4	120.0	107.5	N—C-6—H-6a	109.5	110.0
P—N—C-6	119.9	109.3	N—C-6—H-6b	109.5	109.6
P—O-2—C-3	115.0	112.2	C-7—C-6—H-6a	112.2	111.6
P—O-8—C-7	120.2	116.4	C-7—C-6—H-6b	112.2	110.5
C—N—C	120.1	114.1	H-6a—C-6—H-6b	108.2	108.1
N—C-4—C-3	109.7	113.5	O-8—C-7—H-7a	109.2	109.3
N—C-6—C-7	105.1	107.0	O-8—C-7—H-7b	109.2	107.2
O-2—C-3—C-4	112.6	113.9	C-6—C-7—H-7a	111.6	112.7
O-8—C-7—C-6	106.7	105.6	C-6—C-7—H-7b	111.6	112.5
N—C-4—H-4	121.3	119.9	H-7a—C-7—H-7b	108.5	109.2

planar and folded forms of the half-saturated structure, $H_2 \cdot ADPO$, where one ring is saturated and one is unsaturated. The geometric parameters are given in Table 3.4. The molecule is found to be folded in the ground state. The inversion pathway proceeds through a transition state with a plane of symmetry (C_s) in contrast to the case of the fully saturated structure. The inversion barrier is 28.5 kcal mol^{-1} at the SCF level and is 9.1 kcal mol^{-1} at the MP-2 level. The inversion barrier is clearly lowered relative to $H_4 \cdot ADPO$ by including unsaturation in one ring, as would be expected. The correlation correction for the barrier for $H_2 \cdot ADPO$ is 19.4 kcal mol^{-1}, which is halfway between the correlation corrections found for $H_4 \cdot ADPO$ and ADPO. The difference in energy between the inversion barriers for $H_2 \cdot ADPO$ and $H_4 \cdot ADPO$ is 18.8 kcal mol^{-1}. This difference represents the stabilization of the planar transition state by the introduction of unsaturation in a single ring. Subtracting the 18.8 kcal mol^{-1} from the inversion barrier of $H_2 \cdot ADPO$ gives -9.7, which is an estimate of the stability of the planar form of ADPO (which has two unsaturated rings) over its folded 8-P-3 form. The actual calculated value is -13.9 kcal mol^{-1}, in agreement with this approximate value. This suggests that two unsaturated rings provide even more stabilization than if the effect of each ring was purely additive.

The structure of planar $H_2 \cdot ADPO$ has components of both ADPO and $H_4 \cdot ADPO$. The $P—O_2$ bond is longer in $H_2 \cdot ADPO$ than in ADPO, whereas the $P—O_8$ bond is shorter than the corresponding bond in quasi-planar $H_4 \cdot$ ADPO. Hence, the 3c,4e bond is quite asymmetric in planar $H_2 \cdot$ ADPO. The unsaturated five-membered ring has the longest P—O bond, so that the C—O bond in this ring becomes somewhat carbonyl-like. The P—N bond is like that in quasi-planar $H_4 \cdot$ ADPO, 0.024 Å shorter than the value in 10-P-3 ADPO. The remainder of the bonds are similar to those of the corresponding fragments in 10-P-3 ADPO and quasi-planar $H_4 \cdot$ ADPO. The bond angles behave in a similar fashion. The folded form of $H_2 \cdot$ ADPO behaves as expected from the preceding results for 8-P-3 ADPO and folded $H_4 \cdot$ ADPO. The saturated ring adopts the endo-conformation found in folded $H_4 \cdot$ ADPO. Another bent conformer, which had the ring bent exo, was partially optimized. This structure was not a minimum, as determined by second derivatives, and was 4.5 kcal mol^{-1} higher in energy at the SCF level and 5.7 kcal mol^{-1} higher at the MP-2 level as compared to the ground state. These values are consistent with the difference found between the calculated and experimental barrier heights for saturated ADPO systems and substantiate our steric/conformational model for the differences in the values.

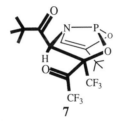

7

Structure **7** is the hexafluorobiacetyl adduct of 10-P-3 ADPO [4] and possesses the basic ring system found in $H_2 \cdot$ ADPO. The X-ray structure of **7** reveals the type of folded geometry depicted. However, because of the rather bulky substituents in the saturated ring of **7**, it may be argued that the observation of a folded 8-P-3 geometry is the result of steric compression that destabilizes the planar 10-P-3 arrangement.

3.6. 10-S-3 Analogs

The thiapentalenes are well-known compounds that are isoelectronic to ADPO. We have considered both the neutral compound (**8**) formed by substituting sulfur for phosphorus and carbon for nitrogen in ADPO as well as the cation (ADSO$^+$) formed by replacement of phosphorus in ADPO with sulfur. The geometrical parameters are given in Table 3.5. In contrast to ADPO, which shows a minimum for the planar C_{2v} structure at the SCF level, neither ADSO$^+$ nor **8** show such a minimum. Indeed, both structures are transition states at the SCF level with one negative direction of curvature. The motion along the negative direction of curvature is not out of the plane but is in the plane and is parallel to the 3c,4e hypervalent bond. Thus, the distortion is side-to-side and corresponds to formation of the resonance structures shown below. Such distortions are indeed well known in the asymmetrically substituted thiapentalenes. At the SCF level the C_{2v} structure for **8** is 3.9 kcal mol^{-1} above the C_s structure, but this preference is reversed at the MP-2 level and the C_{2v} structure is preferred over the C_s structure by 7.4 kcal mol^{-1}, a correlation effect of 11.3 kcal mol^{-1}. The

Table 3.5 ▪ Bond Lengths and Angles for ADSO$^+$ and 8 in C_s and C_{2v} Geometries

Property	8 (C_s)	8 (C_{2v})	9^a (C_{2v})	ADSO$^+$ (C_s)	ADSO$^+$ (C_{2v})
			Bond distances (Å)		
S—O-2	1.696	1.889	1.865	1.659	1.815
S—O-8	2.463	1.889	1.865	2.414	1.815
S—A^b	1.748	1.733	1.702	1.750	1.725
A—C-4	1.447	1.404	1.381	1.418	1.345
A—C-6	1.359	1.404	1.381	1.284	1.345
O-2—C-3	1.335	1.278	1.305	1.334	1.266
O-8—C-7	1.221	1.278	1.305	1.201	1.266
C-4—C-3	1.346	1.382	1.356	1.338	1.393
C-6—C-7	1.443	1.382	1.356	1.482	1.393
C-4—H-4	1.066	1.066	1.064	1.064	
C-6—H-6	1.069	1.066	1.070	1.064	
C-3—H-3	1.070	1.074	1.067	1.071	
C-7—H-7	1.085	1.074	1.078	1.071	
			Bond angles (degrees)		
O-2—S—A	92.0	85.8	87.0	90.4	84.1
O-8—S—A	77.3	85.8	87.0	76.0	84.1
O—S—O	169.4	171.7	173.9	166.4	168.3
S—A—C-4	108.2	114.8	113.9	109.7	116.8
S—A—C-6	123.0	114.8	113.9	124.8	116.8
S—O-2—C-3	111.8	111.4	115.0	113.6	114.9
S—O-8—C-7	101.9	111.4	115.0	105.0	114.9
C—A—C	128.9	130.4	132.2	125.5	126.4
A—C-4—C-3	111.4	110.9	113.4	110.7	110.2
A—C-6—C-7	116.8	110.9	113.4	117.1	110.2
O-2—C-3—C-4	116.7	117.0	115.0	115.6	114.0
O-8—C-7—C-6	121.0	117.0	115.0	117.1	114.0
A—C-4—H-4	123.5	123.9	118.7	120.6	121.7
C-3—C-4—H-4	125.1	125.3	127.9	128.7	128.1
A—C-6—H-6	121.9	123.9	118.7	120.2	121.7
C-7—C-6—H-6	121.3	125.3	127.9	122.6	128.1
O-2—C-3—H-3	115.4	118.2	116.2	116.6	120.3
C-4—C-3—H-3	127.9	124.7	128.8	127.8	125.8
O-8—C-7—H-7	120.2	118.2	116.2	123.7	120.3
C-6—C-7—H-7	118.8	124.7	128.8	119.2	125.8

^{a}Experimental data from the X-ray structure on the 3,7-dimethyl derivative of 8 [30].
^{b}A is used to designate generically the atom (C or N) in the 5 position.

experimentally determined structure [30] for 3,7-dimethyl-2,8-dioxa-1-thiabicyclo [3.3.0]octa-2,4,6-triene 9 is C_{2v}, in agreement with the correlated result. A similar result is found for ADSO$^+$, where the C_s structure is more stable than the C_{2v} structure at the SCF level by 6.6 kcal mol^{-1} but the C_{2v} structure is more stable at the MP-2 level by 14.5 kcal mol^{-1}. Here the effect of correlation is much larger, 20.1 kcal mol^{-1}, comparable to the value found in ADPO, although we note that the C_s structures are different in the two cases (laterally distorted for ADSO$^+$ but folded for ADPO).

8   **9**   **ADSO$^+$**

C_s **distorted 8** C_s **distorted ADSO$^+$**

Comparison of our calculated structure for **8** with the known experimental structure for the dimethyl analog **9** shows a different level of agreement relative to comparisons in the ADPO systems. The bond distances to the sulfur are all calculated to be longer than the experimental distances by less than 0.03 Å in contrast to the comparisons made for ADPO. The calculated C — C bond distances are also longer than the experimental distances by similar amounts, whereas the calculated C — O bond distances are shorter than the experimental values. These differences also show up in the bond angles, as would be expected, with differences between the calculated and experimental values of up to 4°. The structure of the C_s form shows a pronounced shortening of one S—O bond and a lengthening of the other S—O bond. In fact, the latter distance is so long that only a weak interaction exists between O-8 and the sulfur. However, there still is some interaction between O-8 and the sulfur, which constrains the system to remain planar. The bonds in the backbone also remain partially delocalized. The S—C bond length changes only slightly on distortion. The various C—C bond distances change in accord with the valence bond structure; the C-5 — C-4 bond increases and the C-5—C-6 bond decreases. The O-2—C-3 bond in the ring increases, whereas the O-8—C-7 decreases to almost the value of a carbonyl group. The C-4—C-3 bond in the ring decreases toward the value expected for a C=C, whereas the C-6—C-7 bond has significantly more C—C single bond character. The bond angles show the expected changes that correspond to the changes in the bond lengths.

The geometry of C_{2v} ADSO$^+$ is similar to that of ADPO. The S—O distances are comparable to the P—O distances, whereas the S—N bond is shorter than the P—N bond. The C—N and C—O bonds in ADSO$^+$ are shorter than the corresponding bonds in ADPO by 0.05 Å. As a consequence, the C—C bonds are longer in ADSO$^+$ by a comparable amount, 0.045 Å. Clearly, placement of a positive charge in the ring system changes how the backbone interacts with the main group element. The bond angles are quite similar between ADPO and ADSO$^+$. Comparison of the C_{2v} structure of ADSO$^+$ with that of **8** shows that the S—O bond distances are significantly shorter in ADSO$^+$ than in **8**. The S—N and S—C bond distances, however, are quite similar. The C—O and C—N distances are shorter in the cation as compared to **8** whereas the C—C distances are longer in the cation. The C_s distorted structure for ADSO$^+$ follows the distortions found in **8** to give the expected valence bond structure. The S—O-2 interaction is somewhat shorter in ADSO$^+$ as compared to **8**, consistent with the larger positive charge on the sulfur in the former. The C=O carbonyl bond is shorter in ADSO$^+$ as is the C=N bond as compared to the C-5=C-6 bond. The C-6—C-7 single bond is even longer in C_s ADSO$^+$ than in C_s **8**.

3.7. Calculated Charges and Electronic Properties

The atomic charges are given in Table 3.6 together with the molecular dipole moments (D). For ADPO, the charges show that the phosphorus is very positive in both the planar and folded structures. However, there is a significant transfer of negative charge (0.19e) to the phosphorus in the planar 10-P-3 structure as compared to the folded 8-P-3 structure. Most of this charge is transferred from the nitrogen, which loses 0.13e. Care must be taken to realize that the net result of charge transfer from nitrogen to phosphorus does not necessarily imply a P—N double bond. Indeed, similar electronic environments must be involved with the heavier members of the ADPnO series and N=As or N=Sb arrangements are not appealing π-bonds. As shown by the C—H group charges [calculated by adding the charge on the hydrogen(s) to the charge on the carbon], the C—H groups are quite positive with the group bonded to oxygen (more positive than the group bonded to nitrogen). After the phosphorus and nitrogen the largest change in going from the folded to the planar structure is at the carbon attached to nitrogen, which loses 0.03e. This accounts for most of the remaining charge transfer to the phosphorus.

The charge distribution for folded H$_4$·ADPO is very similar to the charge distribution found in 8-P-3 ADPO (C_s), further demonstrating the "normal" type of bonding found in the folded structures. On going to the quasi-planar 'C_{2v}' structure, there is very little change in the populations for H$_4$·ADPO. This is consistent with the inability of the five-membered rings to feed electron density to the phosphorus. Because there is no delocalized π system in the rings, the nitrogen alone cannot feed enough negative charge to the phosphorus to further stabilize the transition state. This again points out that isolated N—P interactions are not sufficient to give planar structures.

As would be expected for H$_2$·ADPO, the behavior of the charges falls between that of ADPO and H$_4$·ADPO. The phosphorus gains 0.10e in the planar transition state and the nitrogen loses 0.08e. Thus, the transition state is partially stabilized by the additional charge transfer and the inversion barrier is lowered.

The dipole moments (Table 3.6) reflect the changes in the charge distributions going from the folded to the planar structures. The dipole moment always increases in the planar form. In the planar form the dipole lies along the nitrogen–pnictogen vector, with its positive end at nitrogen as would be expected from consideration of structure **4a**. The largest dipole moment change is found in ADAsO, where the difference in planar and folded forms is 1.60D. The dipole change in ADPO is 1.23D. The larger change in the dipole of the ADAsO molecule is indicative of the higher degree of charge transfer from the planar ligand to the arsenic center relative to phosphorus. This again indicates that phosphorus returns electron density to the ligand by resonance interactions. For H$_2$·ADPO, the dipole moment change is 1.12D and the smallest change is found for H$_4$·ADPO, 0.65D. These latter changes reflect the poorer electron donating ability of the partially and fully saturated ligand backbones.

Table 3.6 ■ Atomic Charges (e) and Dipole Moments (D) for ADPnO, H$_2$ · ADPO, and H$_4$ · ADPO

Atom	Folded ADPO	Planar ADPO	Folded H$_2$ · ADPO	Planar H$_2$ · ADPO	Folded H$_4$ · ADPO	Quasi-planar H$_4$ · ADPO	Planar ADAsO	Folded ADAsO
Pn	1.15	0.96	1.16	1.06	1.16	1.14	0.93	1.21
O-2	-0.69	-0.68	-0.70	-0.70	-0.70	-0.71	-0.68	-0.72
O-8	-0.69	-0.68	-0.69	-0.71	-0.70	-0.71	-0.68	-0.72
N	-0.71	-0.58	-0.71	-0.63	-0.69	-0.68	-0.54	-0.70
C-3	0.08 (0.29)[a]	0.12 (0.32)	0.08 (0.29)	0.12 (0.31)	-0.03 (0.29)	-0.03 (0.29)	0.15 (0.34)	0.09 (0.30)
C-4	-0.03 (0.18)	-0.05 (0.17)	-0.05 (0.16)	-0.06 (0.15)	-0.19 (0.19)	-0.15 (0.21)	-0.06 (0.16)	-0.04 (0.16)
C-6	-0.03 (0.18)	-0.05 (0.17)	-0.18 (0.20)	-0.17 (0.21)	-0.19 (0.19)	-0.15 (0.21)	-0.06 (0.16)	-0.04 (0.16)
C-7	0.08 (0.29)	0.12 (0.32)	-0.04 (0.31)	-0.03 (0.19)	-0.03 (0.29)	-0.03 (0.29)	0.15 (0.34)	0.09 (0.30)
μ	2.94	4.17	3.75	4.87	4.45	5.10	3.59	1.99

[a] Group charges in parentheses.

The charges for 10-As-3 ADAsO are also shown in Table 3.6. Comparison with the charges of 10-P-3 ADPO shows that the arsenic has even more charge transferred to it from the backbone than does the phosphorus, with this additional charge coming from nitrogen and C-3 (and C-7). This increased charge separation occurs in spite of the electronegativity changes, which should operate in the opposite direction. As we mentioned previously, this effect arises from poorer π overlap in the arsenic case relative to phosphorus—thus diminishing charge return from the arsenic to the ligand. Folding ADAsO causes an even larger change in the charge at the central pnictogen and nitrogen than found in ADPO. Both the oxygens and the carbons bonded to them also show a change in charge. The C—O region becomes more negative on folding.

The charges for the sulfur analogs **8** and ADSO$^+$ are given in Table 3.7. The positive charge on the sulfur is larger in the C_{2v} form as compared to the C_s form. The negative charges on the oxygens increase going from the C_s to the C_{2v} structure. The most striking result in **8** is that C-5 (bonded to the sulfur) is so positive. This is an indication that the electron donation from the ligand π system to sulfur is very significant and reminiscent of effects in the ADPO systems (note that the hypothetical carbon-based analog of ligand **3** would introduce a carbanion line pair in this position rather than a nitrogen lone pair). Carbons C-4 and C-6 are now quite negative and the other carbons C-3 and C-7 are quite positive. In the C_s structure the charges are no longer equal in each "ring" but the pattern of charges is the same. The sulfur is more positive in ADSO$^+$ than in **8**, consistent with the positive molecular charge. However, there is only about one-half of the excess positive charge on the sulfur with the remainder spread over the backbone, mostly on the carbons, as the oxygens and nitrogen are both negative. The dipole moment for **8** in the C_s form is larger than in the C_{2v} form, opposite to the behavior of the structures in ADPO and its saturated analogs.

The electronic structure of the ADPnO molecules is governed by the interaction of the π orbitals on the ligand backbone with the out-of-plane p orbital on the central pnictogen and by the presence of the 3c,4e hypervalent O—Pn—O bond and its interaction with the lone pair on the pnictogen. The orbitals for the C_{2v} structures for

Table 3.7 ▪ Charges (e) and Dipole Moments (D) for 8 and ADSO$^+$

Atom	**8** (C_s)	**8** (C_{2v})	ADSO$^+$ (C_s)	ADSO$^+$ (C_{2v})
S	0.49	0.58	0.80	0.92
O-2	−0.59	−0.63	−0.54	−0.55
O-8	−0.56	−0.63	−0.44	−0.55
A-5	0.21	0.23	−0.35	−0.32
C-3	0.12 (0.37)	0.20 (0.42)	0.17 (0.47)	0.23 (0.52)
C-4	−0.35 (−0.12)	−0.44 (−0.19)	−0.05 (0.24)	−0.07 (0.23)
C-6	−0.45 (−0.22)	−0.44 (−0.19)	0.04 (0.33)	−0.07 (0.23)
C-7	0.23 (0.40)	0.20 (0.42)	0.24 (0.49)	0.23 (0.52)
H-3	0.25	0.22	0.30	0.29
H-4	0.23	0.25	0.29	0.30
H-6	0.23	0.25	0.29	0.30
H-7	0.17	0.22	0.25	0.29
μ	4.32	3.57		

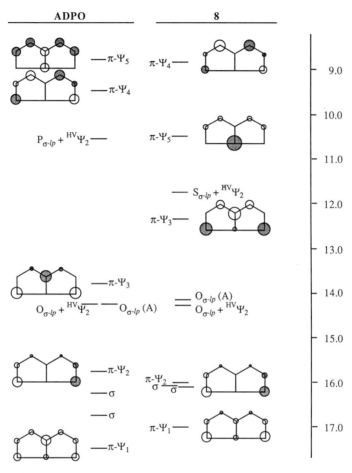

Figure 3.6 ■ The top ten occupied molecular orbitals for ADPO and **8**. The orbital labels represent the contributions from the ligand π system (π-Ψ_n) as given in [4], the σ system, or the hypervalent orbitals ($^{HV}\Psi_n$) as depicted in Figure 3.1. Energy is in eV.

ADPO and **8** are given in Figure 3.6. The orbitals for ADPO are essentially the same as those given previously [4]. The larger basis set makes the HOMO $\Psi_5 + P_p$ instead of Ψ_4 as found with a minimum basis set [4]. However, the two orbitals are very similar in energy. The coefficients of the orbitals in the HOMO have the largest values evenly spread over the phosphorus, nitrogen, and the carbons. This HOMO represents the "delocalized π lone pair" at phosphorus (see summary below). The NHOMO has its largest coefficients on the carbons 3 and 7 followed by the oxygens. The third highest molecular orbital is a σ–lone-pair orbital with the largest component in the s orbital on the phosphorus. There is also some p-orbital character on the phosphorus from the p orbital that lies along the P—N bond. This s–p mixing at the phosphorus center occurs in such a way as to distort this σ lone pair away from the P—N bond. The Ψ_2 level of the hypervalent bond (Figure 3.1) has the proper symmetry to mix with this

σ–lone-pair orbital and does. The next orbital is primarily a ligand-centered π orbital that is symmetric about the P—N axis. This π orbital is followed by two σ orbitals that are predominantly in-plane lone pairs on the oxygens. The lower one of these, which is symmetric about the P—N axis, also mixes a small amount with Ψ_2 of the hypervalent localized molecular orbitals (Figure 3.1). Because of the extensive amount of mixing, we were unable to identify an orbital that shows substantial character from Ψ_1 of the hypervalent set (Figure 3.1). It is important to note that this orbital description is for the SCF wavefunction. Of course there is a low-lying LUMO (-1.58 eV) that is a π orbital with its largest coefficient on the phosphorus. This orbital can mix with the in-plane lone pair in a generalized valence bond or two-configuration sense just as found for the lone pair in ^{1}A CH$_2$ [31]. Thus, it is not surprising that we find a large correlation effect for this system. Indeed, it is the likely interaction of the in-plane lone pair with this π^* orbital that keeps enough density in it (in conjunction with delocalization of the backbone π orbitals) so that this center does not show large reactivity with electron pair donors.

It seems surprising that the phosphorus is so positive in this structure considering the apparent extent of the participation of the phosphorus in the π orbitals. However, as shown in Table 3.8, the phosphorus is clearly electron deficient, with the largest deficiency coming in the p orbital that participates in the hypervalent bond (p_x), only 0.34e. Even though there is as much population in the p π-orbital (p_y) as in the p orbital involved in the in-plane lone pair and the P—N σ-bond (p_z), these orbitals still do not even have one electron each in them.

The top three occupied orbitals of the folded C_s structure map onto the three highest orbitals of planar ADPO. The energies of these orbitals are 9.18, 10.08, and 11.81 eV, for comparison. The LUMO is, of course, much more destabilized in the folded structure at -4.2 eV and the correlating effect of this orbital will be quite different, as found.

The orbitals for **8** are similar to those for ADPO although there are some differences. The HOMO and NHOMO are reversed. The splitting between the HOMO and NHOMO is much larger in **8** than in ADPO. The orbital coefficients are somewhat different in **8** as compared to ADPO. The largest coefficient in the π-HOMO of **8** is clearly on the sulfur whereas in ADPO the density was equally spread over the phosphorus, nitrogen, and carbon centers. The remaining ordering of the orbitals and the basic shapes follow those of ADPO. Clear evidence for the difference between

Table 3.8 ■ Orbital Populations for Planar Phosphorus, Sulfur, and Arsenic Systems (Mulliken Analysis, e)

Orbital	ADPO	$8\,(C_{2v})$	$8\,(C_s)$	ADSO$^+$	H$_2\cdot$ADPO	H$_4\cdot$ADPO	ADAsO
s	1.90	1.94	1.91	1.95	1.87	1.86	
$p_x{}^a$	0.34	0.45	0.54	0.48	0.38	0.39	0.33
$p_y{}^b$	0.80	1.72	1.80	1.67	0.59	0.41	0.95
$p_z{}^c$	0.78	1.13	1.13	0.80	0.83	0.87	0.71
d	0.25	0.14	0.12	0.20	0.29	0.34	

$^a p_x$ is along the 3c,4e bond.
$^b p_y$ is perpendicular to the molecular plane.
$^c p_z$ is along the N—P (S or As) bond.

ADPO and **8** can be seen in the total orbital populations in Table 3.8. The sulfur has significantly more electron density than does the phosphorus. Although there is only 0.11 extra electron density in the p_x orbital, there is almost an additional electron in the sulfur p_y (π orbital). Thus, this orbital has a large lone-pair component to it. The other in-plane p_z orbital has somewhat more density in it on sulfur than does the same orbital on phosphorus. This difference in the orbital populations of the p π-orbital is consistent with the observed distortions at the SCF level. The phosphorus supports only one lone pair so it releases some of its p π-density back to the backbone and folds. For **8**, the sulfur clearly can support two lone pairs (one σ and one π) and it undergoes an in-plane distortion in order to maintain this electron distribution and minimize repulsions in the electron rich σ system. Thus, the orbital populations provide us with some insight into the different distortions. For the planar 10-Pn-3 ADPnO some planar lateral distortion may be observed if unsymmetrically substituted analogs are prepared but sufficient electron density must always be supplied by the ligand to maintain a planar geometry.

The orbitals for ADSO$^+$ are essentially the same as those for **8**. The only real difference besides changes in the coefficients is that there are two σ orbitals between Ψ_3 and Ψ_4. The populations on the sulfur are quite similar to the populations in **8** except for the p_z population, which is the same as that in ADPO. This is due to the fact that the nitrogen in ADSO$^+$ is more electronegative than the carbon (C-5) in **8** and, consequently, the σ bond is more polarized to the nitrogen. The remaining populations are essentially the same as those in **8** and as calculated the distortion in ADSO$^+$ at the SCF level should be the same as that for **8**.

3.8. Summary

With the above transition-state and ground-state structures in mind, it is now clear that the 8-Pn-3 ADPnO molecules would be strongly predisposed to invert through an edge inversion transition state. In fact, the stabilization of the transition state along this pathway is so great that the planar structure actually becomes a minimum (10-Pn-3 ADPnO) on the inversion surface.

Stabilization of the planar 10-Pn-3 ADPnO molecules has occurred by first transferring electron density from the ligand backbone to the out-of-plane orbital at the pnictogen center. Transfer of electron density in this fashion allows the transition state model, **4c**, to more closely approach structure **4a** and allows the planar structures to become minima on the potential surfaces for folding. If even one of the five-membered rings is saturated, the ligand backbone is no longer able to support the planar structure at phosphorus (e.g., folded $H_2 \cdot ADPO$). Although partial or complete saturation of the ADPO backbone does eliminate the minimum on the potential surface at the planar structure, the planar structure (transition state) still enjoys stabilization relative to edge inversion transition states with less effective π-donating substituents (e.g., PF_3 [20]). The unsaturated ligand backbone from **3** is a good enough reducing agent to transfer electron density to arsenic and antimony centers, which are even more electropositive than phosphorus. Thus, 10-As-3 ADAsO and 10-Sb-3 ADSbO are also stable molecules. There is a π-delocalization (resonance) effect that serves to return electron density from the central pnictogen out-of-plane p orbital to the ligand backbone. This back-donation actually has a destabilizing effect on the planar systems. This π-delocalization is most pronounced at phosphorus, which has the best π overlap with its oxygen and nitrogen neighbors. In fact, this effect is so large at phosphorus that planar ADPO is not the global minimum on the folding surface at the SCF level. Fortunately, direct electron transfer from the ligand backbone to the phosphorus center is not the only mechanism by which electron density can be placed in the out-of-plane pnictogen p orbital. Configuration mixing also serves to transfer electron density from the in-plane σ system, which is electron rich, to the out-of-plane pnictogen p orbital. This $\sigma-\pi$ transfer also has a stabilizing effect on the electron-rich σ system that contains the hypervalent bond, the Pn—N σ bond and σ lone pairs. This correlation effect is not sufficient to allow $H_2 \cdot ADPO$ or $H_4 \cdot ADPO$ to be minima at their planar geometries but does reduce greatly the barrier to edge inversion for these molecules. Because arsenic does not suffer the large π-delocalization (destabilizing) effects seen for phosphorus, planar ADAsO is the global minimum on the folding surface even at the SCF level.

Because the planar structure for ADPO was observed as a minimum due to correlation, loss of this effect gives rise to a folding of the structure to a more conventional 8-P-3 derived arrangement. Thus, when phosphorus is coordinated to an electrophilic center so that a phosphorus lone pair becomes involved in a bond to the electrophile, only the tetrahedral 8-P-4 adducts are observed (as with transition metals) [4, 26]. This electronic effect operates in concert with the geometric preference of phosphorus for tetrahedral and TBP geometries over their Ψ-TBP and Ψ-octahedral counterparts (vide supra). By contrast, both ADAsO and ADSbO give adducts with electrophiles in which both central pnictogen lone pairs are evident and Ψ-TBP and Ψ-octahedral geometries are observed [4, 32].

The chemistry of the ADAsO and ADSbO systems clearly demonstrates the presence of two lone pairs of electrons at the central pnictogen. For ADPO the situation is more complicated. We have previously suggested [4] that the phosphorus in ADPO can be best characterized as possessing a σ lone pair and a delocalized π lone pair. From the above discussions, which conceptually were directed at stabilization of **4c**, it is clear that there is some out-of-plane electron density at phosphorus that is necessary to obtain a stable planar structure. It is also evident that this out-of-plane density is extensively delocalized back into the ligand backbone (a smaller effect at arsenic and antimony centers). Thus, the second (π) lone pair is more localized at arsenic than at phosphorus and most localized at antimony. Because the delocalization of the second lone pair at phosphorus is so extensive, it may not be entirely proper to refer to the out-of-plane density as a lone pair. If this "lone pair" π orbital is viewed as being constructed from Ψ_5 of the ligand [4] and the pnictogen out-of-plane p orbital, then one could represent it as $\{\lambda\Psi_5 + (1 - \lambda)Pn_p\}$. When λ assumes a value of zero, then the molecular orbital is accurately characterized as a pnictogen π lone pair (localized). ADSbO is the best example of this extreme model. When λ is 1, we have an extreme in which no electron density is transferred to the pnictogen center and the planar structures are not stable. In this case the ligand folds and 8-Pn-3 structures like $H_2 \cdot ADPO$ and $H_4 \cdot ADPO$ are observed. For all intermediate values of λ there is some degree of mixing or delocalization. For this range of values it is not possible to say at what point the term "lone pair" becomes appropriate. Indeed, this is a continuum and the problem of assigning a label is semantic. We do note, however, that the observation of a stable species like planar 10-P-3 ADPO, which is a global energy minimum on the folding surface, provides a break in the continuum for some value of λ. At this break point, the similarity of the planar structure with the idealized pnictandiide, 4a suggests it may be appropriate to use the label "lone pair." Although, as we have stressed previously, it should be remembered that this is a *delocalized* "lone pair" with significant contributions from the atoms in the ligand backbone.

Isoelectronic molecules like ADSO$^+$ and 8 provide further insight into the bonding and structure of the ADPnO series. Like the ADPnO molecules these sulfur-based systems are planar molecules with the 10-X-3 bonding arrangement. The increased electronegativity of sulfur relative to phosphorus makes it easier to transfer electron density to the out-of-plane sulfur p orbital (and more difficult to remove it by resonance). This makes the planar structures strongly preferred over folded structures. This also reduces the stabilization that can be obtained by configuration mixing. This would be expected because the out-of-plane p_y orbital at the sulfur already bears substantial electron density (Table 3.8). The sulfur analogs, ADSO$^+$ and 8 show a lateral distortion at the SCF level that is small enough to be eliminated by a correlation correction. It is likely that configuration mixing stabilizes the electron-rich σ system as we suggested for the pnictogen molecules.

Appendix: Calculational Details

The calculations were done with the program GRADSCF [33] on CRAY 1A and X-MP computer systems. The geometries were optimized using gradient techniques [34] by minimizing the energy in the appropriate symmetry. If symmetry alone was not good enough to define the transition state, the geometry was optimized using sigma optimization techniques where the sum of the squares of the gradient components was minimized [35]. Force fields were calculated at the optimum SCF geometries, using the rapid analytic second derivatives [36] incorporated in GRADSCF. Correlation corrections were done at the SCF geometries at the MP-2 level [37] incorporating only the valence electrons. The calculations were done with the following basis sets: The basis set for the sulfur is the split valence basis sets of McLean and Chandler [38] augmented by a set of d functions ($\zeta = 0.60$). The basis set for phosphorus is the double-ζ basis set of Dunning and Hay [39] augmented by a set of d functions ($\zeta = 0.50$). The basis set for the arsenic is derived from Dunning's compilation and follows our previous work on AsH_3. The basis set has the form $(14s11p6d)/[8s6p3d]$. The first seven s orbitals are contracted together as are the first six p orbitals and the first four d orbitals. The remaining orbitals are allowed to vary freely. All of the atoms bonded to the main group V element had basis sets of double-ζ quality augmented by a set of d polarization functions. The remaining atoms had double-ζ basis sets. The basis sets for these atoms (O, N, C, H) in the ADPO calculations were taken from Dunning and Hay. The basis sets for O, N, C, and H in the sulfur systems are taken from Dunning [40].

References

1. Culley, S. A.; Arduengo, A. J., III. *J. Am. Chem. Soc.*, **1984**, *106*, 1164.
2. Culley, S. A.; Arduengo, A. J., III. *J. Am. Chem. Soc.*, **1985**, *107*, 1089.
3. Stewart, C. A.; Harlow, R. L.; Arduengo, A. J., III. *J. Am. Chem. Soc.*, **1985**, *107*, 5543.
4. Arduengo, A. J., III; Stewart, C. A.; Davidson, F.; Dixon, D. A.; Becker, J. Y.; Culley, S. A.; Mizen, M. B. *J. Am. Chem. Soc.*, **1987**, *109*, 627.
5. Arduengo, A. J., III. U.S. Patent 4,710,576, Dec. 1, 1987.
6. Pauling, L. *General Chemistry*, 3d ed., W. H. Freeman and Co., San Francisco, 1970, Chapters 6–7, pp. 148–256 (and references therein).
7. Cowley, A. H. *Acc. Chem. Res.*, **1984**, *17*, 386.
8. Holmes, R. R. *Acc. Chem. Res.*, **1979**, *12*, 257.
9. The N-X-L nomenclature system has been previously described: N valence electrons about an atomic center X, with L ligands (Perkins, C. W.; Martin, J. C.; Arduengo, A. J., III; Lau, W.; Algeria, A.; Kochi, J. K. *J. Am. Chem. Soc.*, **1980**, *102*, 7753).
10. (a) Granoth, I.; Martin, J. C. *J. Am. Chem. Soc.*, **1978**, *100*, 7434 (and references therein). (b) *J. Am. Chem. Soc.*, **1979**, *101*, 4623.
11. Lochschmidt, S.; Schmidpeter, A. *Z. Naturforsch*, **1985**, *40b*, 765.
12. (a) Gleiter, R.; Gygax, R. In *Topics in Current Chemistry*, Vol. 63, Springer-Verlag, Berlin, 1976, p. 49. (b) Lozac'h, N. In *Advances in Heterocyclic Chemistry*, Vol. 13, Academic Press, New York, 1971, p. 161.
13. Arduengo, A. J., III; Burgess, E. M. *J. Am. Chem. Soc.*, **1977**, *99*, 2376.
14. (a) Martin, J. C.; Perozzi, E. F. *Science*, **1976**, *191*, 154. (b) Martin, J. C. In *Topics in Organic Sulfur Chemistry*, Tisler, M., ed., Ljubljana, Yugoslavia, 1978, Chapter 8, p. 187.
15. The "ADPO" acronym is used for simplicity in place of the name of the bicyclic ring system it represents: 5-*a*za-2,8-*d*ioxa-1-*p*hosphabicyclo-[3.3.0]octa-2,4,6-triene. See [4] for detailed information on this acronym system.
16. Pauling, L. *The Nature of the Chemical Bond*, 3d ed., Cornell University Press, Ithaca, New York, 1960, pp. 8, 9 (and references therein).
17. Stewart, C. A.; Calabrese, J. C.; Arduengo, A. J., III. *J. Am. Chem. Soc.*, **1985**, *107*, 3397.
18. Bettermann, G.; Arduengo, A. J., III. *J. Am. Chem. Soc.*, **1988**, *110*, 887.
19. Bettermann, G.; Bublak, W.; Breker, J.; Kline, M.; Arduengo, A. J., III, unpublished results.
20. Dixon, D. A.; Arduengo, A. J., III; Fukunaga, T. *J. Am. Chem. Soc.*, **1986**, *108*, 2461.
21. Arduengo, A. J., III; Dixon, D. A.; Roe, D. C. *J. Am. Chem. Soc.*, **1986**, *108*, 6821.
22. Dixon, D. A.; Arduengo, A. J., III *J. Phys. Chem.*, **1987**, *91*, 3195.
23. Dixon, D. A.; Arduengo, A. J., III *J. Am. Chem. Soc.*, **1987**, *109*, 338.
24. Dixon, D. A.; Arduengo, A. J., III *J. Chem. Soc., Chem. Commun.*, **1987**, 498.
25. Arduengo, A. J., III *Pure Appl. Chem.*, **1987**, *59*, 1053.
26. Arduengo, A. J., III; Stewart, C. A., Davidson, F. *J. Am. Chem. Soc.*, **1986**, *108*, 332.
27. Arduengo, A. J., III; Dixon, D. A.; Roe, D. C.; Kline, M. *J. Am. Chem. Soc.*, **1988**, *110*, 4437.
28. Clotet, A.; Rubio, J.; Illas, F. *J. Molec. Struct. (THEOCHEM)*, **1988**, *164*, 351.

29. The structure drawings are made with the KANVAS computer graphics program. This program is based on the program SCHAKAL of E. Keller (Kristallographisches Institut der Universität Freiburg, FRG), which was modified by A. J. Arduengo, III (E. I. du Pont de Nemours & Co., Wilmington, DE) to produce the back and shadowed planes. The planes bear a 50-pm grid and the lighting source is at infinity so that shadow size is meaningful.

30. Dalseng, M.; Hansen, L. K.; Hordvik, A. *Acta Chem. Scand. Ser. A.*, **1981**, *35*, 645.

31. Bauschlicher, C. W., Jr.; Schaefer, H. F., III; Bagus, P. S.; *J. Am. Chem. Soc.*, **1977**, *99*, 7106.

32. Arduengo, A. J., III; Stewart, C. A. *Inorg. Chem.*, **1986**, *25*, 3847.

33. GRADSCF is an ab-initio gradient program system designed and written by A. Komornicki at Polyatomics Research.

34. (a) Komornicki, A.; Ishida, K.; Morokuma, K.; Ditchfield, R.; Conrad, M. *Chem. Phys. Lett.*, **1977**, *45*, 595. (b) McIver, J. W., Jr.; Komornicki, A. *Chem. Phys. Lett.*, **1971**, *10*, 303. (c) Pulay, P., in *Applications of Electronic Structure Theory*, Schaefer, H. F., III, ed., Plenum Press, New York, 1977, p. 153.

35. McIver, J. W., Jr.; Komornicki, A. *J. Am. Chem. Soc.*, **1972**, *94*, 2625.

36. (a) King, H. F.; Komornicki, A. In *Geometrical Derivatives of Energy Surfaces and Molecular Properties*, Jorgenson, P., Simon, S. J., eds., *NATO ASI Series C*, Vol. 166, D. Reidel, Dordrecht, 1986, p. 207. (b) King, H. F.; Komornicki, A. *J. Chem. Phys.*, **1986**, *84*, 5645.

37. (a) Møller, C.; Plesset, M. S. *Phys. Rev.*, **1934**, *46*, 618. (b) Pople, J. A.; Binkley, J. S.; Seeger, R. *Int. J. Quantum Chem. Symp.*, **1976**, *10*, 1.

38. McLean, A. D.; Chandler, G. S. *J. Chem. Phys.*, **1980**, *72*, 5639.

39. Dunning, T. H., Jr.; Hay, P. J. In *Methods of Electronic Structure Theory*, Schaefer, H. F., III, ed.; Plenum Press, New York, 1977, Chapter 1.

40. Dunning, T. H., Jr. *J. Chem. Phys.*, **1970**, *53*, 2823.

New Aspects of Organosulfur and Selenium Chemistry: The Selenosulfonation Reaction

Thomas G. Back, Kurt Brunner, M. Vijaya Krishna, Enoch K. Y. Lai, and K. Raman Muralidharan

Department of Chemistry
University of Calgary
Calgary, Alberta, Canada, T2N 1N4

4.1. Introduction

Synthetic methodology based on unsaturated sulfones has earned a prominent place in the repertoire of most organic chemists [1]. These compounds include vinylic, acetylenic, and allenic sulfones (**1–3**, respectively) that can be used as dienophiles or dipolarophiles in cycloadditions, as Michael acceptors with a wide variety of nucleophiles, and in many other applications. Some examples are shown in Figure 4.1. Furthermore, the sulfone moiety can be reductively removed from the product or employed for the attachment of additional substituents at the α-position via the corresponding sulfone-stabilized carbanion. Consequently, new methods for the preparation of unsaturated sulfones are in continuous demand.

For the past several years we have been studying the use of selenosulfonates (ArSO$_2$SePh, **4**) for this purpose. These compounds are stable, odorless, crystalline solids that were initially reported by Foss [2] in 1947, but were then almost completely ignored until quite recently. They are easily prepared by the oxidation of sulfonhydrazides [3] or sulfinic acids [4a, 4c] with benzeneseleninic acid:

$$\text{ArSO}_2\text{NHNH}_2 + \text{PhSeO}_2\text{H} \longrightarrow \text{ArSO}_2\text{SePh} + \text{H}_2\text{O} + \text{N}_2$$
$$2\text{ArSO}_2\text{H} + \text{PhSeO}_2\text{H} \longrightarrow \text{ArSO}_2\text{SePh} + \text{ArSO}_3\text{H} + \text{H}_2\text{O}$$

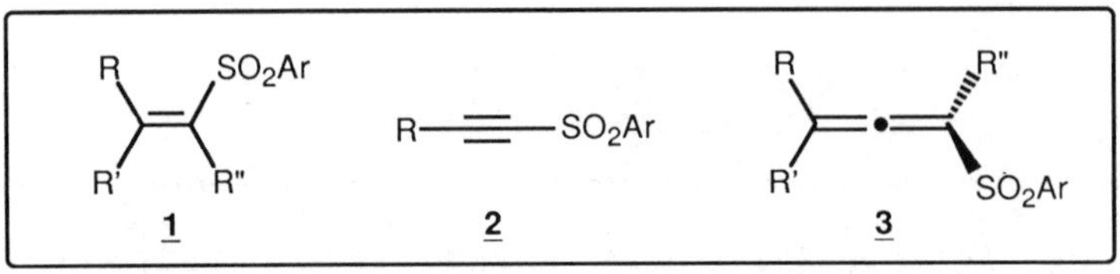

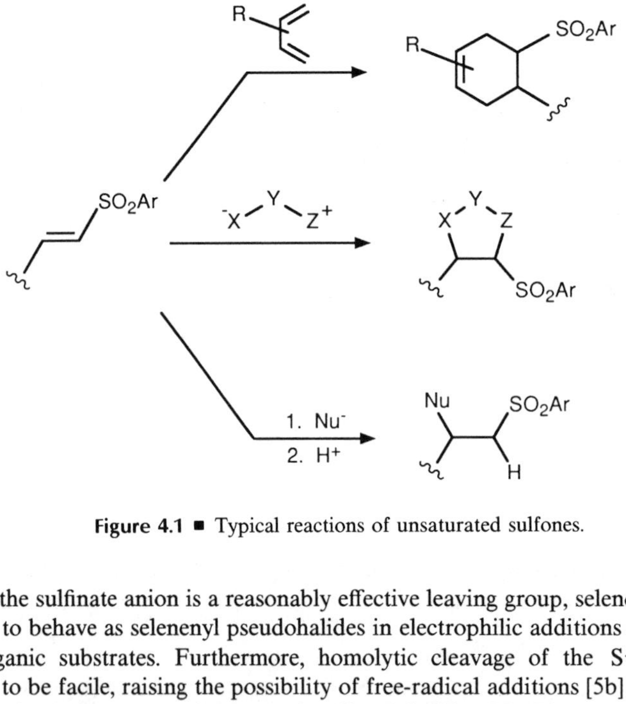

Figure 4.1 ■ Typical reactions of unsaturated sulfones.

Since the sulfinate anion is a reasonably effective leaving group, selenosulfonates are expected to behave as selenenyl pseudohalides in electrophilic additions [5a] to unsaturated organic substrates. Furthermore, homolytic cleavage of the S—Se bond is expected to be facile, raising the possibility of free-radical additions [5b] to appropriate substrates in a manner reminiscent of sulfonyl halides [6] (Figure 4.2). These 1,2-additions, collectively named "selenosulfonations," then afford an efficient and versatile route to various unsaturated sulfones when used in conjunction with selenoxide eliminations.

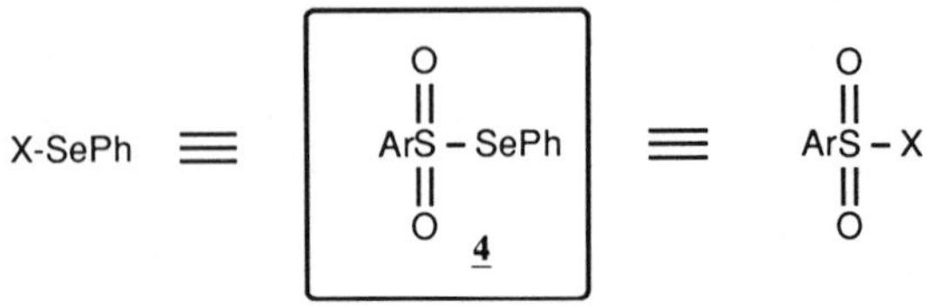

Figure 4.2 ■ Selenosulfonates as selenenyl and sulfonyl pseudohalides.

4.2. Selenosulfonation of Olefins and Allenes

Selenosulfonations of olefins were reported independently by our group [7] and by Kice and co-workers [4b, 4c]. Selenosulfonates add to olefins by an electrophilic mechanism in the presence of boron trifluoride etherate [7]. These reactions are stereospecific (anti) and regioselective, as shown in Figure 4.3, and presumably involve bridged seleniranium ion intermediates. On the other hand, free-radical additions of selenosulfonates can be effected by thermolysis in refluxing chloroform or benzene in the presence of initiators, such as AIBN [7], or by photolysis [4c, 8]. The free-radical processes are nonstereospecific, except with cyclic olefins such as cyclohexene, and afford adducts that are regioisomers of those obtained under electrophilic conditions. Selenoxide eliminations of the respective adducts then afford complementary vinyl sulfone isomers **5** and **6**. A number of synthetic applications of vinyl sulfones obtained in this manner have been reported by other workers [9]. Similar 1,2-additions of selenosulfonates to allenes [10] and to conjugated dienes [7,11] lead to 2-sulfonyl allylic alcohols **7** and 2-sulfonyl-1,3-dienes **8** after [2,3]-sigmatropic rearrangement or syn-elimination, respectively, of the corresponding selenoxides. These processes are also depicted in Figure 4.3.

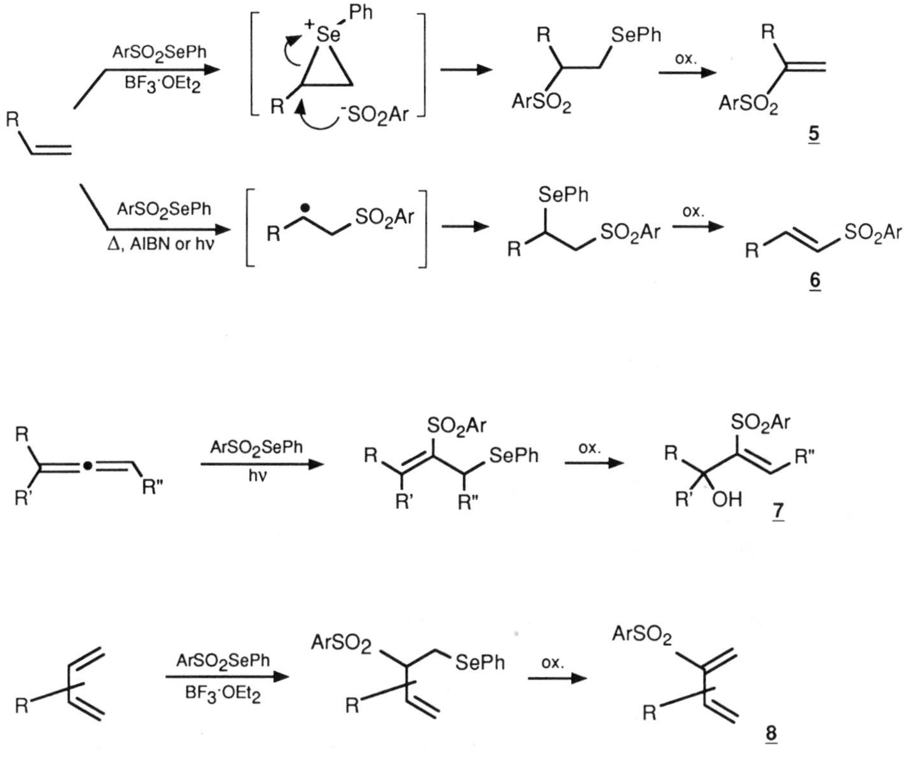

Figure 4.3 ■ Selenosulfonation of olefins and allenes.

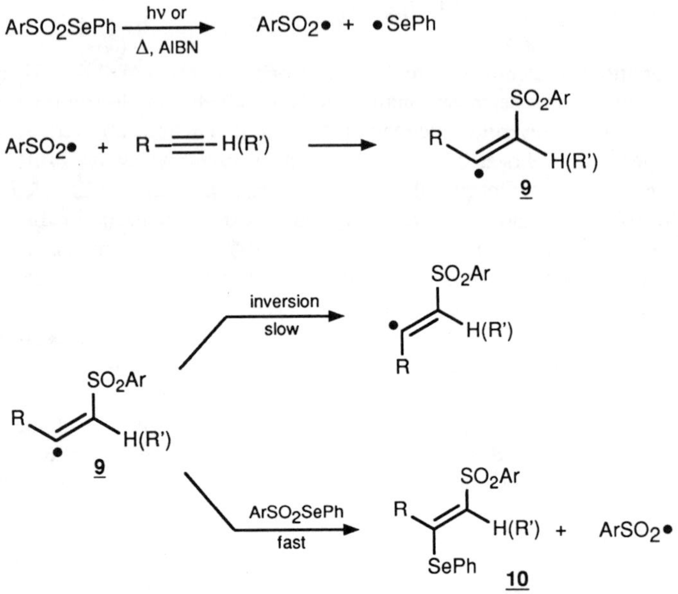

Figure 4.4 ■ Selenosulfonation of acetylenes.

4.3. Selenosulfonation of Acetylenes

Acetylenes fail to react with selenosulfonates under Lewis-acid–catalyzed conditions, but both monosubstituted and disubstituted derivatives undergo efficient free-radical additions [12,13] to afford β-(phenylseleno)vinyl sulfones **10**. These reactions are highly stereoselective and regioselective, and may be carried out thermally in the presence of AIBN or photochemically. Yields are typically in the 80–90 % range for monosubstituted acetylenes and slightly lower for disubstituted ones. The process proceeds by the chain mechanism shown in Figure 4.4 [12] and is initiated by homolysis of the S—Se bond of the selenosulfonate, followed by addition of the resulting sulfonyl radical to the acetylene. In the case of monosubstituted acetylenes, attack occurs only at the terminal carbon atom, producing the more substituted vinyl radical intermediate. The exclusive formation of the anti adduct **10** can be explained [12] by a chain-transfer step that is fast compared to relatively slow inversion in the intermediate vinyl radical **9**.

4.4. Substitution Reactions of β-(Phenylseleno)vinyl Sulfones

As indicated above, the adducts **10** are readily available from the selenosulfonation of acetylenes. They are capable of acting as Michael acceptors by virtue of the vinyl sulfone moiety. Furthermore, the phenylseleno group can function as a leaving group, which can be activated by oxidation to the corresponding selenoxide. Thus, selenides

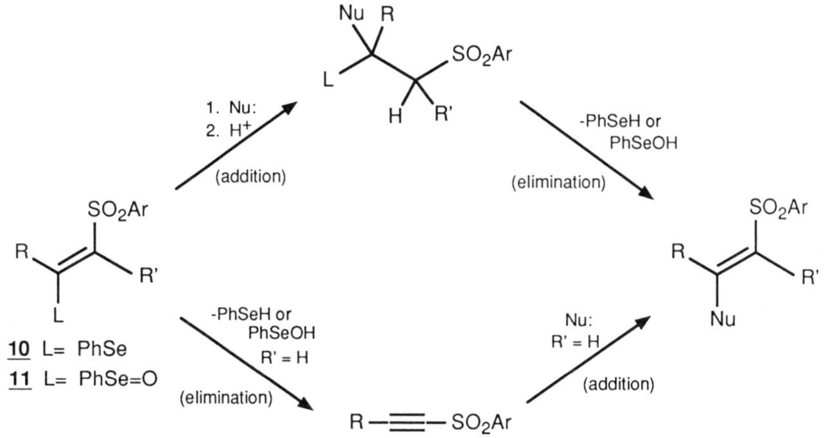

Figure 4.5 ▪ Addition–elimination and elimination–addition reactions of selenosulfonation adducts **10** and selenoxides **11**.

10 or selenoxides **11** react with nucleophiles by addition–elimination or, alternatively, by elimination–addition sequences (Figure 4.5), which depend upon the precise reactants and conditions. In either case, overall substitution of the phenylseleno group by the nucleophile results.

4.4.1. With N and O Nucleophiles

Selenoxides **11** react with amines to afford enamine sulfones [14]. The products undergo further selenenylation unless the by-product PhSeOH is removed by an appropriate scavenger such as ethyl vinyl ether. Alcohols produce enol ethers or ketals (by a second Michael addition to the enol ethers) of β-keto sulfones, and hydrolysis of these products, or of selenides **10**, affords the parent β-keto sulfones [12a]. Representative examples are shown in Figure 4.6.

4.4.2. With Carbon Nucleophiles

A wide variety of carbon nucleophiles effect substitution with selenoxides **11** [15]. These include cyanide ion, active methylene compounds, and the lithium derivatives of dithiane and trimethylsilylpropyne. In most cases, allylic sulfones rather than vinyl sulfones are produced. Examples of these reactions are also shown in Figure 4.6. They demonstrate that diverse types of synthetically useful sulfones are readily available from common acetylenes via the selenosulfonation approach.

4.4.3. With Organocuprates

An especially useful variation of this theme involves the reaction of selenides **10** with organocuprates [16]. Stereospecific substitution of the selenium moiety by the cuprate alkyl group occurs in high yield, with retention of configuration. This provides an

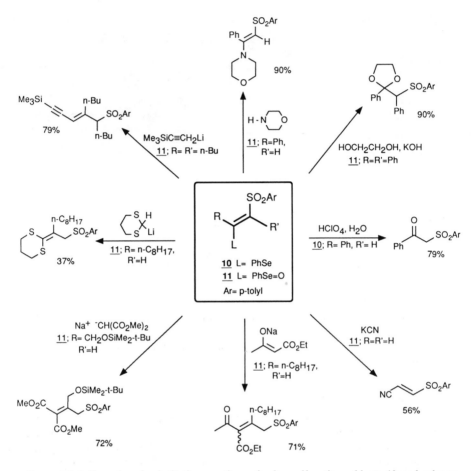

Figure 4.6 ▪ Examples of substitution reactions of selenosulfonation adducts **10** and selenoxides **11**.

efficient connective route to variously substituted vinyl sulfones, and also to olefins after reductive desulfonylation. The novel mixed cuprates RCu(SePh)Li proved to be particularly efficacious for this purpose, as they avoid further Michael addition to the vinyl sulfone products, a side reaction observed with conventional cuprates. This procedure is depicted in Figure 4.7 and permits acetylenes to act as synthetic equivalents of vinyl cations **12**.

4.5. Synthesis of Marine Sterols

A number of marine sterols, such as ostreasterol (**13**), dehydroaplysterol (**14**), and xestosterol (**15**), share the same steroid nucleus, as well as a 24-methylene group, and differ from each other only with respect to the branched substituent attached at C-24.

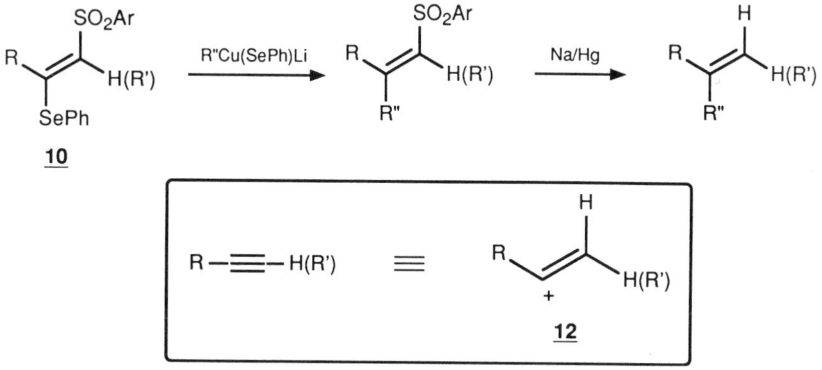

Figure 4.7 ■ Substitution reactions of selenosulfonation adducts **10** with cuprates.

Retrosynthetic analysis suggested that all such compounds could be obtained from the common acetylenic precursor (**16**) by selenosulfonation, substitution with the appropriate organocuprate, and reductive desulfonylation. When this strategy was applied to the suitably protected steroidal iodide **17**, which is itself readily available from inexpensive stigmasterol, the desired marine sterols were obtained without difficulty [17]. This is illustrated in Figure 4.8.

4.6. Acetylenic and Allenic Sulfones

4.6.1. Preparation of Acetylenic Sulfones

Terminal acetylenes can be smoothly converted to acetylenic sulfones **2** by selenosulfonation and selenoxide elimination, as shown in Figure 4.9 [12,13]. This is possible because the high regioselectivity and stereoselectivity of the selenosulfonation step ensures both that the sulfone group is incorporated at the 1-position and that the vinylic hydrogen atom is cis to the selenium moiety, providing for facile selenoxide syn-elimination toward that site. Vinyl selenoxides normally produce mixtures of acetylenes and allenes [18], and the exclusive formation of acetylenic (instead of allenic) sulfones in the present case is attributed to the activating effect of the sulfone group upon the vinylic hydrogen atom.

4.6.2. Preparation of Allenic Sulfones

Although the convenient preparation of synthetically useful acetylenic sulfones by the above method was gratifying, we also desired an alternative protocol that would permit the selective preparation of their allenic counterparts, as the latter are of increasing importance in many types of synthetic applications [19]. A solution to this problem is also shown in Figure 4.9. The adducts **10** isomerize readily and usually quantitatively to allylic sulfones **18** [20] when treated with bases ranging in strength from triethylamine to LDA. In contrast to the original adducts **10**, it is the allylic (instead of vinylic) hydrogen atoms in compounds **18** that are activated by the sulfone group. Conse-

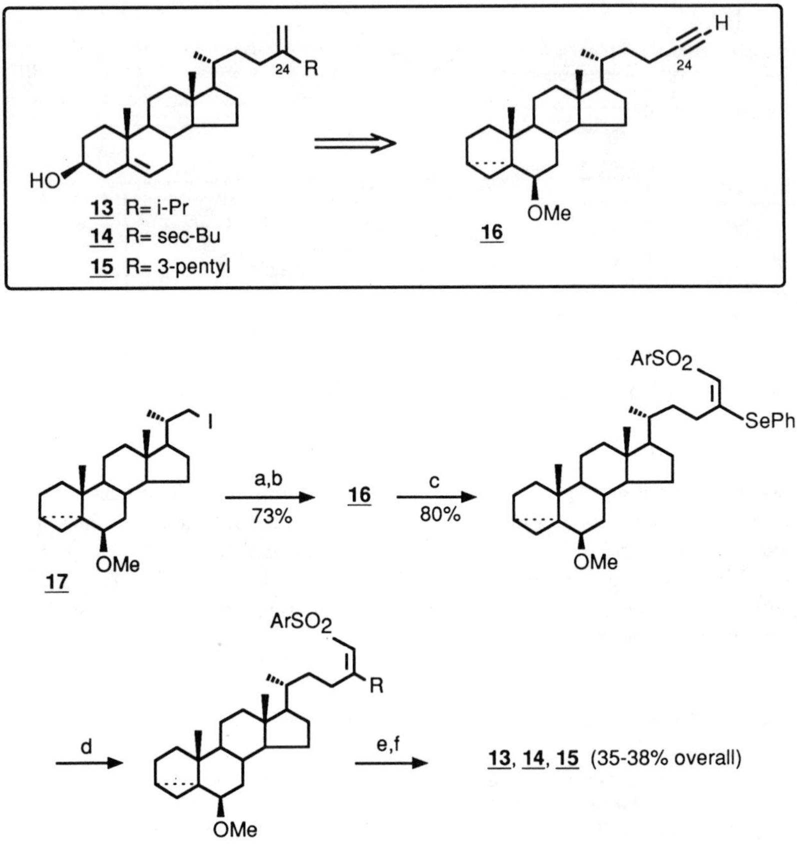

a) $Me_3SiC{\equiv}CCH_2Li$ b) $n\text{-}Bu_4N^+\ F^-$ c) $ArSO_2SePh$ (Ar= p-tolyl); C_6H_6, Δ, AIBN
d) RCu(SePh)Li e) Na/Hg f) H_3O^+

Figure 4.8 ■ Synthesis of marine sterols.

quently, these compounds produce the desired γ-substituted allenic sulfones **3** instead of the corresponding propargylic isomers when subjected to oxidation and elimination [21].

It is interesting to note that the base-catalyzed isomerization step affords the Z isomers of the allylic sulfones **18** with high selectivity, as established by Nuclear Overhauser Effect (NOE) experiments. The Z configuration further ensures that selenoxide elimination will be directed towards the allylic hydrogens, as the vinylic hydrogen atom is now trans to the selenium residue.

Block and Aslam [20a] studied the stereochemistry of the vinylogous Ramberg–Bäcklund reaction, which also requires the γ-deprotonation of a vinyl sulfone. They demonstrated that the "syn effect" controls the stereochemical outcome of the reaction unless it is overridden by steric hindrance between a γ-substituent and a *cis*-sulfone

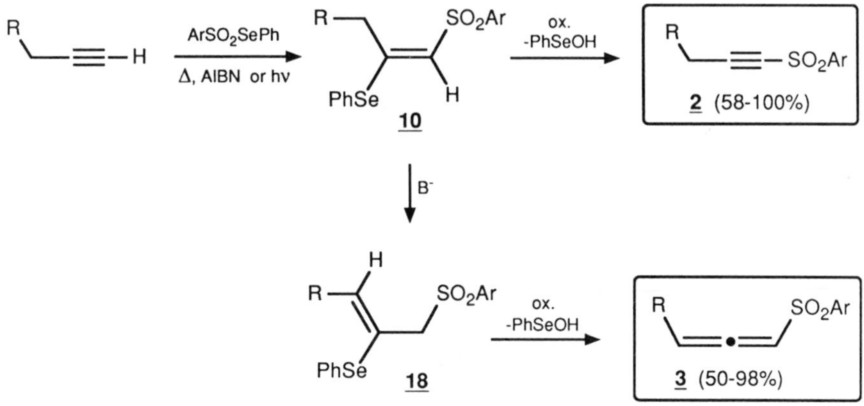

Figure 4.9 ■ Preparation of acetylenic and γ-substituted allenic sulfones.

group. In our case, the syn effect would be expected to favor the E isomer of **18** and so we conclude that the steric effects depicted in Figure 4.10 account for the observed stereochemistry in the present case.

The isomerization of **10** to **18** proceeds via the sulfone-stabilized allyl anion **19**. This presents the opportunity to introduce an additional substituent at the α-position of the product allenic sulfone by means of a suitable electrophile prior to selenoxide elimination (see Figure 4.11). Thus, anion **19**, generated quantitatively with LDA in THF at −78 °C, was subjected to alkylation with electrophiles (such as methyl, ethyl, or allyl iodide), deuteration with D$_2$O, or silylation with Me$_3$SiCl [21]. In some cases, spontaneous elimination of PhSeH occurred, thereby obviating the need for a subsequent oxidation step. These procedures therefore provide a versatile and convenient route to γ-substituted and α,γ-disubstituted allenic sulfones **3**, as well as to acetylenic sulfones **2**.

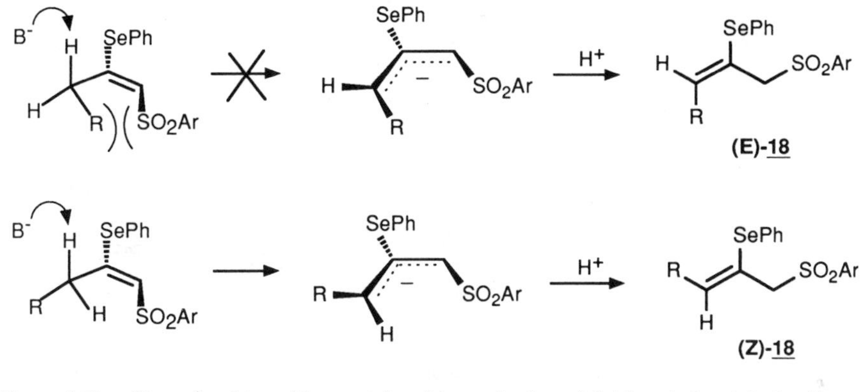

Figure 4.10 ■ Stereochemistry of base-catalyzed isomerization of β-(phenylseleno)vinyl sulfones.

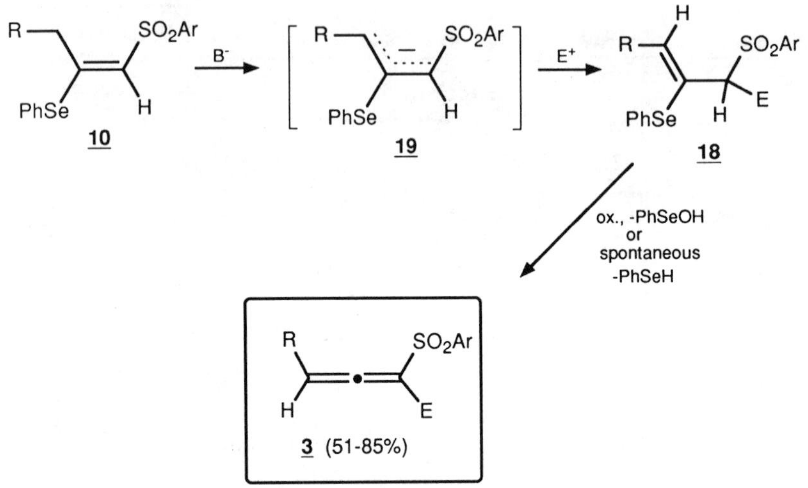

Figure 4.11 ▪ Preparation of α,γ-disubstituted allenic sulfones.

4.6.3. Preparation of Allenes

Reductive desulfonylation of allenic sulfones would permit the synthetically useful overall transformation of acetylenes to allenes. Unfortunately, all such attempts with conventional reagents [22] resulted in incomplete desulfonylation or in overreduction of the product. Alternatively, desulfonylation prior to selenoxide elimination was investigated, producing the corresponding allenes as the principal products [21a]. An example is displayed in Figure 4.12. In some cases, however, allene formation is accompanied by smaller amounts of acetylene isomers.

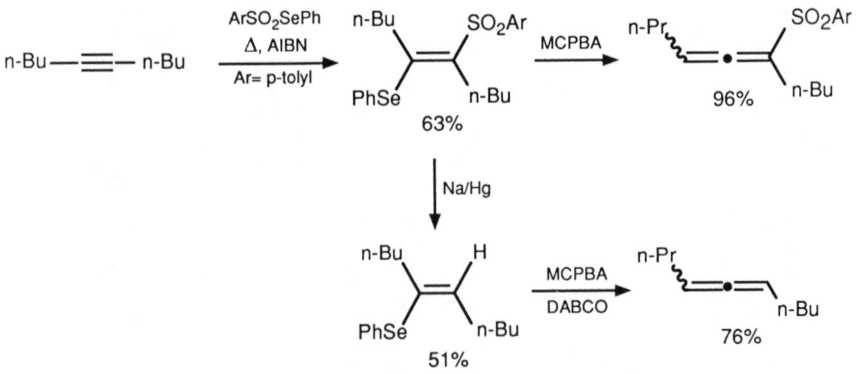

Figure 4.12 ▪ Preparation of an allene from an acetylene.

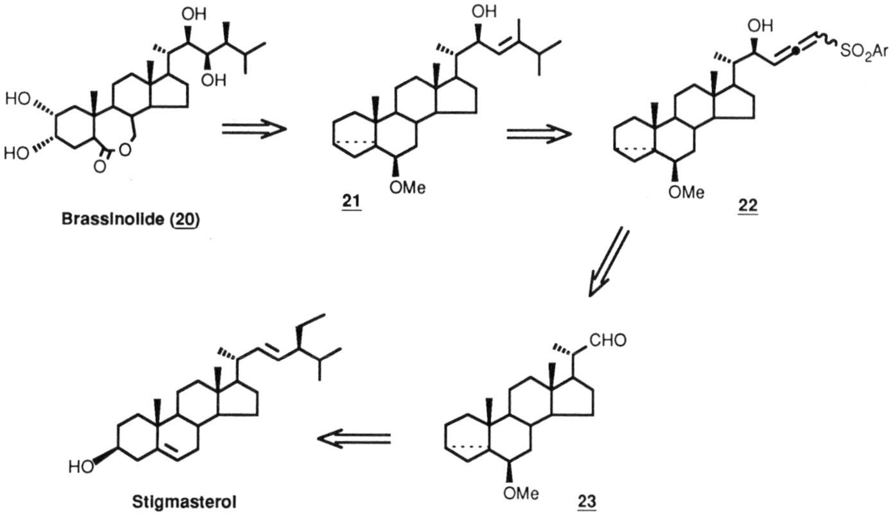

Figure 4.13 ■ Retrosynthetic analysis of brassinolide.

4.7. Synthesis of Brassinosteroids

Brassinolide (**20**) is a powerful plant growth-promoter that was originally isolated from the pollen of *Brassica napus* L. [23] in minute quantities. The allylic alcohol **21** served as a key intermediate in the first synthesis of brassinolide by Fung and Siddall [24], and also comprises a potentially useful intermediate for the synthesis of other related sterols. A new synthesis of **21** (and, therefore, formally of brassinolide) was developed from the allenic sulfone **22**, in turn obtained from aldehyde **23** using selenosulfonation-based methodology [25]. This is illustrated in the retrosynthetic scheme shown in Figure 4.13.

Thus, aldehyde **23**, which is readily obtained from stigmasterol, was treated with 3-lithio-1-trimethylsilylpropyne. This afforded the Cram addition product **24a** as the major C-22 epimer. Selenosulfonation of the latter compound, followed by base-catalyzed isomerization and selenoxide elimination in the usual manner, then afforded the required allenic sulfone **22** as a mixture of diastereomers. Fortunately, their separation was not required, as both stereoisomers could be converted to the allylic alcohol **21**. Thus, the addition of i-Pr$_2$CuLi—SMe$_2$ to **22**, followed by desulfonylation of **25**, produced the desired allylic alcohol along with smaller amounts of the double-bond isomer **26** and the elimination product **27**. These transformations are shown in Figure 4.14. It is interesting to note that the cuprate addition step occurred with high stereoselectivity, affording only the required Z isomer of **25**. This suggests that the direction of addition is determined by steric factors and that complexation of the cuprate with the unmasked C-22 hydroxyl group does not play a significant role (Figure 4.15).

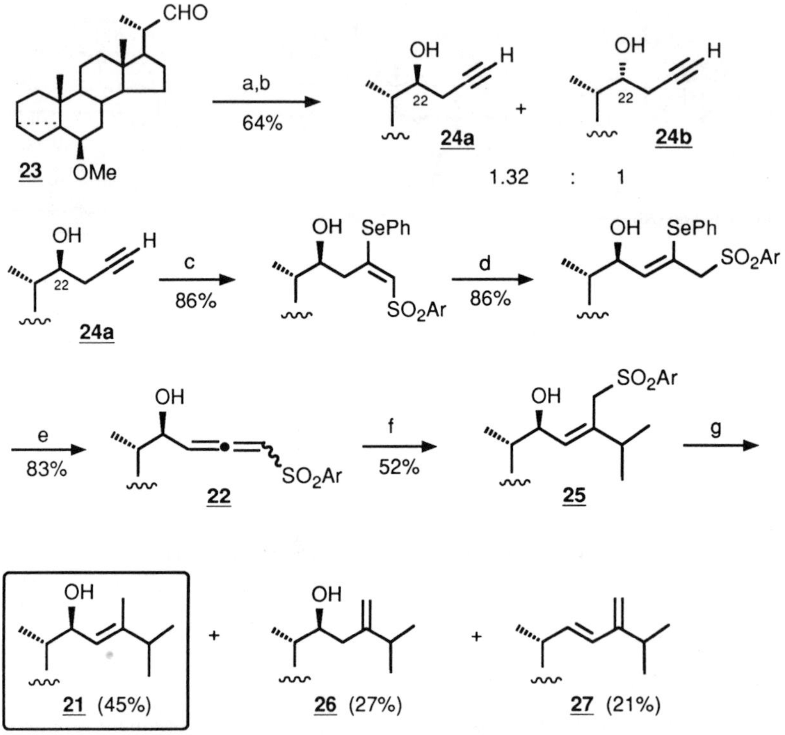

a) LiCH$_2$C≡CSiMe$_3$ b) n-Bu$_4$N$^+$ F$^-$ c) ArSO$_2$SePh, hv (Ar= p-tolyl)

d) Et$_3$N, toluene, Δ   e) t-BuOOH   f) i-Pr$_2$CuLi·SMe$_2$ g)Mg, MeOH

Figure 4.14 ▪ Synthesis of key intermediate **21**.

4.8. Selenosulfonation of Vinyl and Acetylenic Cyclopropanes

The selenosulfonation of vinyl cyclopropanes is accompanied by ring-opening of the intermediate cyclopropylcarbinyl radicals **28** to afford high yields of the corresponding 1,5-adducts **29** [26] (Figure 4.16). Selenoxide elimination of the latter produces dienyl sulfones **30** that are in turn synthetically useful in Julia olefination reactions [27]. Furthermore, the ring-opening of cyclopropylcarbinyl radicals can be employed as a "free-radical clock" [28], against which the rates of competing radical reactions can be measured. On this basis we conclude that chain-transfer is substantially slower than ring-opening (ca. 1.3×10^8 s^{-1}) in **28**. In contrast, cyclopropylacetylene affords chiefly the 1,2-adduct **32**, formed exclusively with the E-configuration [26] (Figure 4.16). This indicates that chain-transfer is considerably faster than either ring-opening or inversion in the cyclopropylvinyl radical **31**. Finally, selenoxide elimination of adduct **32** produced the novel acetylenic sulfone **33**, which may have interesting synthetic applications of its own.

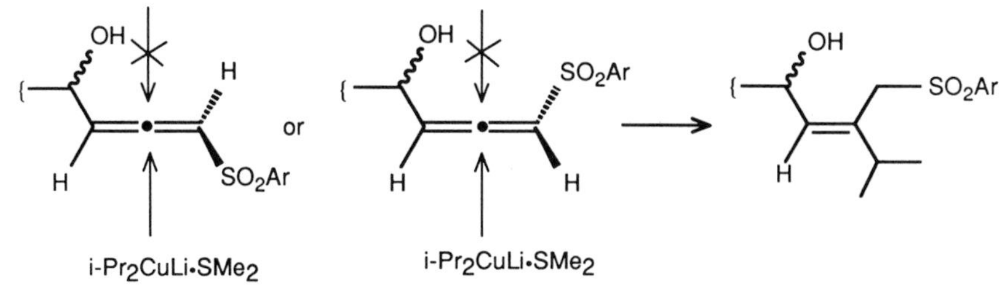

Figure 4.15 ■ Organocuprate addition to allenic sulfone **22**.

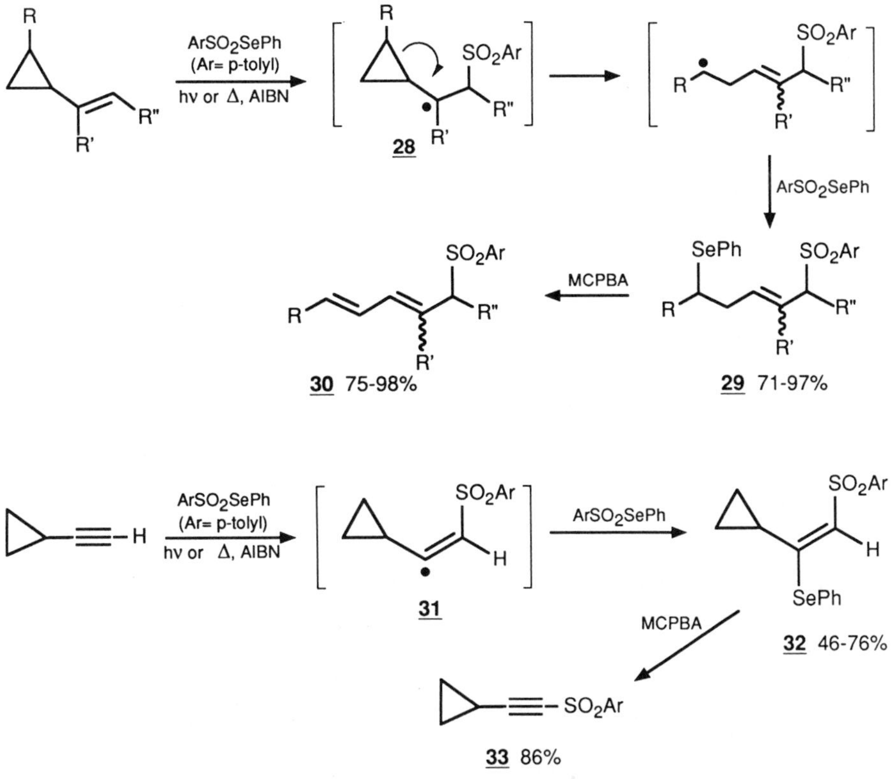

Figure 4.16 ■ Selenosulfonation of vinyl and acetylenic cyclopropanes.

4.9. Conclusion

Selenosulfonates are readily available and easily handled reagents that can be efficiently employed in the selenosulfonation of olefins and acetylenes. These reactions provide β-seleno sulfone adducts that in turn afford access to a variety of useful unsaturated sulfone derivatives by means of selenoxide eliminations or via substitution reactions of the selenium moiety. High regioselectivity and stereoselectivity is generally observed. This methodology has now been successfully applied to the synthesis of several types of steroid side chains and offers promise for many future applications.

Acknowledgments. We wish to acknowledge the excellent contributions of Dr. S. Collins, Dr. R. G. Kerr, U. Gokhale, and K.-W. Law to some of the early aspects of this work. We are also grateful to the Natural Sciences and Engineering Research Council of Canada for financial support.

References

1. For reviews, see the following: (a) De Lucchi, O.; Pasquato, L. *Tetrahedron*, **1988**, *44*, 6755. (b) Trost, B. M. *Bull. Chem. Soc. Jpn.*, **1988**, *61*, 107. (c) Block, E.; Aslam, M. *Tetrahedron*, **1988**, *44*, 281. (d) Fuchs, P. L.; Braish, T. F. *Chem. Rev.* **1988**, *86*, 903. (e) Durst, T., in *Comprehensive Organic Chemistry*, Barton, D. H. R.; Ollis, W. D., eds., Vol. 3, Pergamon Press, New York, 1979, Chapter 11.9. (f) Magnus, P. D. *Tetrahedron*, **1977**, *33*, 2019.

2. Foss, O. *J. Am. Chem. Soc*, **1947**, *69*, 2236.

3. Back, T. G.; Collins, S.; Krishna, M. V. *Can. J. Chem.*, **1987**, *65*, 38.

4. (a) Gancarz, R. A.; Kice, J. L. *Tetrahedron Lett.*, **1980**, *21*, 1697. (b) Gancarz, R. A.; Kice, J. L. *Tetrahedron Lett.*, **1980**, *21*, 4155. (c) Gancarz, R. A.; Kice, J. L. *J. Org. Chem.*, **1981**, *46*, 4899.

5. (a) For a review of electrophilic selenium reactions, see Back, T. G., in *Organoselenium Chemistry*, Liotta, D., ed., Wiley, New York, 1987, Chapter 1. (b) For a review of radical reactions of selenium compounds, see ibid., Chapter 7.

6. For some examples, see (a) Liu, L. K.; Chi, Y.; Jen, K.-Y. *J. Org. Chem.*, **1980**, *45*, 406. (b) Amiel, Y. *J. Org. Chem.*, **1974**, *39*, 3867. (c) Truce, W. E., Wolf, G. C. *J. Org. Chem.*, **1971**, *36*, 1727. (d) Truce, W. E.; Gorlaski, C. T. *J. Org. Chem.*, **1970**, *35*, 4220.

7. (a) Back, T. G.; Collins, S. *Tetrahedron Lett.*, **1980**, *21*, 2215. (b) Back, T. G.; Collins, S. *J. Org. Chem.*, **1981**, *46*, 3249.

8. Kang, Y.-H.; Kice, J. L. *J. Org. Chem.*, **1984**, *49*, 1507.

9. (a) Paquette, L. A.; Lin, H.-S.; Gunn, B. P.; Coghlan, M. J. *J. Am. Chem. Soc.*, **1988**, *110*, 5818. (b) Kinney, W. A.; Crouse, G. D.; Paquette, L. A. *J. Org. Chem.*, **1983**, *48*, 4986. (c) Paquette, L. A.; Crouse, G. D. *J. Org. Chem.*, **1983**, *48*, 141. (d) Paquette, L. A.; Kinney, W. A. *Tetrahedron Lett.*, **1982**, *23*, 5127. (e) Black, K. A.; Vogel, P. *J. Org. Chem.*, **1986**, *51*, 5341.

10. Kice, J. L.; Kang, Y.-H. *Tetrahedron*, **1985**, *41*, 4739.

11. Bäckvall, J.-E.; Nájera, C.; Yus, M. *Tetrahedron Lett.*, **1988**, *29*, 1445.

12. (a) Back, T. G.; Collins, S.; Kerr, R. G. *J. Org. Chem.*, **1983**, *48*, 3077. (b) Back, T. G.; Collins, S.; Gokhale, U.; Law, K.-W. *J. Org. Chem.*, **1983**, *48*, 4776.

13. Miura, T.; Kobayashi, M. *J. Chem. Soc., Chem. Commun.*, **1982**, 438.

14. Back, T. G.; Collins, S.; Law, K.-W. *Can. J. Chem.*, **1985**, *63*, 2313.

15. Back, T. G.; Krishna, M. V. *J. Org. Chem.*, **1987**, *52*, 4265.

16. Back, T. G.; Collins, S.; Krishna, M. V.; Law, K.-W. *J. Org. Chem.*, **1987**, *52*, 4258.

17. Back, T. G.; Proudfoot, J. R.; Djerassi, C. *Tetrahedron Lett.*, **1986**, *27*, 2187.

18. Reich, H. J.; Willis, W. W., Jr. *J. Am. Chem. Soc.*, **1980**, *102*, 5967.

19. For some lead references, see the following: (a) Hayakawa, K.; Nishiyama, H.; Kanematsu, K. *J. Org. Chem.*, **1985**, *50*, 512. (b) Barbarella, G.; Cinquini, M.; Colonna, S. *J. Chem. Soc., Perkin Trans. 1*, **1980**, 1646. (c) Padwa, A.; Kline, D. N.; Norman, B. H. *Tetrahedron Lett.*, **1988**, *29*, 265. (d) Padwa, A.; Yeske, P. E. *J. Am. Chem. Soc.*, **1988**, *110*, 1617. (e) Parpani, P.; Zecchi, G. *J. Org. Chem.*, **1987**, *52*, 1417. (f) Denmark, S. E.; Harmata, M. A.; White, K. S. *J. Org. Chem.*, **1987**, *52*, 4031.

20. For examples of isomerizations of vinyl to allyl sulfones, see the following: (a) Block, E.; Aslam, M. *J. Am. Chem. Soc.*, **1983**, *105*, 6164. (b) Block, E.; Eswarakrishnan, V.; Gebreyes, K. *Tetrahedron Lett.*, **1984**, *25*, 5469. (c) Eisch, J. J.; Galle, J. E. *J. Org. Chem.*, **1979**, *44*, 3279. (d) Baldwin, J. E.; Adlington, R. M.; Ichikawa, Y.; Kneale, C. J. *J. Chem. Soc., Chem. Commun.*, **1988**, 702. (e) Sváta, V.; Procházka, M.; Bakos, V. *Coll. Czech. Chem. Commun.*, **1978**, *43*, 2619. (f) O'Connor, D. E.; Lyness, W. I. *J. Am. Chem. Soc.*, **1964**, *86*, 3840. (g) Broaddus, C. D. *J. Am. Chem. Soc.*, **1966**, *88*, 3863. (h) Broaddus, C. D. *J. Am. Chem. Soc.*, **1968**, *90*, 5504.

21. (a) Back, T. G.; Krishna, M. V.; Muralidharan, K. R. *J. Org. Chem.*, **1989**, *54*, 4146. (b) Back, T. G.; Krishna, M. V.; Muralidharan, K. R. *Tetrahedron Lett.*, **1987**, *28*, 1737.

22. (a) Trost, B. M.; Arndt, H. C.; Strege, P. E.; Verhoeven, T. R. *Tetrahedron Lett.*, **1976**, 3477. (b) Trost, B. M.; Verhoeven, T. R. *J. Am. Chem. Soc.*, **1978**, *100*, 3435. (c) Pascali, V.; Umani-Ronchi, A. *J. Chem. Soc., Chem. Commun.*, **1973**, 351. (d) Bremner, J.; Julia, M.; Launay, M.; Stacino, J.-P. *Tetrahedron Lett.*, **1982**, *23*, 3265. (e) Cuvigny, T.; Herve du Penhoat, C.; Julia, M. *Tetrahedron*, **1987**, *43*, 859. (f) Inomata, K.; Igarashi, S.; Mohri, M.; Yamamoto, T.; Kinoshita, H.; Kotake, H. *Chem. Lett.*, **1987**, 707. (g) Brown, A. C.; Carpino, L. A. *J. Org. Chem.*, **1985**, *50*, 1749.

23. Grove, M. D.; Spencer, G. F.; Rohwedder, W. K.; Mandava, N.; Worley, J. F.; Warthen, J. D., Jr.; Steffens, G. L.; Flippen-Anderson, J. L.; Cook, J. C., Jr. *Nature (London)*, **1979**, *281*, 216.

24. Fung, S.; Siddall, J. B. *J. Am. Chem. Soc.*, **1980**, *102*, 6580.

25. Back, T. G.; Brunner, K.; Krishna, M. V.; Lai, E. K. Y. *Can. J. Chem.*, **1989**, *67*, 1032.

26. Back, T. G.; Muralidharan, K. R. *J. Org. Chem.*, **1989**, *54*, 121.

27. (a) Julia, M.; Paris, J.-M. *Tetrahedron Lett.*, **1973**, 4833. (b) Arnould, D.; Chabardes, P.; Farge, G.; Julia, M. *Bull. Soc. Chim. Fr.*, **1985**, 130.

28. (a) Griller, D.; Ingold, K. U. *Acc. Chem. Res.*, **1980**, *13*, 317. (b) Mathew, L.; Warkentin, J. *J. Am. Chem. Soc.*, **1986**, *108*, 7981.

Ligand Coupling in Bismuth and Lead Derivatives

Derek H. R. Barton

Department of Chemistry
Texas A&M University
College Station, Texas 77843

The application of Bi(V) reagents in the arylation of a wide range of nucleophiles to give (even very hindered) compounds in good yield has been reviewed. The use of copper catalysis enables a range of oxygen and nitrogen functions, including hindered systems, to be arylated under very mild conditions.

Similar chemistry has been developed by J. T. Pinhey (Sydney) using aryllead and alkenyllead compounds. Bismuth reagents are more reactive than their lead analogues. However, the lead derivatives are more efficient (utilization of the single aryl residue) and are more easy to prepare when electron-rich aryl group transfer is required.

In recent work, bismuth and lead derivatives have been compared in the arylation of various 4-hydroxycoumarins. We have also compared bismuth and lead derivatives in the arylation of amines with copper catalysis. We find that bismuth compounds are more reactive and can be applied to phenols and other oxygen nucleophiles where lead reagents are inactive.

We consider that the chemistry of bismuth and lead is dominated by ligand coupling. This explains why very hindered compounds can be prepared in good yield.

The concept of ligand coupling was originally due to Trost [1]. It implies that two groups attached to an atom couple in a concerted manner without separating into radicals or ions. The expression "reductive elimination," much used in organometallic chemistry, designates the same overall transformation, but without specifying the mechanism involved. Oae and his collaborators [2] have recently presented important evidence for ligand coupling in S(IV) and P(V) compounds. In particular, the reaction of alkyl or aryl Grignard reagents with 1-phenylethyl-1-(2'-pyridyl) sulfoxide induces ligand coupling of the 1-phenylethyl group with the 2'-pyridyl group with complete retention of configuration.

We first met ligand coupling in the pyrolysis of tetraphenyltellurium to furnish diphenyltellurium and diphenyl [3]. A careful mechanistic study showed that phenyl radicals were not involved.

Our next encounter [4] came while studying the use of Bi(V) reagents for the oxidation of alcohols under very mild conditions. The alkaloid quinine (1) gave, with triphenylbismuth carbonate, the corresponding ketone and then, with excess of reagent, the phenylated quininone 2 in high yield [eq. (5.1)].

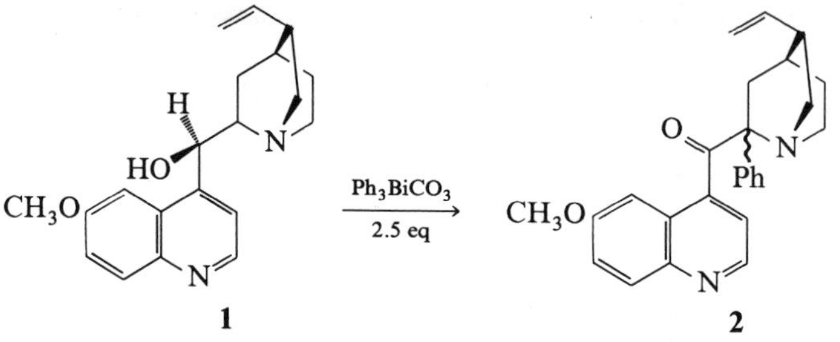

$$(5.1)$$

We were impressed by the formation of such a hindered compound in good yield under mild conditions. We quickly showed that under basic conditions, ketones, phenols, enols, and nitro-compounds could all be converted in high yield into very hindered arylated derivatives. For example, cyclohexanone (3) afforded 2,2,6,6-tetraphenyl-cyclohexanone (4) [eq. (5.2)] in 93% yield and 3,5-di-t-butylphenol (5) gave the 2,6-diphenyl derivative 6 in 77% yield [eq. (5.3)].

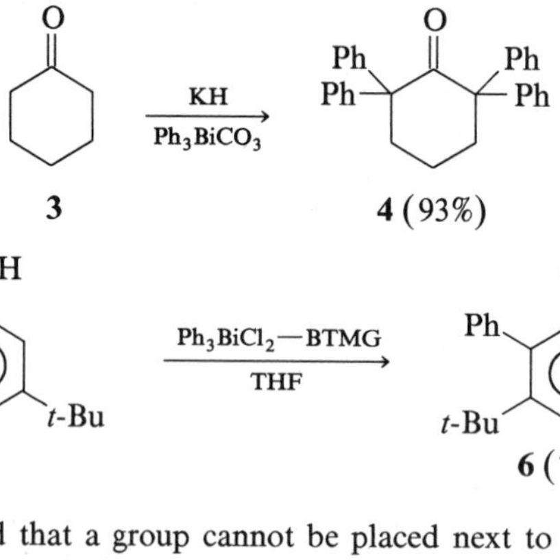

$$(5.2)$$

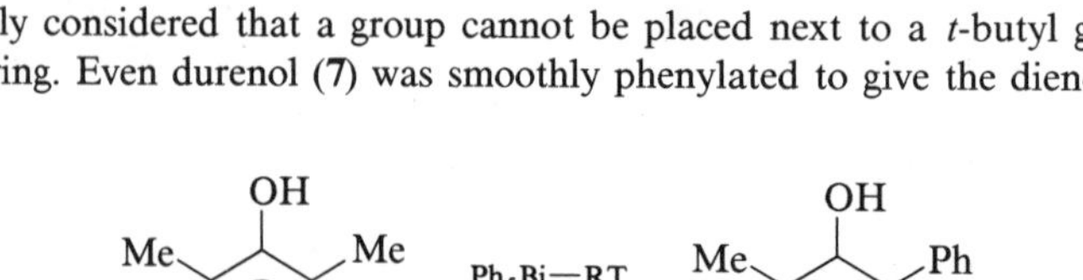

$$(5.3)$$

It is usually considered that a group cannot be placed next to a t-butyl group on an aromatic ring. Even durenol (7) was smoothly phenylated to give the dienone 8 (83%) [eq. (5.4)].

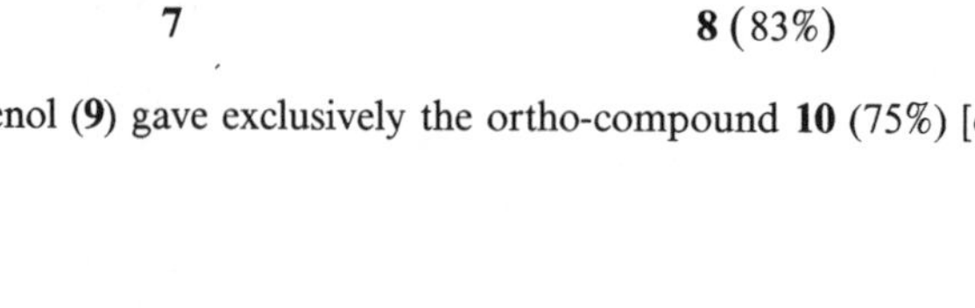

$$(5.4)$$

2,6-Dimethylphenol (9) gave exclusively the ortho-compound 10 (75%) [eq. (5.5)].

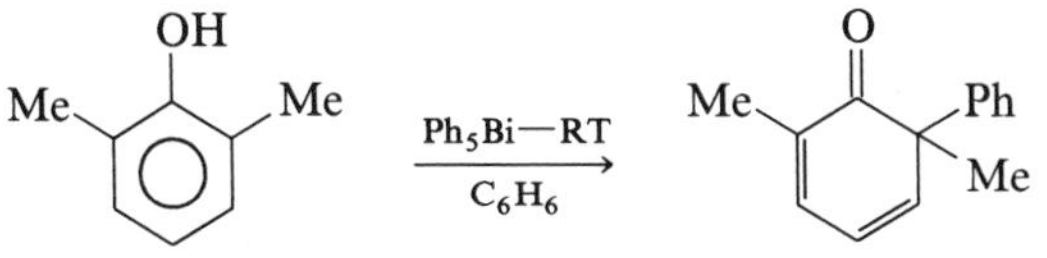

$$\mathbf{9} \xrightarrow[\text{C}_6\text{H}_6]{\text{Ph}_5\text{Bi—RT}} \mathbf{10\ (75\%)} \qquad (5.5)$$

Enols reacted readily as in **11** giving **12** and **13**, furnishing **14** [eqs. (5.6) and (5.7)].

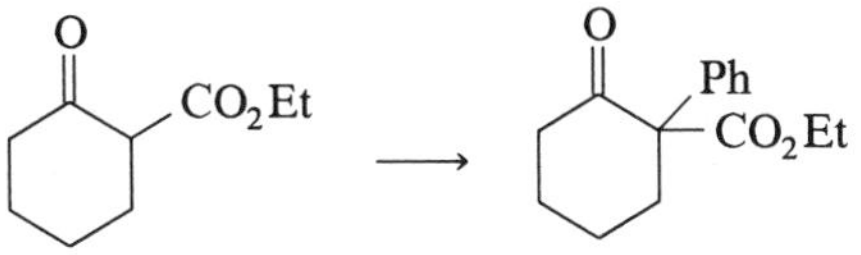

$$\begin{cases} \text{Ph}_3\text{BiCl}_2 & 75\% \\ \text{Ph}_4\text{BiOCOCF}_3/\text{BTMG} & 91\% \\ \text{Ph}_5\text{Bi} & 57\% \end{cases} \qquad (5.6)$$

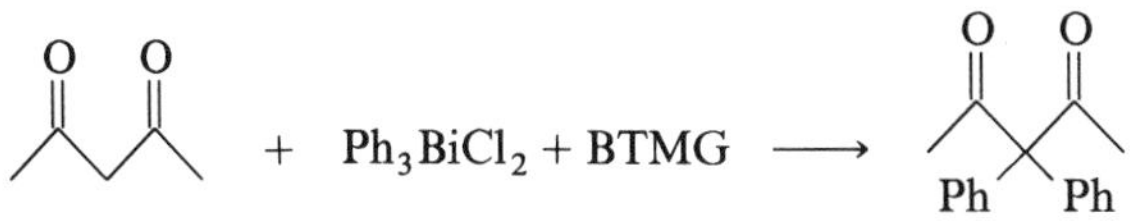

$$\mathbf{13} + \text{Ph}_3\text{BiCl}_2 + \text{BTMG} \longrightarrow \mathbf{14\ (74\%)} \qquad (5.7)$$

Similarly, 2-nitropropane (**15**) phenylated readily to give **16**, while **17** afforded the hindered **18** (81%) [eqs. (5.8) and (5.9)].

$$(\text{CH}_3)_2\text{CHNO}_2 + \text{Ph}_3\text{BiCl}_2 + \text{BTMG} \longrightarrow \text{Ph}\!-\!\overset{\text{CH}_3}{\underset{\text{CH}_3}{\text{C}}}\!-\!\text{NO}_2 \qquad (5.8)$$

$$\mathbf{15} \qquad\qquad\qquad\qquad \mathbf{16\ (77\%)}$$

$$\text{CH}_3\text{CH}(\text{NO}_2)\text{CO}_2\text{C}_4\text{H}_9 + \text{Ph}_3\text{BiCl}_2 + \text{TMG} \longrightarrow \text{Ph}\!-\!\overset{\text{NO}_2}{\underset{\text{CH}_3}{\text{C}}}\!-\!\text{CO}_2\text{C}_4\text{H}_9 \qquad (5.9)$$

$$\mathbf{17} \qquad\qquad\qquad\qquad\qquad\qquad \mathbf{18\ (81\%)}$$

In the formulas, BTMG is *t*-butyltetramethylguanidine and TMG is tetramethylguanidine.

Two other phenylations that proceed in good yield are that of sulfinate **19** to afford sulfone **20** and that of skatole (**21**) to give 9-phenylskatole (**22**) [eqs. (5.10) and (5.11)].

$$p\text{-CH}_3\text{C}_6\text{H}_4\text{SO}_2\text{Na} + \text{Ph}_4\text{BiOCOCF}_3 \longrightarrow p\text{-CH}_3\text{C}_6\text{H}_4\text{SO}_2\text{Ph} \qquad (5.10)$$

$$\mathbf{19} \qquad\qquad\qquad\qquad\qquad \mathbf{20\ (86\%)}$$

$$\mathbf{21} \qquad\qquad\qquad\qquad\qquad \mathbf{22\ (95\%)}$$

It was clear from the experiments recorded above that one had an unusual arylation system with a remarkable capacity to make hindered compounds. We studied the

Table 5.1 ■ Changes in the ^{1}H NMR Chemical Shifts of 3,5-Di-t-butylphenol (5) upon Addition of Bi(V) Reagents

	δ (H ortho)	δ (H para)
Phenol 5	6.66	6.97
Phenol 5 + TMG	6.67	6.82
Phenol 5 + TMG + Ph$_3$BiCl$_2$	6.10	6.50
Phenol 5 + TMG + Ph$_3$Bi(OCOCF$_3$)$_2$	6.13	6.46
Phenol 5 + Ph$_5$Bi	6.15	6.47

ortho-arylation reaction of phenol **5** with special attention. As shown in Table 5.1, mixing phenol **5** with various Bi(V) reagents caused a marked upfield shift of both ortho and para protons in the ^{1}H NMR spectrum as well as the appearance of a typical yellow color. Using triphenylbismuth dichloride with tetramethylguanidine as base, we were able to isolate a crystalline intermediate **23** [eq. (5.12)].

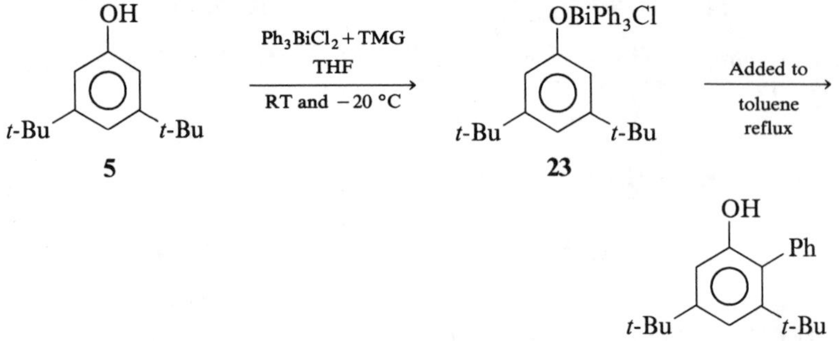

(5.12)

This was fully characterized spectroscopically and by microanalysis. It had the same shifted NMR spectrum. On adding **23** to refluxing toluene, 2-phenyl-3,5-di-t-butylphenol (**24**) was formed (82%). A further 14% of starting phenol **5** was recovered.

The intermediate **23** might decompose by cationic or radical mechanisms or by ligand coupling. The relative migratory aptitudes of aryl groups can easily distinguish between cationic and radical migration. The usual spectrum from *para*-nitrophenyl to *para*-methoxyphenyl showed that the mechanism was not cationic and also probably not radical. We proposed the ligand coupling process shown in **25** [eq. (5.13)].

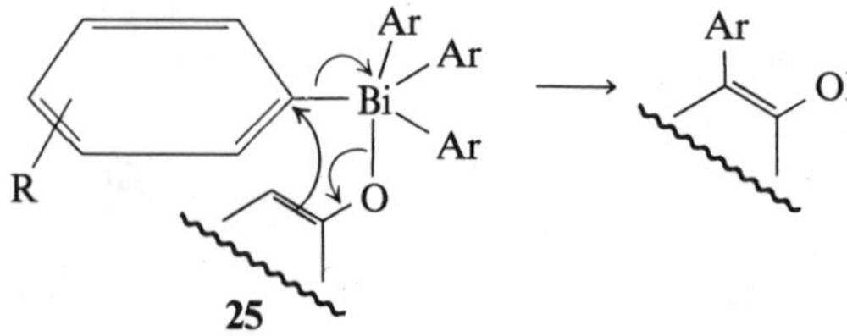

(5.13)

In order to exclude a radical mechanism, we resorted [9] to quantitative radical trapping experiments. Using nitrosobenzene as the trap, t-butyltetramethylguanidine as the base, tetraphenylbismuth tosylate as the phenylating agent, and symmetrical tetraphenylacetone **26** as substrate, phenyl radicals, trapped as diphenylnitroxide, could be easily identified in the ESR spectrum. However, the same radical trapping was seen in the absence of the substrate **26**. This ketone is convenient because it phenylates once in a high yield, but not a second time. Pentaphenylacetone (**27**) is a nicely crystalline compound that is easily isolated [eq. (5.14)].

$$Ph_2CHCOCHPh_2 + Ph_4BiOTos + PhNO \xrightarrow[\text{(2) Fe, AcOH, Ac}_2\text{O}]{\text{(1) BTMG, THF, RT}}$$

26   Reflux

$$Ph_3CCOCHPh_2 + PhNHAc + Ph_2NAc \quad (5.14)$$

27 (90%)   83%   3%

We then devised a method to quantitate radical formation. Reduction of the end product with acetic acid–iron powder in the presence of acetic anhydride gave a mixture of acetanilide (from the nitrobenzene) and N-acetyldiphenylamine (from the trapped phenyl radicals). The nitrosobenzene was used in stoichiometric amounts. The product **27** was formed in high yield with or without the nitrosobenzene. Only 3% of phenyl radicals were trapped with or without the substrate **26**. This experiment clearly shows that when radicals are present they are trapped, but that they have nothing to do with the mechanism of phenylation.

A different radical trap, 1,1-diphenylethylene, was used in a number of other arylation reactions. The results were the same with or without a stoichiometric amount of the trap.

David and Thieffry [5] discovered an unusual reaction while experimenting with triphenylbismuth diacetate as a glycol cleaving reagent. Instead of glycol cleavage, clean monophenylation of the substrate was observed after a two-hour induction period. Eventually, we showed that this reaction was specific to methylene dichloride, that laboratory light was necessary, and that a second phenylation could occur, but it was 10 times slower than the first.

Following an observation of Dodonov, Gushchin, and Brilkina [6] that copper salts catalyze the phenylation of isopropanol, we were able to show that copper salts or metallic copper had a dramatic effect on the reaction of David and Thieffry [5]. The solvent was no longer restricted to methylene dichloride, there was no longer an induction period, and light was not needed. The reaction was fast even at room temperature.

Subsequent work showed that many alcohols and phenols could be O-phenylated very efficiently at room temperature using triphenylbismuth diacetate and catalysis by cupric acetate or metallic copper.

The conditions of these phenylations are so mild that they can easily be applied to natural product chemistry. Thus, a number of amino acid esters (Scheme 5.1) were smoothly phenylated in moderate to good yield under the same conditions as above [7]. Amino acids, themselves, did not react. A second phenyl residue could only be added with difficulty. The mild conditions of this reaction should allow phenylation of terminal amino residues in amino acid esters and peptides and probably of side chain amino-functions and of tyrosine phenolic groups.

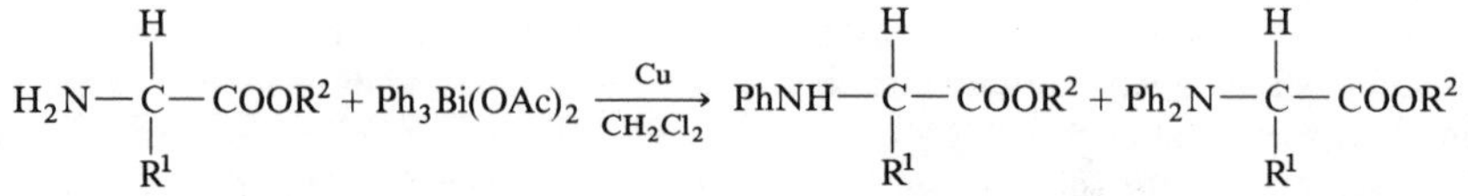

$$\underset{R^1}{\overset{H}{H_2N-\overset{|}{\underset{|}{C}}-COOR^2}} + Ph_3Bi(OAc)_2 \xrightarrow[\text{CH}_2\text{Cl}_2]{\text{Cu}} \underset{R^1}{\overset{H}{PhNH-\overset{|}{\underset{|}{C}}-COOR^2}} + \underset{R^1}{\overset{H}{Ph_2N-\overset{|}{\underset{|}{C}}-COOR^2}}$$

R^1	R^2	%	%
H	Et	81	4
PhCH$_2$ (Bzl)	Bzl	80	
BzlO$_2$C—CH$_2$	Bzl	50	
BzlO$_2$C—(CH$_2$)$_2$	Bzl	58	
Indolylmethyl	Me	66	

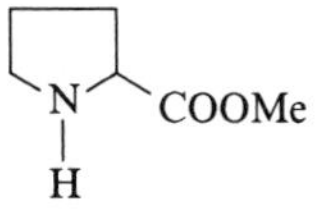

61

(structure: pyrrolidine ring with N—H and COOMe)

Scheme 5.1

A comparison of triphenylbismuth diacetate and bis-trifluoroacetate showed that the latter was an even faster phenylating reagent than the former. For example, a number of indoles with a free β-position could be smoothly phenylated at room temperature. Examples are indole (**28**) itself, 2-methylindole (**29**), and 2,N-dimethylindole (**30**) [eq. (5.15)].

C-Phenylation:

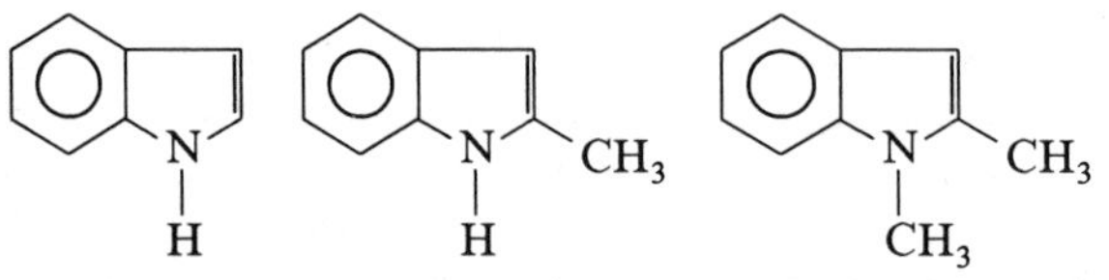

28 (50%)   **29** (60–95%)   **30** (84–94%)   (5.15)

In the presence of a stoichiometric amount of cupric acetate, Bi(V) reagents were not needed and triphenylbismuth sufficed. Yields were satisfactory for basic amino-functions, but poor for less basic amines like 4-nitroaniline.

We consider that the mechanism of all this copper-catalyzed chemistry also involves ligand coupling on copper(III). Thus, Cu(I) is considered to react with triphenylbismuth diacetate to add a phenyl and an acetate (or, if appropriate, trifluoroacetate) group. Scheme 5.2 summarizes our views on the N-phenylation process and Scheme 5.3 applies to glycol O-phenylation and to other O-phenylation processes. Other copper-catalyzed reactions involving diaryliodonium salts have been interpreted similarly [8]. In any case, these copper-catalyzed reactions were not affected by massive amounts of 1,1-diphenylethylene. In a separate experiment we showed that this was an efficient trap for phenyl radicals.

How many other elements will show the phenomenon of ligand coupling? We think that many other elements will show the same behavior and have interpreted I(III) and Sb(V) reactions as also involving ligand coupling [9].

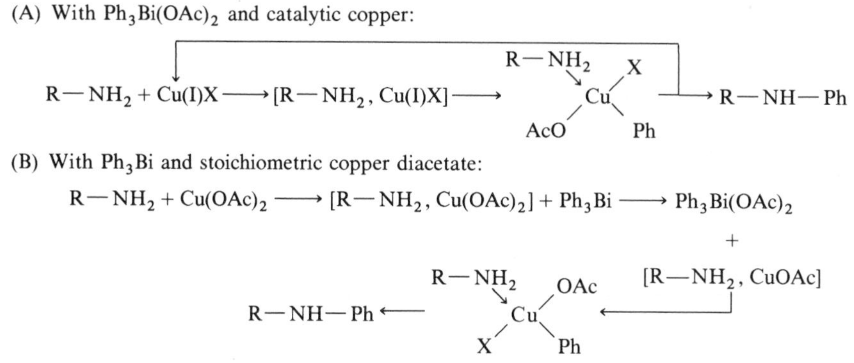

(A) With $Ph_3Bi(OAc)_2$ and catalytic copper:

$$R\!-\!NH_2 + Cu(I)X \longrightarrow [R\!-\!NH_2, Cu(I)X] \longrightarrow \underset{\underset{AcO}{}}{\overset{\overset{R\!-\!NH_2}{\diagdown}}{\underset{\diagup}{Cu}}}\!\!\!\overset{X}{\underset{Ph}{\diagup}} \longrightarrow R\!-\!NH\!-\!Ph$$

(B) With Ph_3Bi and stoichiometric copper diacetate:

$$R\!-\!NH_2 + Cu(OAc)_2 \longrightarrow [R\!-\!NH_2, Cu(OAc)_2] + Ph_3Bi \longrightarrow Ph_3Bi(OAc)_2$$

$$+$$

$$R\!-\!NH\!-\!Ph \longleftarrow \underset{\underset{X}{}}{\overset{\overset{R\!-\!NH_2}{\diagdown}}{\underset{\diagup}{Cu}}}\!\!\!\overset{OAc}{\underset{Ph}{\diagup}} \longleftarrow [R\!-\!NH_2, CuOAc]$$

Scheme 5.2 ■ Amine *N*-Phenylation

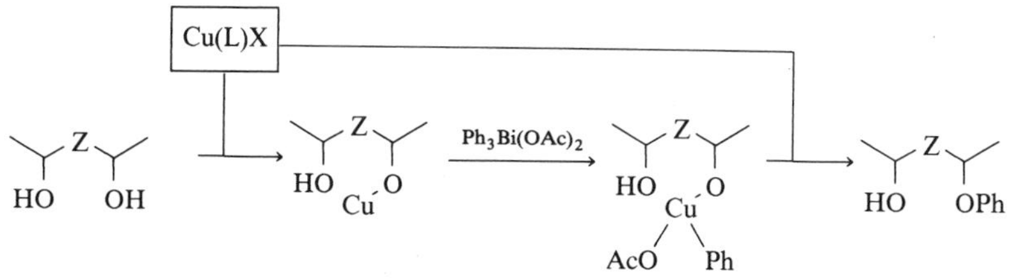

Scheme 5.3 ■ Glycol *O*-Phenylation

J. T. Pinhey and his colleagues [10] have carried out elegant and extensive studies on the use of Pb(IV) reagents for the arylation and alkenylation of anions of the type that react also with Bi(V) reagents.

The work on the Pb(IV) reagents was, of course, independent of knowledge of the Bi(V) reactions and essentially simultaneous in time. A climax to the work has led to the synthesis of important natural products [11].

The Pb(IV) reagents have the advantage of using their sole aryl group in the formation of the desired carbon–carbon bond. In Bi(V) chemistry only one out of the three aryl groups is used in the desired way.

Recently, in collaboration with D. M. X. Donnelly (Dublin), we have had occasion to compare Bi(V) and Pb(IV) chemistry in the synthesis of 3-aryl-4-hydroxycoumarins [12]. The results with the Bi(V) reagents are given in Scheme 5.4. They show the usual high yields in the arylation. It is important that the reaction stops after the transfer of one aryl group. We presume that the enolic hydroxyl group has become too hindered to react further. The reagent C of Scheme 5.4 is of interest in that it permits the transfer of the *m*-nitrophenyl group. We could manipulate easily the nitro group to phenol, methoxyl, etc.

The lead reagents (Schemes 5.5 and 5.6) have the advantage that they are easier to prepare when the aryl groups are electron rich. The kind of substituents needed for isoflavone synthesis can thus be introduced without difficulty. Indeed, the electron rich, but sterically hindered, 2,4,6-trimethoxyphenyl group gives high yields of product.

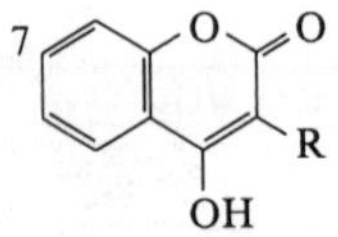

A = $Ph_3Bi(OAc)_2$
B = $(p\text{-Tol})_3Bi(OAc)_2$
C = $(m\text{-}NO_2C_6H_4)_3Bi(ONO_2)_2$

7 (H) A ⟶ R = Ph (92%)
7 (H) B ⟶ R = p-Tol (85%)
7 (H) C ⟶ R = m-$NO_2C_6H_4$ (83%)
7 (OMe) A ⟶ R = Ph (82%)
7 (OMe) B ⟶ R = p-Tol (80%)

Scheme 5.4 ■

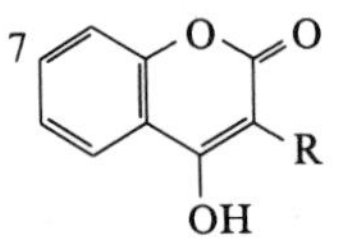

D = $PhPb(OAc)_3$
E = $4\text{-}MeOC_6H_4Pb(OAc)_3$
F = $2,4\text{-}(MeO)_2C_6H_3Pb(OAc)_3$
G = $2,4,6\text{-}(MeO)_3C_6H_2Pb(OAc)_3$

7 (H) D ⟶ R = Ph (49%)
7 (H) E ⟶ R = 4-$MeOC_6H_4$ (47%)
7 (H) F ⟶ R = 2,4-$(MeO)_2C_6H_3$ (95%)
7 (H) G ⟶ R = 2,4,6-$(MeO)_3C_6H_2$ (87%)
7 (OMe) E ⟶ R = 4-$MeOC_6H_4$ (44%)
7 (OMe) F ⟶ R = 2,4-$(MeO)_2C_6H_3$ (85%)

Scheme 5.5 ■

The lead reagents react slowly (16 h) in refluxing chloroform containing three equivalents of pyridine. These are the usual reaction conditions of Pinhey and his colleagues [10].

We conclude that for the synthesis of, say, 3',4'-dioxygenated isoflavonoids, lead reagents will be superior because of the easy synthesis of the reagent from lead tetraacetate.

The interpretation of all the lead(IV)-induced arylation reactions as being ligand coupling was supported by our demonstration, in a typical example, that radicals are

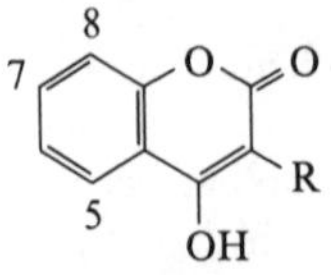

H = $3,4\text{-}(MeO)_2C_6H_3Pb(OAc)_3$ (59%)
I = $3,4\text{-}Methylenedioxy\text{-}C_6H_3Pb(OAc)_3$

7 = 8 (OMe) E ⟶ R = 4-$MeOC_6H_4$ (59%)
7 = 8 (OMe) G ⟶ R = 2,4,6-$(MeO)_3C_6H_2$ (94%)
5 (OMe) E ⟶ R = 4-$MeOC_6H_4$ (75%)
5 (OMe) F ⟶ R = 2,4-$(MeO)_2C_6H_3$ (95%)
5 (OMe) G ⟶ R = 2,4,6-$(MeO)_3C_6H_2$ (97%)
5 (OMe) H ⟶ R = 3,4-$(MeO)_2C_6H_3$ (68%)

Scheme 5.6 ■

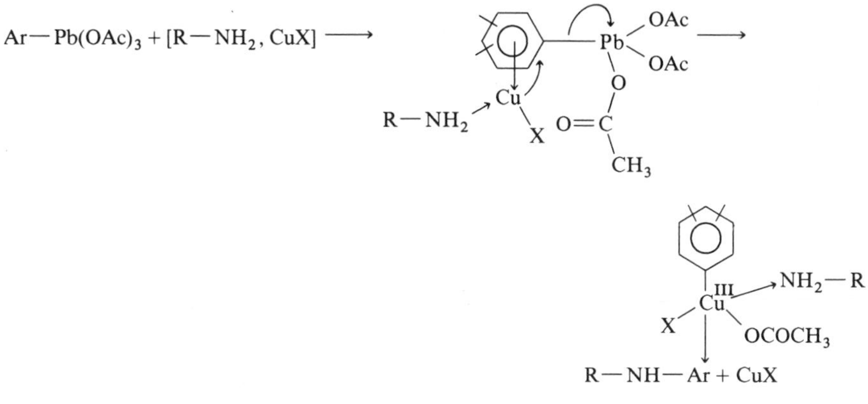

Scheme 5.7 ■

not present [9]. In a recent article Pinhey and his colleagues are not in disagreement with this proposal [13].

It was of interest to see if aryllead compounds would arylate amines using copper catalysis, as works so well for arylbismuth compounds (see above). Of the family of reagents $PbPh_4$, Ph_3PbOAc, $Ph_2Pb(OAc)_2$, and $PhPb(OAc)_3$, the last named was the most efficient. Electron-rich aromatic amines were converted in high yield to the corresponding monophenylated derivative [14]. Even the hindered 2,4,6-trimethyl-aniline (mesidine) gave the N-phenyl derivative in 94% yield. Similar results were obtained with either cupric acetate or with cupric trifluoroacetate. However, phenols, indoles, and aliphatic and alicyclic amines were phenylated in poor yield or not at all. The family of bismuth reagents is therefore more suitable than lead reagents for O- or N-phenylation.

Recently, in collaboration with D. M. X. Donnelly (Dublin), we have examined the applicability of a range of electron-rich aryllead reagents for the arylation of aromatic and other amines [15]. The arylation of aromatic amines mostly proceeded in moderate to good yield. In the extreme case of mesidine the p-methoxyphenyl, the 2,4-dime-thoxyphenyl, and the 2,4,6-trimethoxyphenyl derivatives were obtained in yields of 65, 74, and 18%, respectively. The reaction conditions were with cupric acetate catalysis at 0 °C to room temperature. Dimesidine is a very hindered compound so that the formation of even 18% yield of product is remarkable.

Again, aliphatic and alicyclic amines were not, in general, efficiently arylated. Indole, carbazole, and phenols were not transformed at all. A suitable mechanism is presented in Scheme 5.7.

In summary, ligand coupling promises to be an important branch of organic synthesis, particularly for the preparation of hindered compounds [16].

References

1. Trost, B. M.; LaRochelle, R. W.; Atkins, R. C. *J. Am. Chem. Soc.*, **1969**, *91*, 2175. LaRochelle, R. W.; Trost, B. M. *J. Am. Chem. Soc.*, **1971**, *93*, 6077. Trost, B. M.; Arndt, H. C. *J. Am. Chem. Soc.*, **1973**, 95, 5288.
2. Oae, S. *Croat. Chem. Acta*, **1986**, *59*, 129. Oae, S. *Phosphorus and Sulfur*, **1986**, *27*, 13.

Oae, S.; Kawai, T.; Furukawa, N. *Tetrahedron Lett.*, **1984**, *25*, 69. Oae, S.; Kawai, T.; Furukawa, N.; Iwasaki, J. *J. Chem. Soc., Perkin Trans. II*, **1987**, 405. Uchida, Y.; Onoue, K.; Tada, N.; Nagao, F.; Oae, S. *Tetrahedron Lett.*, **1989**, *30*, 567.

3. Barton, D. H. R.; Glover, S. A.; Ley, S. V. *J. Chem. Soc., Chem. Commun.*, **1977**, 266.

4. The text that follows is an abstract of two review articles in which full references are given: Abramovitch, R. A.; Barton, D. H. R.; Finet, J.-P. *Tetrahedron*, **1988**, *44*, 3039. Finet, J.-P. *Chem. Rev.*, **1989**, *89*, 1487.

5. David, S.; Thieffry, A. *Tetrahedron Lett.*, **1981**, *22*, 5063. David, S.; Thieffry, A. *J. Org. Chem.*, **1983**, *48*, 441.

6. Dodonov, V. A.; Gushchin, A. V.; Brilkina, T. G. *Zh. Obshch. Khim.*, **1984**, *54*, 2157 and **1985**, *55*, 2514.

7. Barton, D. H. R.; Finet, J.-P.; Khamsi, J. *Tetrahedron Lett.*, **1989**, *30*, 937.

8. Caserio, M. C.; Glusher, D. L.; Roberts, J. D. *J. Am. Chem. Soc.*, **1959**, *81*, 336. Scherrer, R. A.; Beatty, H. R. *J. Org. Chem.*, **1980**, *45*, 2127. Lockardt, T. P. *J. Am. Chem. Soc.*, **1983**, *105*, 1940.

9. Barton, D. H. R.; Finet, J.-P.; Giannotti, C.; Halley, F. *J. Chem. Soc., Perkin Trans. I*, **1987**, 241.

10. Bell, H. C.; Pinhey, J. T.; Sternhell, S. *Aust. J. Chem.*, **1979**, *32*, 1551. Pinhey, J. T.; Rowe, B. A. *Aust. J. Chem.*, **1979**, *32*, 1561. Pinhey, J. T.; Rowe, B. A. *Tetrahedron Lett.*, **1980**, *21*, 965. Pinhey, J. T.; Rowe, B. A. *Aust. J. Chem.*, **1980**, *33*, 113. May, G. L.; Pinhey, J. T. *Aust. J. Chem.*, **1982**, *35*, 1859. Pinhey, J. T.; Rowe, B. A. *Aust. J. Chem.*, **1983**, *36*, 789. Kopinski, R. P.; Pinhey, J. T. *Aust. J. Chem.*, **1983**, *36*, 311. Kopinski, R. P.; Pinhey, J. T.; Rowe, B. A. *Aust. J. Chem.*, **1984**, *37*, 1245. Moloney, M. G.; Pinhey, J. T. *J. Chem. Soc., Chem. Commun.*, **1984**, 965. Kozyrod, R. P.; Pinhey, J. T. *J. Chem. Soc., Chem. Commun.*, **1985**, *38*, 713. Kozyrod, R. P.; Pinhey, J. T. *J. Chem. Soc., Chem. Commun.*, **1985**, *38*, 1155. Ackland, D. J.; Pinhey, J. T. *J. Chem. Soc., Perkin Trans. I*, **1987**, 2689.

11. Ackland, D. J.; Pinhey, J. T. *Tetrahedron Lett.*, **1985**, *26*, 5331. Ackland, D. J.; Pinhey, J. T. *J. Chem. Soc., Perkin Trans. I*, **1987**, 2695.

12. Barton, D. H. R.; Donnelly, D. M. X.; Finet, J.-P.; Guiry, P. J. *Tetrahedron Lett.*, **1989**, *30*, 1539. Barton, D. H. R.; Donnelly, D. M. X.; Finet, J.-P.; Stenson, P. H. *Tetrahedron*, **1988**, *44*, 6387.

13. Moloney, M. G.; Pinhey, J. T.; Roche, E. 'G. *J. Chem. Soc., Perkin Trans. I*, **1989**, 333.

14. Barton, D. H. R.; Yadav-Bhatnagar, N.; Finet, J.-P.; Khamsi, J. *Tetrahedron Lett.*, **1987**, *28*, 3111.

15. Barton, D. H. R.; Donnelly, D. M. X.; Finet, J.-P.; Guiry, P. J. *Tetrahedron Lett.*, **1989**, *30*, 1377.

16. For an elegant application of alkenyllead chemistry see Hashimoto, S.; Shirioda, T.; Ikegami, S. *J. Chem. Soc., Chem. Commun.*, **1988**, 1137.

Cycloaddition Reactions and Rearrangements of Stable Silenes and their Derivatives

Adrian G. Brook

Lash Miller Chemical Laboratories
University of Toronto
Toronto, M5S 1A1, Canada

6.1. Introduction

After decades of unsuccessful efforts spent in their attempted syntheses, Gusel'nikov and Flowers in 1967 reported [1] the first compelling evidence for the existence of silenes, compounds containing a double bond between silicon and carbon. This initiated a renewed interest in the synthesis and behavior of silenes. During the 1960s we had been studying the photochemical behavior of acylsilanes [2], which resulted in a 1,2-shift of a silyl group from carbon to oxygen forming a siloxycarbene, as shown in Scheme 6.1.

This study led to the investigation of acyldisilanes [3], which showed evidence for both a 1,2-silyl shift to form a disilyloxycarbene and also for a 1,3-shift of a silyl group from silicon to oxygen, forming a silene. Although the silene could not be observed directly, it could be trapped by methanol (Scheme 6.1). This novel low-energy route to

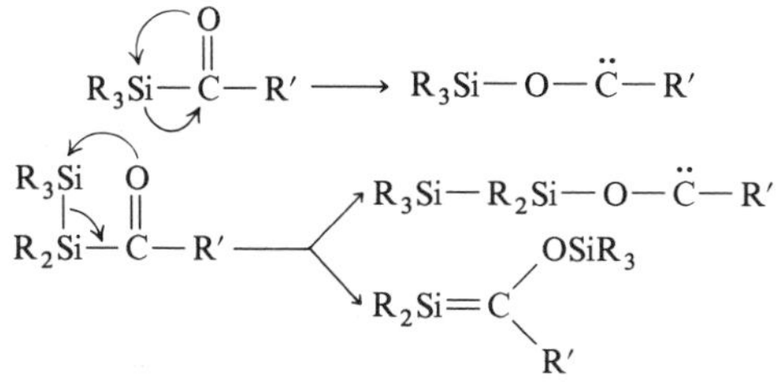

Scheme 6.1

silenes led us ultimately to study the photochemistry of acylpolysilanes, which when photolyzed at about 360 nm gave clear evidence for the exclusive 1,3-shift of a trimethylsilyl group from silicon to oxygen, leading only to silenes. The silenes could be trapped by methanol or by 1,3-dienes [3], then known as efficient traps for silenes [4]. In addition, in the absence of an added trapping reagent, many of the simpler silenes dimerized to give 1,2-disilacyclobutanes, as shown in eq. (6.1), a most unusual mode of dimerization, because many highly reactive, unstable silenes produced by other workers were known to dimerize to yield 1,3-disilacyclobutanes [4].

Stable for R = Ad
CEt_3, Mesityl, (t-Bu)

Ad =

It was recognized that the unusual head-to-head dimerization that these species underwent might be inhibited if the groups on the ends of the double bond were sufficiently large, and hence a program was begun to establish whether this could be accomplished. Ultimately, it was found that, with R = t-Bu [eq. (6.1)], a dynamic equilibrium existed between the silene and its dimer, which was very temperature sensitive [5]. When R was changed to 1-adamantyl, triethylmethyl, or mesityl, no dimer was formed and only the silene could be observed in solution, using NMR methods [6].

Ultimately, it was possible to isolate the adamantylsilene as a solid whose crystal structure, shown in Figure 6.1, was of some interest [7]. Thus, the silicon–carbon double bond had a length of 1.764 Å, rather longer than the 1.70 Å predicted by the best calculations [4], and the molecule was twisted by 14° about the double bond, presumably because of steric interactions of adjacent bulky groups attached to the ends of the double bond. Several relatively stable silenes of this family are now available for use in studying the chemistry of the silicon–carbon double bond.

Some time later the crystal structure of a second crystalline silene and of its THF complex were reported by Wiberg and co-workers [8]. This synthesis, shown in eq. (6.2), was remarkable because of the 1,3-shift of a methyl group from an sp^3- to an sp^2-hybridized silicon atom that occurred spontaneously during the synthesis. The nearly planar silene had a silicon–carbon bond length of 1.70 Å, the value predicted by calculations.

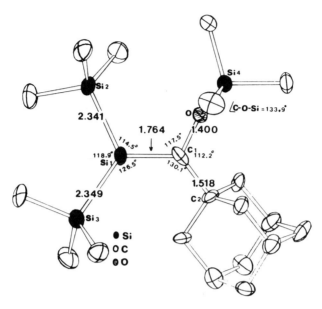

Figure 6.1 ■ ORTEP diagram for adamantylsilene $(Me_3Si)_2Si{=}CAd(OSiMe_3)$.

$$(t\text{-Bu})_2\underset{\underset{F}{|}}{Si}\!-\!\underset{\underset{Li}{|}}{C}(SiMe_3)_2 \longrightarrow [\,t\text{-Bu}_2Si{=}C(SiMe_3)_2\,] \xrightarrow[\text{shift}]{1,3\text{-Me}} t\text{-Bu}_2\underset{\underset{Me}{|}}{Si}\!-\!\overset{\overset{Me_2Si}{\|}}{C}\!-\!SiMe_3$$

$$(6.2)$$

The availability of silenes that were stable in solution over long periods of time (in the absence of oxygen or protic solvents with which they rapidly react) allowed for the first time a study of the chemistry of the silicon–carbon double bond under the conventional conditions familiar to an organic chemist—e.g., in solution at or near room temperature. Much of what is described below concerns our studies of these systems, but others, particularly Wiberg [9], have made a number of important contributions to the understanding of the chemistry of these highly reactive species. Where our results have overlapped, there generally has been close agreement.

6.2. Reactions of Silenes with Dienes and Alkenes

Based on the analogy of the carbon–carbon double bond, whose chemical behavior is well known, it might be expected that silenes could act as dienophiles in the Diels–Alder reaction. This had been found to be true for relatively unstable silenes produced by high temperature thermolyses, and was found to be true for these stable silenes also [3,10]. Thus, all members of our family of silenes (i.e., those silenes that have silyl groups on the sp^2-hybridized silicon atom and a siloxy group on the sp^2-hybridized carbon atom) underwent [2 + 4] cycloadditions with 2,3-dimethylbutadiene to give the expected silacyclohexene, accompanied by the product of an "ene" reaction as a minor byproduct, as shown in eq. (6.3) (R = Ad, *t*-Bu, and Mes).

$$(6.3)$$

In the light of this clean and conventional "organic" behavior it was surprising to find that with butadiene itself the major products of the reactions were silacyclobutanes, often in yields as high as 80%, as the result of a [2 + 2] cycloaddition pathway, with the [2 + 4] cycloaddition product being formed in lesser amounts [10] [eq. (6.4), R = Ad, *t*-Bu, and Mes].

$$(6.4)$$

It was shown that the [2 + 2] cycloaddition was a dark (nonphotochemical) reaction, a process that would be "disallowed" in carbon chemistry, based on the Woodward–Hoffmann rules. A related departure from conventional alkene behavior was also found when the silenes were allowed to react with arylalkenes such as styrene, as shown in Scheme 6.2, or vinylnaphthalene, and so on, where again high yields of [2 + 2] cycloaddition products were obtained. However, if the arylalkenes had allylic hydrogen, or if simple alkenes with allylic hydrogen were employed, the only products from these reactions were those resulting from ene type reactions (Scheme 6.2).

   This raised a question about the mechanism of these cylcoaddition reactions of silenes. It was inferred that radical processes were not involved, based on the inability to affect the nature of proportions of products formed during the course of the reactions by having a large excess of tributyltin hydride, an efficient radical trap, present during the reaction (although if any radical intermediates formed were very short-lived this might not be an effective test). Also, the reactions showed almost no change in rates or proportions of products when the solvents were changed from nonpolar hydrocarbons to ether or THF, suggesting that dipolar intermediates were not being formed in the course of the reactions. (It is not easy to study this mechanistic problem because any reagent that might be employed to trap a dipolar intermediate will almost certainly react immediately with the silene present.) Based on this negative evidence, together with the facts that no polystyrene was formed during the reaction

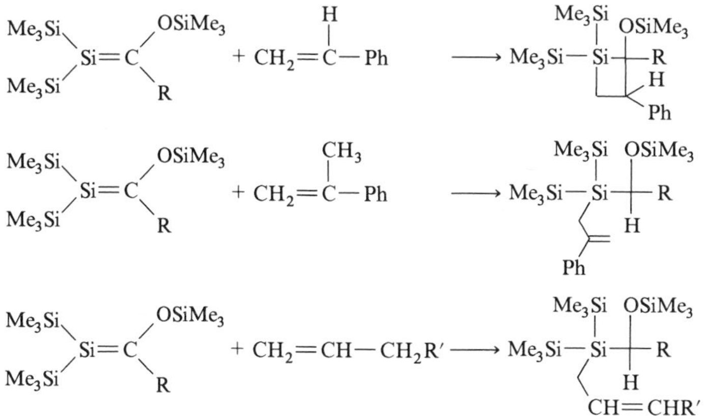

$$\underset{Me_3Si}{\overset{Me_3Si}{>}}Si=C\underset{R}{\overset{OSiMe_3}{<}} \;+\; CH_2=\underset{|}{\overset{H}{C}}-Ph \;\longrightarrow\; Me_3Si-\underset{\substack{| \\ Ph}}{\overset{\substack{Me_3Si \\ |}}{Si}}\underset{H}{\overset{OSiMe_3}{\rceil}}R$$

$$\underset{Me_3Si}{\overset{Me_3Si}{>}}Si=C\underset{R}{\overset{OSiMe_3}{<}} \;+\; CH_2=\underset{|}{\overset{CH_3}{C}}-Ph \;\longrightarrow\; Me_3Si-\underset{Ph}{\overset{Me_3Si}{Si}}\underset{H}{\overset{OSiMe_3}{\rceil}}R$$

$$\underset{Me_3Si}{\overset{Me_3Si}{>}}Si=C\underset{R}{\overset{OSiMe_3}{<}} \;+\; CH_2=CH-CH_2R' \;\longrightarrow\; Me_3Si-\underset{\substack{H \\ CH=CHR'}}{\overset{Me_3Si}{Si}}\overset{OSiMe_3}{\rceil}R$$

R = Ad, *t*-Bu, and Mes

Scheme 6.2

with styrene and that one geometric isomer of a silene capable of geometrical isomerism formed only a single diastereomeric product on reaction with dimethylbuta-diene [11], it was concluded that these cycloaddition reactions occur by a concerted process, Woodward–Hoffman rules not being applicable to these reactions.

6.3. Reactions of Silenes with Carbonyl Compounds: Siloxetanes

It has long been suggested and widely accepted that silenes react with carbonyl compounds to yield cyclic siloxetanes [4] and that these latter species, being quite unstable, spontaneously decompose into silanones and alkenes by a retro-[2 + 2] process as shown in eq. (6.5). Although the expected alkenes are readily isolated, often in high yield, the proposed silanones have not been observed, their cyclic oligomers, particularly the trimer, being the other product normally observed from the reaction.

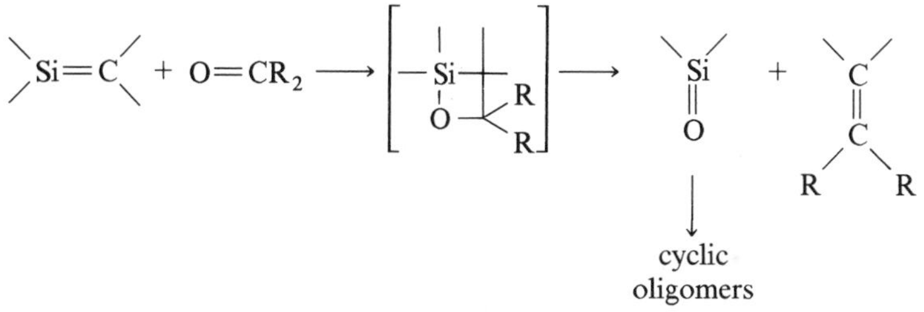

$$\underset{/}{\overset{\backslash}{>}}Si=C\underset{\backslash}{\overset{/}{<}} \;+\; O=CR_2 \;\longrightarrow\; \left[\begin{array}{c} -\underset{|}{\overset{|}{Si}}-\\ \underset{O-}{\overset{}{}} \end{array} \underset{R}{\overset{R}{\rceil}} \right] \;\longrightarrow\; \underset{O}{\overset{\backslash /}{Si}} \;+\; \underset{R\;\;\;R}{\overset{\backslash /}{C}} \qquad (6.5)$$

cyclic oligomers

Each of the silenes in our family reacted cleanly with a wide variety of aldehydes or ketones to yield stable siloxetanes as shown in Scheme 6.3, one of which has been characterized by an X-ray crystal structure [12]. Many of these siloxetanes were quite stable at room temperature in solution or as crystalline solids. However, some of them, having an aryl group attached either to the carbon of the Si = C bond, or to the carbonyl group of the starting ketone, underwent spontaneous rearrangement involving

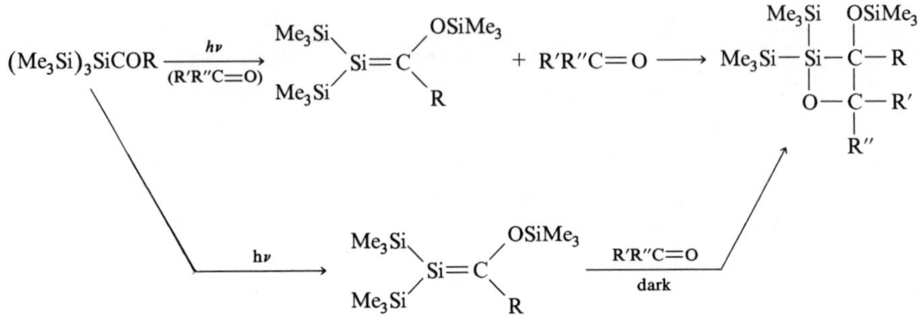

R = *t*-Bu, Ad, Ph, and Mes

R'R''C=O = Ph_2CO, $(p\text{-}Tol)_2CO$, Fluorenone, PhCHO, Me_3CCHO, Me_2CO

One diastereomer for R' = R''
Two diastereomers for R' = R''

Scheme 6.3

substitution into the aromatic ring yielding a bicyclo[4.4.0] product as the major component of an equilibrium mixture with the starting siloxetane, as shown in eq. (6.6).

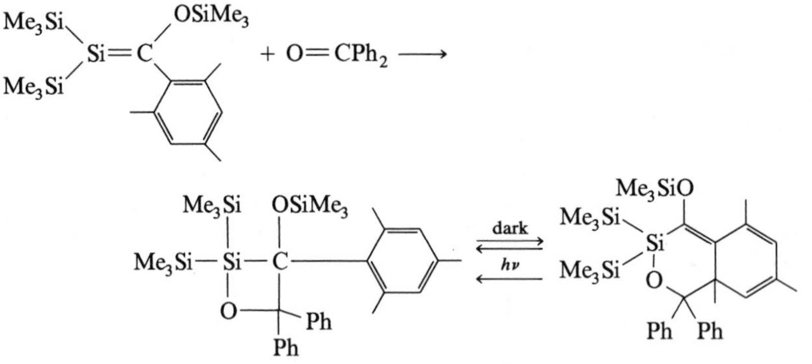

(6.6)

This equilibrium has been approached from both directions. The presence in the reaction mixture of excess tributyltin hydride, as a possible radical trap, or of methanol, as a trap for a dipolar intermediate, had absolutely no influence on the course of the rearrangements, which thus appear to be concerted 1,3-processes. In addition, photolysis of the bicyclo[4.4.0] compound with 360-nm radiation rapidly converted it back to the siloxetane, and the two species could be cycled cleanly back and forth with alternating periods of dark and light without loss of material. In some cases the first detected product of these nonphotochemical reactions between silenes and carbonyl compounds was the bicyclo[4.4.0] species shown in eq. (6.7), which usually subsequently isomerized spontaneously to the siloxetane.

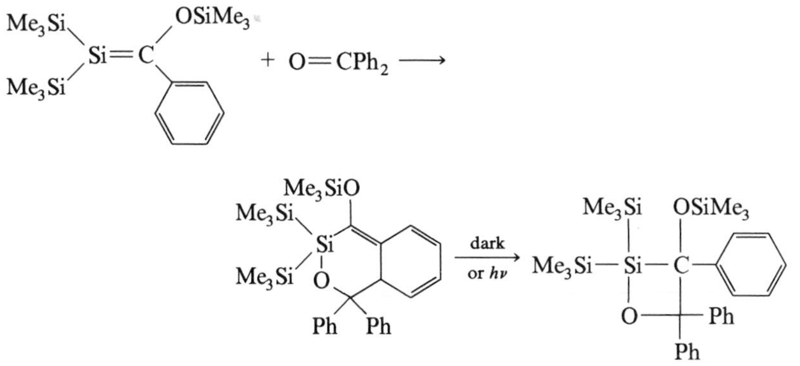

$$(6.7)$$

It is obvious that the siloxetane and bicyclo[4.4.0] ring systems are nearly equal in energies (the equilibrium constant in one case was about 2), suggesting that the ring strain and other strains in the crowded siloxetane ring system are about equal to the aromatic resonance energy lost in forming the bicyclo[4.4.0] ring system.

6.3.1. Thermal Behavior of the Siloxetanes

The isolation of stable siloxetanes allowed testing of the hypothesis that these species thermally break down into a silanone and an alkene. In fact, very few of the siloxetanes prepared decomposed thermally by this pathway. The kinetics of decomposition of those that did, such as the fluorenyl derivative shown in eq. (6.8), were very complex and totally incompatible with the anticipated simple first-order kinetics expected if the reaction was the simple retro-[2 + 2] reaction depicted in eq. (6.5).

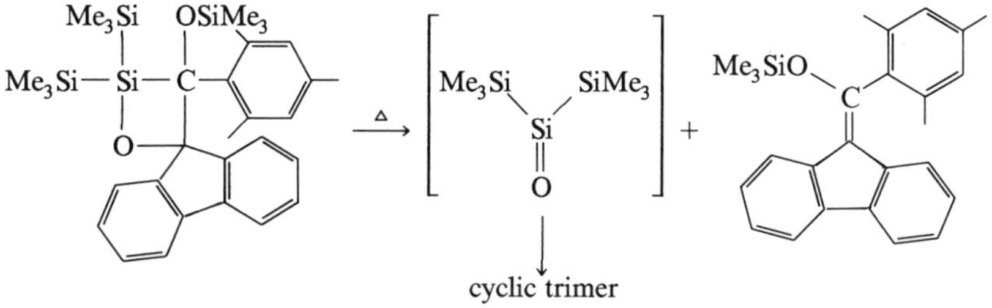

$$(6.8)$$

The kinetics, followed by NMR techniques, showed the formation and disappearance of two or more intermediates and were more consistent with the proposals of Barton [13] and Bachrach and Streitwieser [14], who suggested that bimolecular reactions between siloxetane molecules could lead to the expulsion of the alkenes observed and the ultimate formation of the silanone oligomers, without requiring the formation of intermediate silanones, a process calculated to be highly endothermic.

Most of the siloxetanes, when thermolyzed in the temperature range 75–110 °C, underwent clean intramolecular rearrangements, depicted in eq. (6.9), leading to the formation of siloxysiloxyalkenes, which have been isolated and characterized.

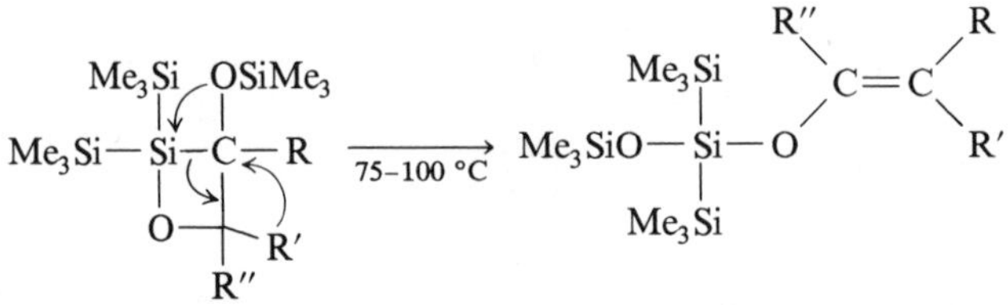

$$(6.9)$$

This appears to be a concerted process because attempts to intercept possible carbene intermediates failed completely. A related thermal rearrangement of this type has also been proposed by Märkl and Horn [15] to explain the results of thermolysis of a 1,2-disilacyclobutane with ketones.

6.4. Reactions of Silenes with Imines

As might have been expected, the reactions of the stable silenes with imines were similar to those with carbonyl compounds [11]. For example, the mesitylsilene shown in eq. (6.10) reacted with the fluorenylimine in the dark to give the [2 + 2] adduct shown, and this behavior was found with some other silenes and imines. In the above case, the 1-sila-2-azacyclobutane was rather unstable and slowly decomposed by a retro-[2 + 2] process, since the reaction products were the siloxyalkene and the 1,3-disila-2,4-diazacyclobutane, the product expected from head-to-tail dimerization of an intermediate silaimine. The silaimine was not observed directly, and its dimer may have been formed directly from a bimolecular process.

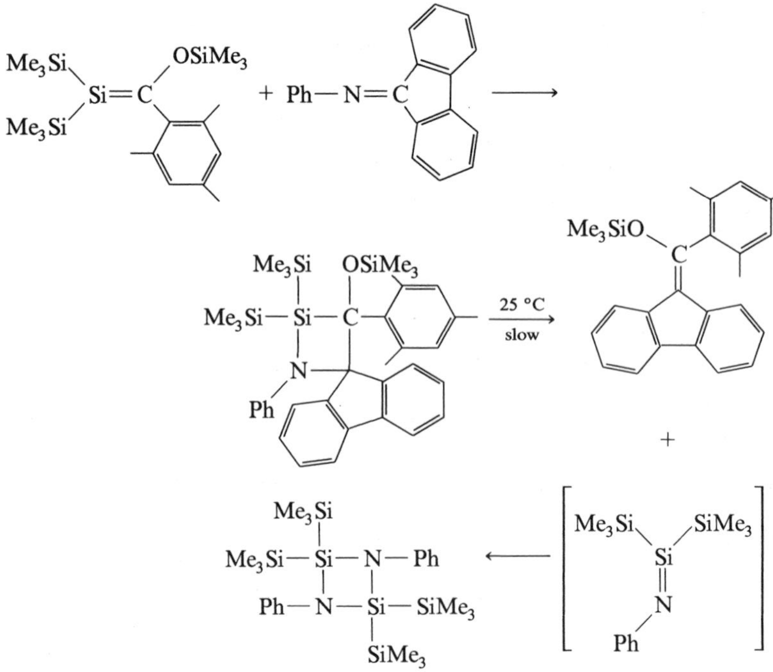

$$(6.10)$$

In other cases, as illustrated in eq. (6.11) (where Ad = 1-adamantyl), the silene reacted with *C*-arylimines to give a bicyclo[4.4.0] product, the nominal [2 + 4] adduct of the silene (2π) reacting with the imine (4π). Generally, the bicyclo[4.4.0] species were not very stable and slowly rearranged in solution, either photochemically or in the dark when warmed, to the more stable silaazacyclobutane.

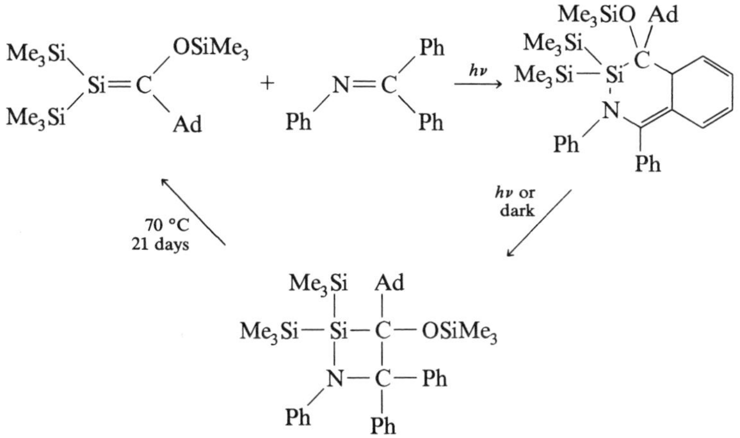

(6.11)

Wiberg studied the reactions of his stable silenes with *N*-silylimines and also observed both the [2 + 2]- and [2 + 4]-type products [9]. Whereas Wiberg found his adducts to be useful sources of his stable silenes under mild warming, this was not generally true for the adducts shown in eqs. (6.10) and (6.11). The former slowly decomposed by a retro-[2 + 2] process of the opposite sense to give an alkene and silaimine dimer, and the latter reverted to silene and imine only very slowly over 21 days when heated at 70 °C. At this temperature our silenes rapidly rearrange back to their parent acylsilanes, so this reaction is not a useful thermal route to our silenes.

6.5. Cycloadditions Leading to Three-Membered Rings

6.5.1. Disilacyclopropanes

Recently, the reactions of silenes with reagents that lead to the formation of three-membered rings have been investigated [11]. In an attempt to form disilacyclopropanes, the reaction of our family of silenes with hexamethylsilirane, a known source of dimethylsilylene when heated to about 60 °C, was explored. Because heating our silenes above room temperature greatly accelerates their retrorearrangement back to the parent acylsilane, it was necessary to allow the silene to react with hexamethylsilirane at room temperature. Under these conditions a slow reaction ensued, which probably involved a radical process, as has been noted by Seyferth and co-workers [16] under similar conditions, and the products of the reaction included the desired disilacyclopropanes accompanied by products in which all the atoms of the hexamethylsilirane became attached to the silene, as shown in eq. (6.12) (R = Ad and *t*-Bu).

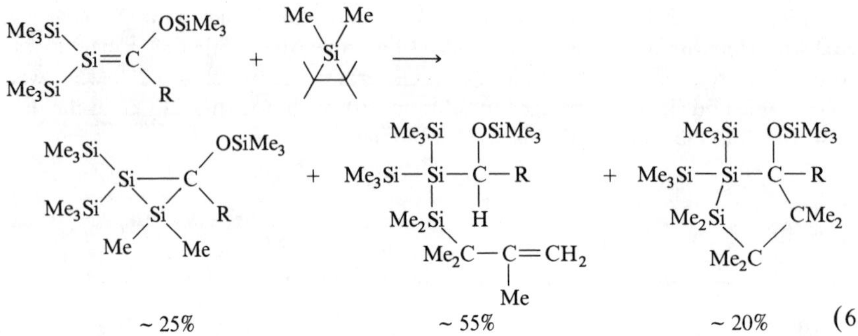

$$\sim 25\% \qquad\qquad \sim 55\% \qquad\qquad \sim 20\% \qquad (6.12)$$

The reaction of the mesitylsilene with hexamethylsilirane, shown in eq. (6.13) was much cleaner, with the disilacyclopropane being the major product, accompanied by only small amounts of other byproducts. Surprisingly, but in direct analogy with the previously mentioned behavior of the siloxetanes, this disilacyclopropane slowly isomerized over two days to a bicyclo[4.3.0] species, which could be photochemically reconverted to the disilacyclopropane on brief photolysis at 360 nm [17]. These species could be cycled back and forth cleanly, using alternating periods of dark and light.

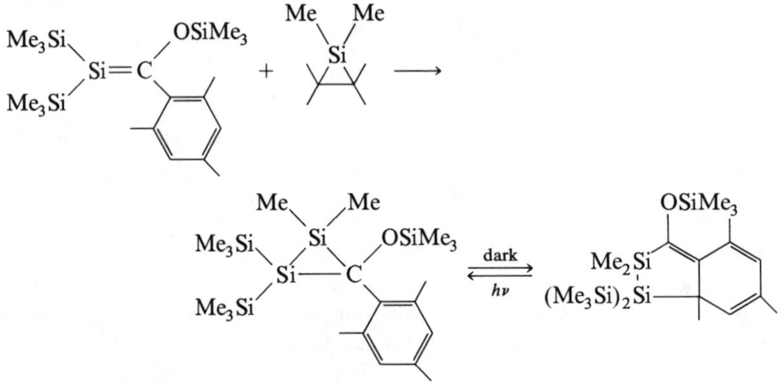

$$(6.13)$$

6.5.2. Silaaziridines and 1-Sila-3-azacyclobutanes

A second example of cycloaddition leading to the formation of a three-membered ring came from a study of the reactions of the silenes with isonitriles. West and co-workers [18] had reported that one of his stable disilenes reacted cleanly with 2,6-dimethylphenylisocyanide to give the expected disilacyclopropanimine, as shown in eq. (6.14) (Xyl = 2,6-dimethylphenyl).

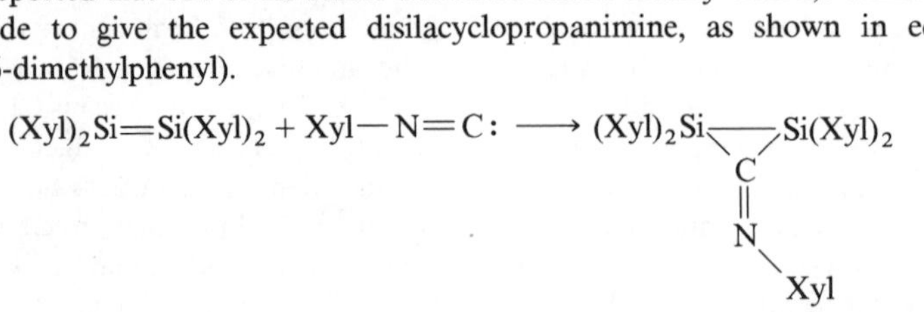

It would be expected that our silenes would react to give the analogous silacyclopropanimines, as shown in eq. (6.15), but this was found not to be the structure of the family of compounds formed from the reaction of several silenes with two different alkyl isocyanides.

$$(Me_3Si)_2Si\!=\!C\!\!\begin{array}{c}R\\ \\OSiMe_3\end{array} + R'N\!=\!C\!: \longrightarrow (Me_3Si)_2Si\!\!\begin{array}{c}R\\ \diagdown\\C\!\!-\!\!C\\ \parallel\quad OSiMe_3\\N\\ \diagdown R'\end{array} \quad (6.15)$$

The crystal structure of one of the compounds, shown in Figure 6.2, indicated that the species formed were the isomeric silaaziridines [19], which presumably had arisen by rearrangement of the anticipated silacyclopropanimines. This was confirmed from studies carried out at -70 °C. At this temperature the products formed from the reaction of the alkylsilenes with t-butyl isocyanide showed totally different physical properties from those of the silaaziridines, particularly the NMR data which were fully consistent with the properties expected of silacyclopropanimines. However, when the temperature was then raised to -40 °C the signals of the silacyclopropanimines disappeared as the signals of the silaaziridines grew in. This sequence is shown in eq. (6.16) (R, R' aliphatic).

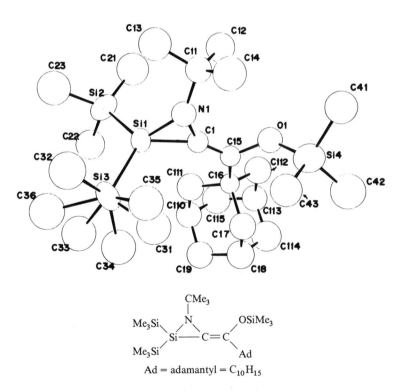

$$Me_3Si\!\!\begin{array}{c}\quad CMe_3\\\quad |\\\diagdown \quad N \qquad OSiMe_3\\ \diagdown \quad \diagup\\Si\!\!-\!\!C\!=\!C\\ \diagup \qquad\qquad \diagdown\\Me_3Si \qquad\qquad Ad\end{array}$$

$$Ad = adamantyl = C_{10}H_{15}$$

Figure 6.2 ■ ORTEP diagram for silaaziridine.

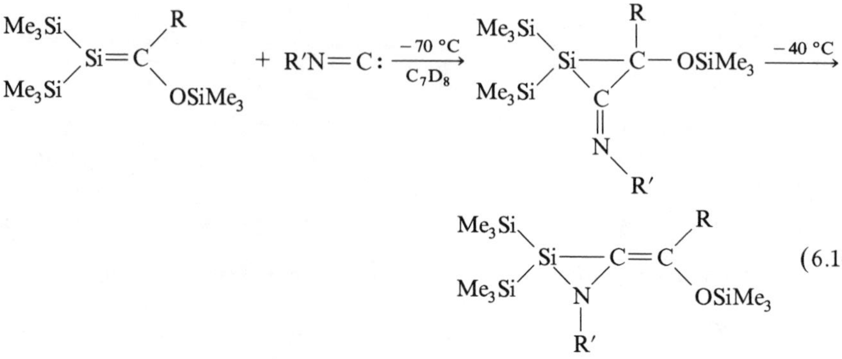

$$(6.16)$$

The mechanism by which this rearrangement occurs is of some interest. Although methylenecyclopropanes are known to rearrange by a process involving homolysis of the ring bond opposite the exocyclic double bond, this process occurs only at relatively high temperatures (above 100 °C) whereas the present rearrangement occurs at −40 °C or lower. A more reasonable mechanism would appear to be the heterolysis of the silicon–carbon ring bond to give a 1,3-dipolar species that contains a silyl-stabilized silanion and a delocalized cation, stabilized by the adjacent oxygen and the exocyclic double bond, as shown in eq. (6.17), which can then rearrange to yield the silaaziridine. Attempts are currently underway to trap the proposed intermediate 1,3-dipolar species with appropriate reagents.

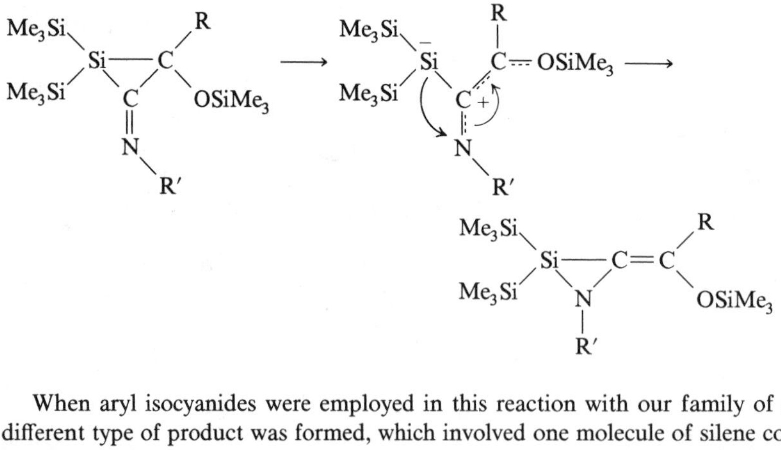

$$(6.17)$$

When aryl isocyanides were employed in this reaction with our family of silenes a different type of product was formed, which involved one molecule of silene combining with two molecules of isonitrile. A crystal structure of one of these products showed that they were 1-sila-3-azacyclobutanes [20], having the structure shown in Figure 6.3.

The atoms comprising the ring and the exocyclic double bonds lie essentially in a plane and the compound can be visualized as arising from insertion of a second molecule of aryl isocyanide into the ring Si—N bond of an initially formed silaaziridine. Evidence that this occurred was obtained in an experiment at −70 °C involving equimolar amounts of silene and 2,6-dimethylphenyl isocyanide that gave rise to a mixture of about one part of silaaziridine and four parts of 1-sila-3-azacyclobutane (together with unconsumed silene), as shown in eq. (6.18).

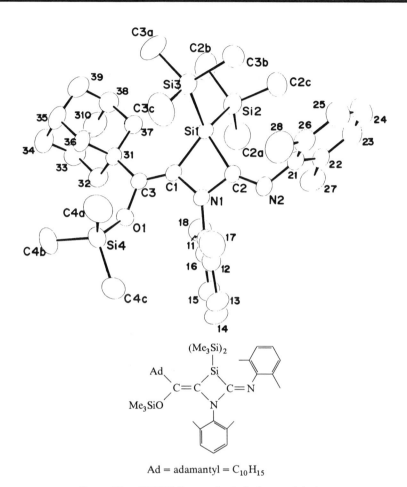

Ad = adamantyl = $C_{10}H_{15}$

Figure 6.3 ■ ORTEP diagram for 1-sila-3-azacyclobutane.

$$(6.18)$$

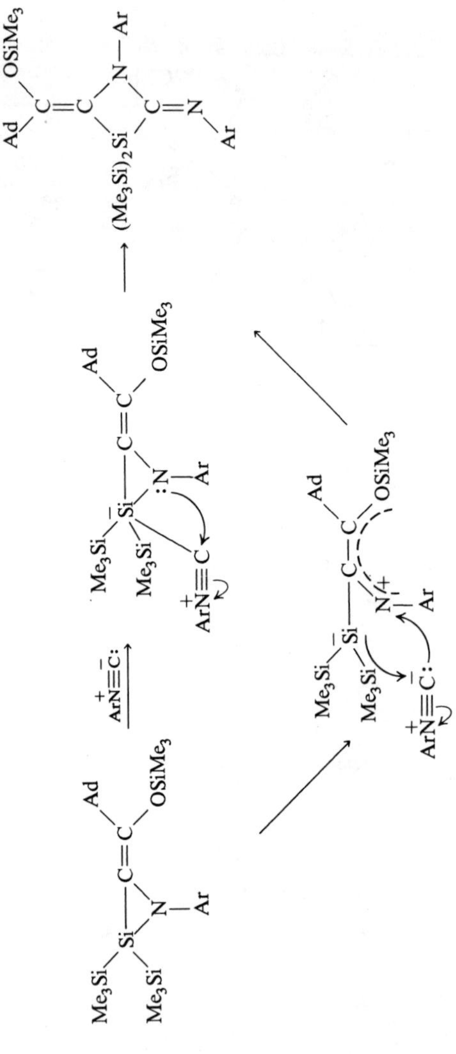

Scheme 6.4

If more isonitrile was then added to the mixture, the silaaziridine was converted to the four-membered ring compound. The mechanism of this ring expansion is as yet unknown: It could be conceived as involving nucleophilic attack by the isonitrile on the ring silicon atom, followed by rearrangement, or alternatively, as a heterolysis of the ring Si—N bond to give a 1,3-dipolar species involving a silyl-stabilized carbanion and a delocalized cation that reacted with the isonitrile. These pathways are shown in Scheme 6.4. In either case, it is clear that the aryl group must lower the energy barrier for this ring expansion, because N-alkyl-silaaziridines do not undergo ring expansion when treated with either alkyl or aryl isocyanides in the dark.

Overall, the reaction between silenes and isonitriles leading to the formation of these new silaheterocycles is the result of a series of very facile molecular rearrangements that occur at quite low temperatures: the sequence of reactions is summarized in eq. (6.19) (where Ar = 2,6-dimethylphenyl). We are continuing to explore the chemistry of these new silaheterocycles.

$$(Me_3Si)_2Si{=}C\!\!\begin{array}{c}R\\\\OSiMe_3\end{array} + R'N{=}C: \longrightarrow (Me_3Si)_2Si\!\!\begin{array}{c}R\\|\\\\C-OSiMe_3\\\parallel\\N\\\backslash R'\end{array} \longrightarrow$$

$$(Me_3Si)_2Si\!\!\begin{array}{c}R\\|\\C{=}C\\N\;\;\;\;OSiMe_3\\|\\R'\end{array} \xrightarrow[\substack{\text{if } R' \text{ or}\\ R = Ar}]{Ar-N{=}C:} (Me_3Si)_2Si\!\!\begin{array}{c}R\;\;\;OSiMe_3\\\backslash\;/\\C\\\parallel\\C\\/\;\;\;\backslash\\\;\;\;\;N-R'\\C\\\parallel\\NAr\end{array} \quad (6.19)$$

6.6. Photochemical Rearrangements of Alkyl- or Aryl-substituted Silenes

All of the cycloadditions described above involved silenes of the family that carry a siloxy group on sp^2-hybridized carbon and two trimethylsilyl groups on sp^2-hybridized silicon. There is no doubt that these substituents, electron-releasing at carbon and electron-withdrawing at silicon, have an effect on the behavior of the silenes. Calculations by Apeloig and Karni [21] suggest that groups of this type would cause the length of the silicon–carbon double bond to be longer than normal, as observed from the crystal structure, and that the silene would be less polarized than a "simple" and more typical silene that lacks these groups, which may help to explain the unusual head-to-head dimerization. Although the chemistry described above is typical of silenes in general, as far as it has been investigated by other workers using "simple" silenes, we have been concerned to determine the behavior of our family of silenes when one or more of the trimethylsilyl groups was replaced by more conventional hydrocarbon substituents.

The synthesis of the necessary polysilylacylsilanes was carried out in the conventional manner by coupling of the appropriate silyllithium reagents [prepared by cleavage of tris(trimethylsilyl)SiR with methyl lithium], with the appropriate acid chloride in an inert solvent, as shown in eq. (6.20) (where R = Me, t-Bu, and Ph and R' = Ad, CEt$_3$, and Mes).

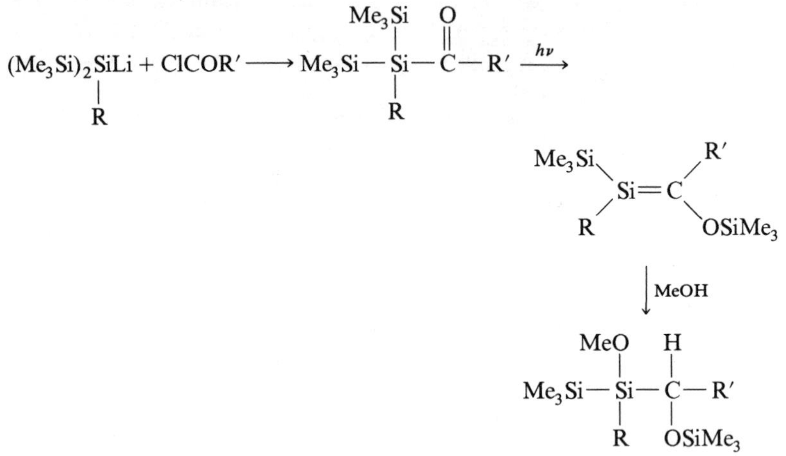

$$(Me_3Si)_2SiLi + ClCOR' \longrightarrow Me_3Si-\underset{\underset{R}{|}}{\overset{\overset{Me_3Si}{|}}{Si}}-\overset{\overset{O}{||}}{C}-R' \xrightarrow{h\nu}$$

$$\underset{R}{\overset{Me_3Si}{\diagdown}}Si{=}C\underset{OSiMe_3}{\overset{R'}{\diagup}} \quad (6.20)$$

$$\downarrow \text{MeOH}$$

$$Me_3Si-\underset{\underset{R}{|}}{\overset{\overset{MeO}{|}}{Si}}-\underset{\underset{OSiMe_3}{|}}{\overset{\overset{H}{|}}{C}}-R'$$

The resulting acylsilanes differed little from the previous tris(trimethyl)silyl analogs: Each when photolyzed in methanol gave a diastereomeric mixture of methanol adducts of the anticipated silene, as confirmed by NMR and other spectroscopic methods [22].

Three different systems have been studied in depth and will be described below. Several further systems have been investigated that confirm the generality of the reactions.

6.6.1. The Methylsilene (Me$_3$Si)MeSi = CAd(OSiMe$_3$)

The behavior of the methyl-substituted acylsilane is shown in Scheme 6.5. When photolyzed in inert solvent, the anticipated silene was evidently rather unstable because it could not be observed: It underwent the typical head-to-head dimerization to a 1,2-disilacyclobutane that molecular modeling predicts should have the all-trans

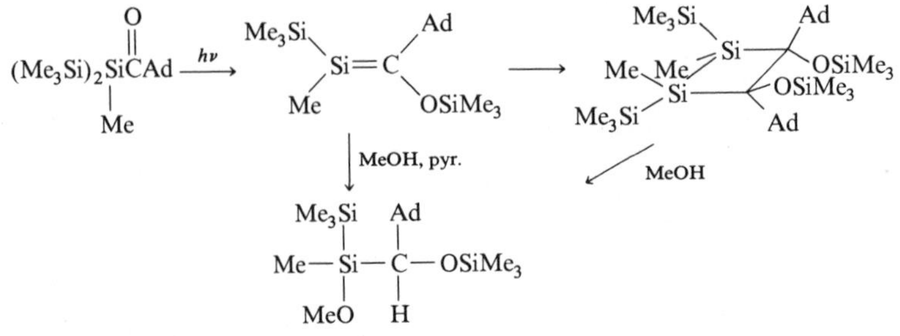

Scheme 6.5

structure shown [23]. Initially, it seemed surprising that a dimer with adjacent adamantyl groups could be formed, in light of the fact that the much-studied silene $(Me_3Si)_2Si = CAd(OSiMe_3)$ did not dimerize, which was attributed to the unfavorable steric interaction of neighboring bulky groups, especially the adamantyl groups on adjacent carbon atoms. In the case of the methyl-substituted silene, minor changes in bond angle appear to be possible because of the smaller size of methyl relative to trimethylsilyl, reducing the strain of the adjacent adamantyl groups. However, the head-to-head dimer was not very stable, and simply placing it in methanol soon led to its disappearance and the formation of the same methanol adducts of the silene precursor as obtained by photolysis of the acylsilane in methanol. Thus, replacement of a trimethylsilyl group by a methyl group led to no new chemistry but only the formation of a silene that readily dimerized, presumably because, having somewhat less bulky groups, its dimerization was less sterically hindered.

6.6.2. The *t*-Butylsilene 2, $(Me_3Si)t$-BuSi $= CAd(OSiMe_3)$

Replacement of a trimethylsilyl group on the acylsilane by a *t*-butyl group led to much more unusual behavior. Photolysis of the acylsilane **1** led to a quite stable silene **2** having the *t*-butyl group on sp^2-hybridized silicon atom (shown in Scheme 6.6), which was studied by ^1H, ^{13}C, and ^{29}Si NMR spectroscopy. This is the first reported case of a stable silene that can exist as a pair of cis–trans isomers: Spectroscopic investigation indicated that only one geometric isomer was present (the configuration is unknown but is predicted by molecular modeling to be the E isomer [23]). That only one isomer was present was supported by the finding that its Diels–Alder reaction with 2,3-dimethylbutadiene gave only a single diastereomer where two should have been obtained if two isomeric silenes were present. When the silene was added to methanol the same mixture of diastereomeric methanol adducts was isolated as was obtained when the parent acylsilane was photolyzed in methanol, as shown in Scheme 6.6. This finding, coupled with the observed single product from the reaction with dimethylbutadiene indicates that the reaction of methanol with these silenes is not a stereospecific process, but must be a two-step process initiated by nucleophilic attack of methanol on the sp^2-hybridized silicon atom of the silene, followed by a proton transfer from oxygen to carbon.

During the course of the complete photolysis of the *t*-butyl acylsilane **1** to its silene isomer **2**, it was observed that the silene itself began to disappear, with the formation of new NMR signals attributable to the formation of a second silene **3**. This new dimethylsilene was sufficiently long-lived that it also could be studied by NMR spectroscopy. Its structure was shown to have three methyl groups on one silicon atom, two methyls on a second silicon atom, and one methyl group on a third silicon atom, a structure totally different from its precursor silene, which had both trimethylsilyl and trimethysiloxy groups. It was evident that a major photochemical rearrangement had taken place.

The new silene when treated with methanol gave a different pair of diastereomeric adducts, one of which was crystalline so that its crystal structure could be obtained, as shown in Figure 6.4. This indicated that the new silene had the structure **3**, $Me_2Si = CAd(SiMe(t$-Bu$) — OSiMe_3)$, because methanol addition to silenes is known to be a reliable reaction occurring without molecular rearrangement. Also supportive of the structure of the new silene was the fact that when this "simple" silene

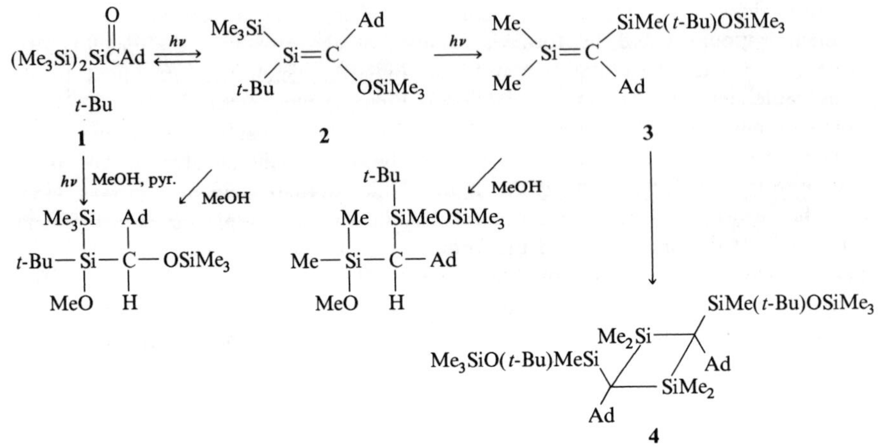

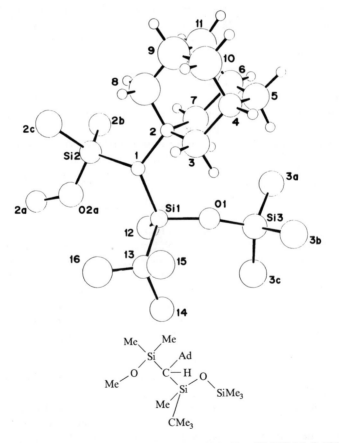

Scheme 6.6

Figure 6.4 ■ ORTEP diagram for methanol adduct of Me$_2$Si=CAd(SiMe(t-Bu)OSiMe$_3$).

dimerized, it gave as a major product the expected head-to-tail dimer **4** of the new silene, shown in Scheme 6.6, as confirmed by its crystal structure [24]. These two crystal structures and the additional ^{1}H, ^{13}C, and ^{29}Si NMR data that were obtained on the silene clearly establish its structure. It is evident that a remarkable rearrangement has occurred during the photolysis of **2** to **3**, since the new silene lacks the original trimethylsilyl group attached to silicon (there are now no Si—Si bonds), it lacks the trimethylsiloxy group attached to carbon, which is now attached to silicon, and it is evident that a methyl group has moved from one silicon atom to a silicon atom that originally lacked a methyl group. Before attempting to explain a possible pathway for these processes, it is useful to consider the behavior of an acylsilane where a phenyl group replaces a trimethylsilyl group.

6.6.3. The Phenylsilene 6, (Me₃Si)PhSi=CAd(OSiMe₃)

The acylsilane **5** in which a phenyl group replaces one of the trimethylsilyl groups of the original acylsilane, shown in Scheme 6.7, when photolyzed in methanol gave the expected pair of diastereomeric methanol adducts, indicating that the anticipated silene **6** had been formed.

However, the silene was evidently unstable, because NMR data indicated that a pair of head-to-head dimers **7** were present as the major initial products of the photolysis. These were probably derived from a single silene geometric isomer. Thus, although the same two diastereomeric methanol adducts were obtained when methanol was added to the dimers (which presumably dissociate back to parent silene and may even be in equilibrium with a small and undetected amount of silene in solution at room temperature), the addition of dimethylbutadiene gave rise to only a single [2 + 4] product, as would be expected of a stereoselective concerted reaction with a single silene isomer.

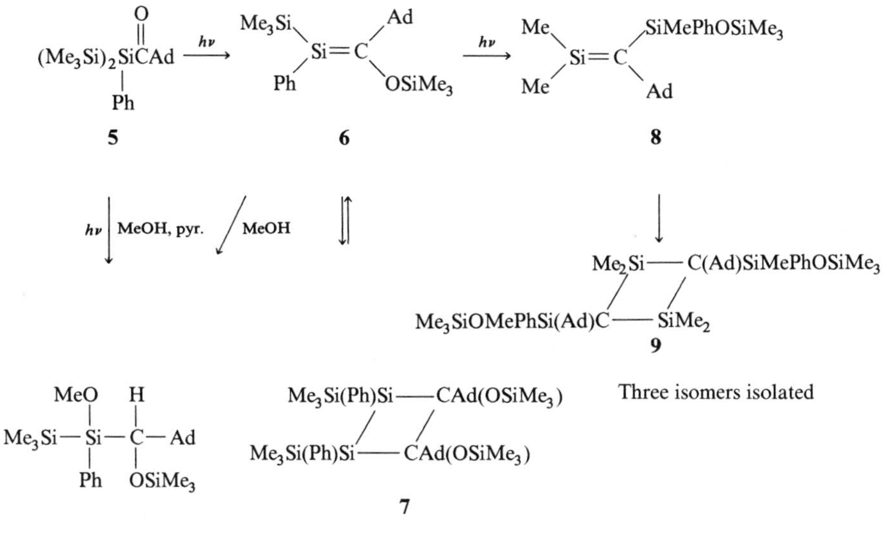

Scheme 6.7

Analogous to the behavior of the *t*-butyl silene described previously, further photolysis of the solution containing the phenyl silene dimers **7** led to their disappearance and the formation of new crystalline head-to-tail dimers **9**, each derived from the same silene parent **8**, namely, $Me_2Si=CAd(SiMePh-OSiMe_3)$. This silene can exist as a pair of enantiomers and can combine to form dimers having *cis*-RR,SS, *cis*-RS,SR, and *trans*-RR,SS or *trans*-RS,SR configurations. Three of these dimers were isolated, and the structures of two of them, both *cis*-isomers, were established by X-ray methods [24]. The structure of the third isomer was shown to be the *trans*-RS,SR compound, based on NMR studies. The crystal structure of the crowded *cis*-RS,SR dimer is shown in Figure 6.5. The very long silicon–carbon ring bonds of 1.96 Å (normal value 1.89 Å) reflects the crowding in these compounds.

The photoisomerization of the original phenylsilene **6** to the monomeric precursor **8** of the dimers involved exactly the same rearrangement process enumerated previously for the *t*-butyl silene, i.e., migration of the trimethylsilyl group from silicon to carbon, migration of the trimethylsiloxy group from carbon to silicon, and a methyl migration from one silicon atom to a different silicon atom, as shown in eq. (6.21) (where R = *t*-Bu and Ph).

$$Me_3Si \diagdown \underset{R}{\overset{}{Si}}=C\underset{OSiMe_3}{\overset{Ad}{}} \longrightarrow Me \diagdown \underset{Me}{\overset{}{Si}}=C\underset{Ad}{\overset{Si-OSiMe_3}{\overset{Me \quad R}{}}} \qquad (6.21)$$

Studies with other systems, in particular where the R group on the carbonyl carbon of the initial acylsilane was changed from adamantyl to triethylmethyl or mesityl, indicated that similar photochemical rearrangements occurred with these molecules

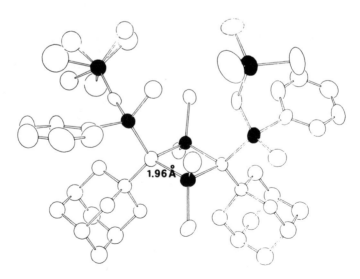

Figure 6.5 ■ ORTEP diagram of (RS,SR) cis dimer of $Me_2Si=CAd(SiMePhO-SiMe_3)$.

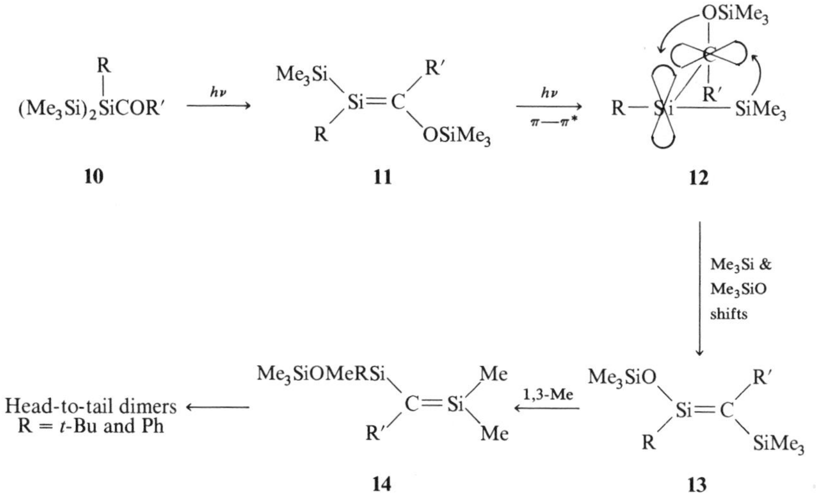

Scheme 6.8

also, so that the photochemical silene-to-silene isomerization shown in eq. (6.21) is a fairly general process.

The simplest interpretation consistent with the available facts of what is occurring is summarized in Scheme 6.8. Photolysis of the general acylsilane **10** causes the usual 1,3-trimethylsilyl shift from silicon to oxygen to form only one of the two possible silene geometric isomers **11**. Which isomer is formed is not known with certainty but molecular modeling [23] indicates that the E isomer is the more stable, regardless of which group, Me, *t*-Bu, or Ph, replaced the trimethylsilyl group on silicon in the parent acylsilane. The photochemistry of the silene **11** by a π-to-π^*-process (silenes are known to absorb in a broad band at about 340 nm that encompasses the wavelength used to irradiate the acylsilane [6]), will be expected to involve a twisted excited state **12** as shown. In this conformation the empty p orbital on silicon is favorably disposed to interact with the migrating trimethylsiloxy group, as is the empty p orbital on carbon toward migration of the trimethylsilyl group from silicon. Both of these processes could occur simultaneously, that is, by a dyotropic rearrangement [25]. Whereas dyotropic rearrangements across single bonds are well established, dyotropic rearrangements across double bonds are not well known, although West has recently suggested this mechanism in explaining a rearrangement of one of his stable disilenes [26]. Because of the excited state, the present case seems best considered as a dyotropic rearrangement across a single bond.

The possibility that migration of the two groups occurred in a stepwise manner was investigated. If the siloxy group migrated first, a carbene would be formed as an intermediate, but attempts to trap such a carbene failed. Similarly, if the trimethylsilyl group migrated first, a silylene would be formed as an intermediate, but the reaction occurred unaffected by the presence of a large excess of any of several silanes (Si — H) known to be very effective silylene traps. Hence, it was concluded that the double rearrangement probably occurred simultaneously. The rearranged species is then presumed to collapse to the ground state to give the new silene **13**, which is not detected. This suggests that the subsequent 1,3-methyl shift from sp^3-hybridized silicon to sp^2-hybridized silicon is a rapid process, as has been observed previously both by Wiberg [9] and by Eaborn [27] with rather different systems. This methyl migration leads to the silene **14**, which has been trapped in various cases by methanol, or as head-to-tail dimers with known crystal structures, as described previously. This rather complex process adequately accounts for the wide variety of data that have been obtained, some of which have been given above.

6.7. Summary

From the previous results it is clear that the replacement of one of the trimethylsilyl groups on silicon in the Brook family of stable silenes by a simple hydrocarbon group has led to new, unusual, and exciting chemistry by these compounds. Although some of the chemistry of silenes obviously parallels that of alkenes, much of the chemistry of the silicon–carbon double bond is unique and sometimes highly dependent on the substituents attached thereto. The future promises further interesting results.

Acknowledgments. The author is indebted to numerous talented, dedicated, and hard-working graduate students and postdoctoral fellows over the years for the success of this work, and to the Natural Science and Engineering Research Council of Canada for the funding of this research.

References

1. Gusel'nikov, L. E.; Flowers, M. C. *J. Chem. Soc., Chem. Commun.*, **1967**, 864.
2. Duff, J. M.; Brook, A. G. *Can. J. Chem.*, **1973**, *53*, 2893.
3. Brook. A. G.; Harris, J. W.; Lennon, J.; El Sheikh, M. *J. Am. Chem. Soc.*, **1979**, *101*, 83.
4. For recent reviews on the chemistry of silenes see the following: Brook, A. G., Baines, K. M. *Adv. Organomet. Chem.*, **1986**, *25*, 1. Raabe, G.; Michl, J. *Chem. Rev.*, **1985**, *85*, 419.
5. Brook, A. G.; Nyburg, S. C.; Reynolds, W. F.; Poon, Y. C.; Chang, Y.-M.; Lee, J.-S.; Picard, J.-P. *J. Am. Chem. Soc.*, **1979**, *101*, 6750.
6. Brook, A. G.; Nyburg, S. C.; Abdesaken, F.; Gutekunst, B.; Gutekunst, G.; Kallury, R. K. M. R.; Poon, Y. C.; Chang, Y.-M.; Wong-Ng, W. *J. Am. Chem. Soc.*, **1982**, *104*, 5667.
7. Nyburg, S. C.; Brook, A. G.; Abdesaken, F.; Gutekunst, G.; Wong-Ng, W. *Acta. Cryst.*, **1985**, *C41*, 1632.
8. Wiberg, N.; Wagner, G.; Reide, J.; Müller, G. *Organometallics*, **1987**, *6*, 32.
9. Wiberg, N. *J. Organomet. Chem.*, **1984**, *272*, 141.
10. Brook, A. G.; Vorspohl, K.; Ford, R. R.; Hesse, M.; Chatterton, W. *J. Organometallics*, **1987**, *6*, 2128.
11. Unpublished research by A. G. Brook and W. J. Chatterton.
12. Brook, A. G.; Chatterton, W. J.; Sawyer, J. F.; Hughes, D. W.; Vorspohl, K. *Organometallics*, **1987**, *6*, 1246.
13. Barton, T. J., personal communication.
14. Bachrach, S. M.; Streitwieser, A., Jr. *J. Am. Chem. Soc.*, **1985**, *107*, 1186.
15. Märkl, G.; Horn, M. *Tetrahedron Lett.*, **1983**, *24*, 1477.
16. Seyferth, D.; Vick, S. C.; Shannon, M. L.; Lim, T. F. O.; Duncan, D. P. *J. Organomet. Chem.*, **1977**, *C37*, 135.

17. Brook, A. G.; Wessely, H.-J. *Organometallics*, **1985**, *4*, 1487.
18. Yokelson, H. B.; Millevolte, A. J.; Haller, K. J.; West, R. *J. Chem. Soc., Chem. Commun.*, **1987**, 1605.
19. Brook, A. G.; Kong, Y. K.; Saxena, A. K.; Sawyer, J. F. *Organometallics*, **1988**, *7*, 2245.
20. Brook, A. G.; Saxena, A. K.; Sawyer, J. F. *Organometallics*, **1989**, *8*, 850.
21. Apeloig, Y.; Karni, M. *J. Am. Chem. Soc.*, **1984**, *106*, 6676.
22. Baines, K. M.; Brook, A. G.; Ford, R. R.; Lickiss, P. D.; Saxena, A. K.; Chatterton, W. J.; Sawyer, J. F.; Behnam, B. *Organometallics*, **1989**, *8*, 693.
23. Brook, A. G.; Gajewski, J. G. *Heteroatom Chem.*, **1990**, *1*, 57.
24. Baines, K. M.; Brook, A. G.; Lickiss, P. D.; Sawyer, J. F. *Organometallics*, **1989**, *8*, 709.
25. Reetz, M. T. *Adv. Organomet. Chem.*, **1977**, *16*, 33.
26. Yokelson, H. B.; Maxka, J.; Siegel, D. A.; West, R. *J. Am. Chem. Soc.*, **1986**, *108*, 4239.
27. Eaborn, C.; Happer, D. A. R.; Hitchcock, P. B.; Hopper, S. P.; Safa, K. D.; Washburne, S. S.; Walton, D. M. *J. Organomet. Chem.*, **1980**, *186*, 309.

Carbanion Production by Reductive Cleavage of Carbon–Sulfur and Carbon–Oxygen Bonds

Theodore Cohen

Department of Chemistry
University of Pittsburgh
Pittsburgh, Pennsylvania 15260

7.1. Reductive Lithiation of Phenylthio Compounds

7.1.1. Introduction

Synthetic organic chemistry relies heavily upon the use of carbanions, usually organo-lithium compounds. Until recently, synthetic chemists nearly always have prepared organoalkali compounds by removal of a proton or other electrophile from a suitable precursor by the use of a strong base [1]. In the 1970s, publications from our laboratory [2] and that of Screttas [3] reported that certain organolithium compounds can be prepared by the action of lithium naphthalenide on phenylthio ethers or phenylthio acetals, a process referred to as reductive lithiation (e.g., eq. (7.1) [2a]). During the past decade, reductive metalation of phenylthio compounds by radical anions has gradually gained acceptance as a useful method of preparation of organoalkali compounds [4]. We have found that the two most useful aromatic radical-anions are lithium 1-(dimethylamino)naphthalenide (**1**, LDMAN) [5] and lithium p,p'-di-*tert*-butylbiphenylide (**2**, LDBB) [6]. The latter is the most powerful radical-anion known but the former has the advantage in many cases of allowing ready removal of the by-product, 1-(dimethylamino)naphthalene, by a dilute acid wash. This advantage becomes especially pronounced when the ultimate product is nonpolar and thus separable with difficulty by chromatography from a by-product such as p,p'-di-*tert*-butylbiphenyl or in large-scale preparations where mammoth chromatographies would otherwise be required.

$$\text{(structure)} + 2\left[Li^+ \text{(naphthalene radical anion)}\right] \longrightarrow \text{(structure)} \begin{matrix} SPh \\ Li \end{matrix} + PhSLi \qquad (7.1)$$

1   **2**

Reductive lithiation of phenylthio compounds by the use of aromatic radical-anions has some very attractive features for the generation of organolithium compounds. Because of the great versatility of divalent sulfur, substrates required for the preparation of most anions are readily available and, in contrast to the situation with alkyl halides that are occasionally used in synthetic chemistry as substrates for aromatic radical anions [7], the substrates are usually stable in the presence of the generated anions. Furthermore, reductive metalation is quite complementary to the conventional method, electrophile removal. In the latter case, it is likely that the rate-determining step is formation of the anion itself. The latter may be generated either in a one-step process, as appears likely in the case of proton removal, or, as may be the case with halogen, tin, and so on as the electrophile, in a two-step process in which the slow step is rupture of the bond between the carbon atom and the electrophile [eq. (7.2); E = electrophile and Nu = nucleophile]. There is thus a direct relationship between the stability of the anion and its ease of production. On the other hand, reductive metalation is believed to involve a rate-determining homolytic cleavage of the weak carbon–sulfur bond of a radical-anion, produced when an electron is absorbed by the substrate, followed by rapid electron transfer to the generated carbon radical. The carbon–sulfur bond cleavage could be concerted with electron transfer to the substrate or it could follow a rapid, reversible electron transfer as depicted in eq. (7.3) [4]. Since the stability order of radicals is frequently opposite to that of anions in solution, the rate-determining radical formation in reductive metalation has profound consequences with regard to the ease of anion production. Whereas sp^3-hybridized and more highly substituted anions, respectively, are more difficult to generate by electrophile removal than sp^2-hybridized and less-substituted anions, exactly the opposite is true for reductive metalation [4].

$$RE + Nu^- \rightleftharpoons \left[R\dot{E}Nu \right] \xrightarrow[slow]{} R^- + ENu \qquad (7.2)$$

$$RSPh \overset{e^-}{\rightleftharpoons} [R\dot{-}SPh] \xrightarrow[slow]{-SPh^-} R\cdot \xrightarrow[fast]{e^-} R^- \qquad (7.3)$$

7.1.2. α-Lithiotetrahydropyrans

The existence of a radical intermediate has interesting and useful stereochemical consequences as well. As shown in Scheme (7.1), reductive lithiation of 2-(phenylthio)tetrahydropyrans, prepared in a one-pot procedure from the lactone [eq. (7.4)], results in the

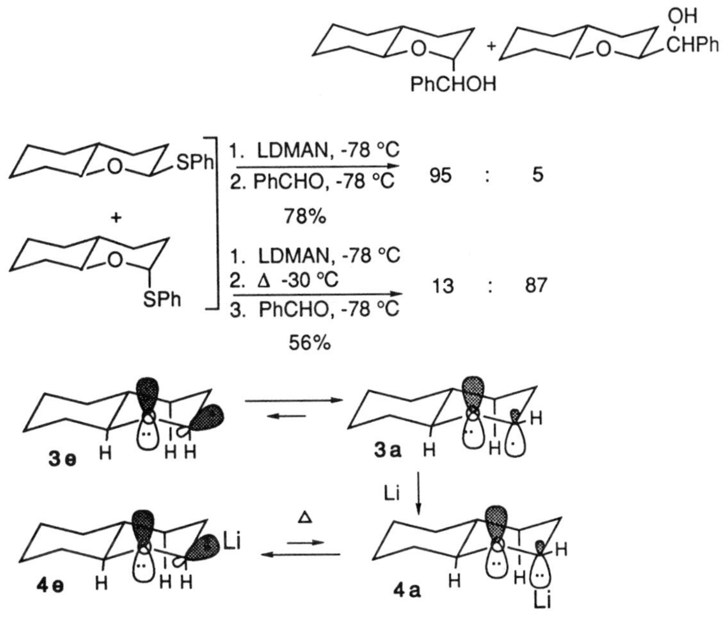

Scheme 7.1

proximate production of an axial 2-lithiotetrahydropyran (**4a**) [8]. It is believed that this product arises because of the configurational flexibility at $-78\ °C$ of the intermediate radical **3** and the thermodynamic preference for **3a** due to favorable overlap between the half-filled orbital and one of those bearing the nonbonding electrons on the oxygen atom. Rapid transfer of an electron to **3a** produces the axial α-lithioether that, at $-78\ °C$, is not configurationally mobile. Fortunately, the equatorial α-lithioether **4e** is the thermodynamically favored epimer and it can be generated with moderate selectivity by simply warming the axial epimer **4a** from -78 to $-27\ °C$. Thus, one can exert considerable stereochemical control.

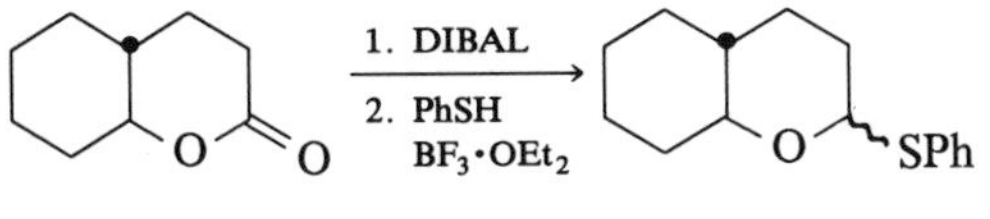

$$(7.4)$$

DIBAL = diisobutylaluminum hydride

An application of this discovery to highly stereoselective syntheses of *cis*- and *trans*-rosoxides, naturally occurring substances that are used as perfume components, is shown in Scheme 7.2. Boeckman and co-workers [9] have elegantly applied this method of producing axial α-lithiotetrahydropyrans in the key step of their synthesis of (−)-X-14547A (indanomycin); the pertinent segment of the synthesis is shown in Scheme 7.3.

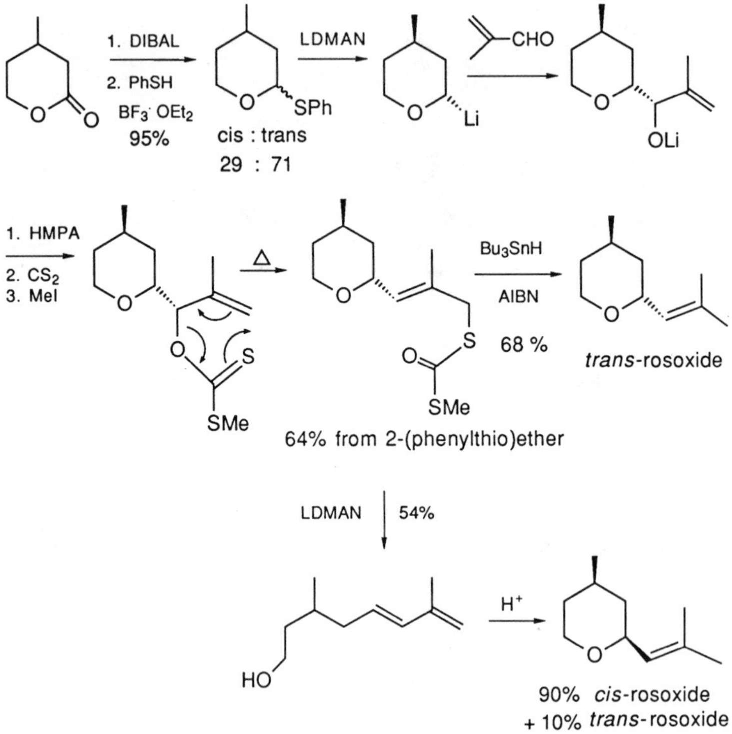

Scheme 7.2

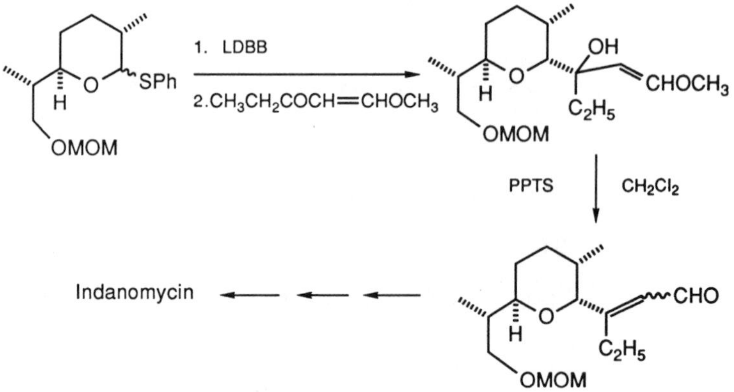

Scheme 7.3

7.1.3. Hydrocarbon Allyl Anions

Until very recently, reductive lithiation was the only reported method for producing α-lithio cyclic ethers [10]. The ease with which their precursors can be prepared [8] provides promise that these organolithium compounds will become widely useful synthetically. Reductive lithiation of phenylthio compounds has proven to be a general method not only for the production of α-lithioethers [11] but also for the generation of α-lithiosilanes, useful in the Peterson olefination reaction, and of hydrocarbon allyl anions [4, 12]. Indeed, it is apparently the only general method for preparing the latter class of compounds. It is also a very convenient procedure because the substrates are readily available by a variety of methods [4, 12]. One example of the preparation of an allyl anion that could not be produced by conventional methods, but which was required for NMR studies [13], is shown in eq. (7.5), [14]. Two steps were required starting from mesityl oxide, and the tertiary nature of the target anion was a help rather than a hindrance, the reduction step being extremely rapid at $-78\ ^\circ$C.

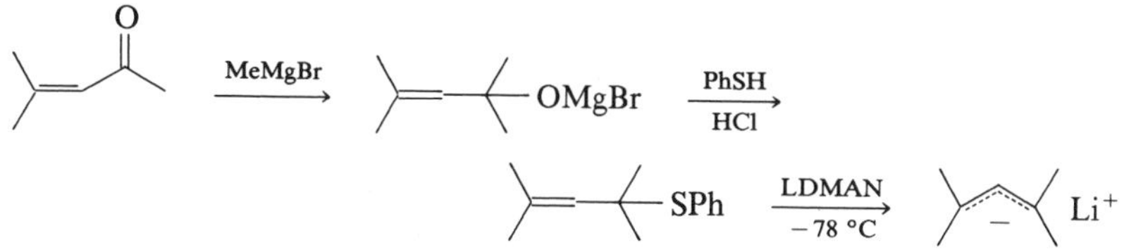

$$(7.5)$$

A crucial factor in the synthetic promise of this method of production of allyllithiums is that regiochemical control can be exerted in the case of unsymmetrical allyl anions. The addition of titanium tetraisopropoxide results in rather clean attack of the carbonyl partner at the most substituted terminus [eq. (7.6)], presumably via the cyclic Zimmerman–Traxler [15] type transition state 5 [eq. (7.7)] [12]. However, the use of cerium(III) chloride [16] results in predominant reaction at the least-substituted terminus (eq. 7.8), apparently via a π-allylcerium complex [4, 12].

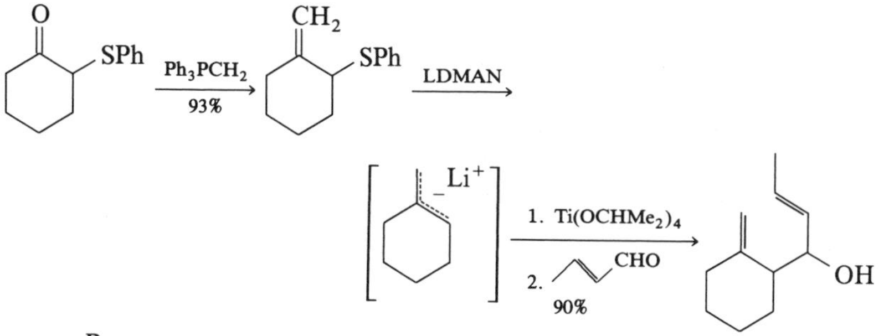

$$(7.6)$$

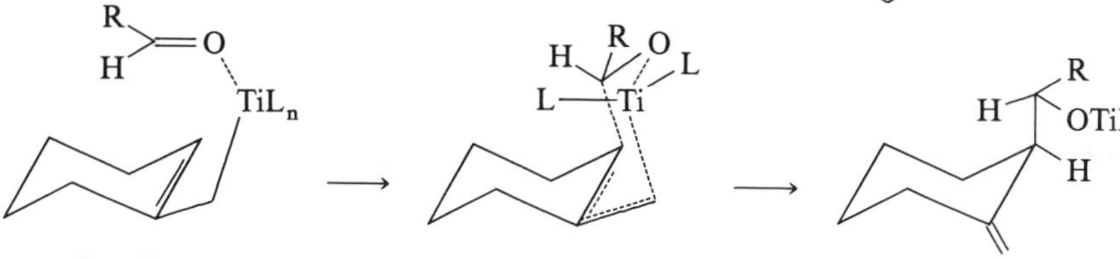

L = ligand   5

$$(7.7)$$

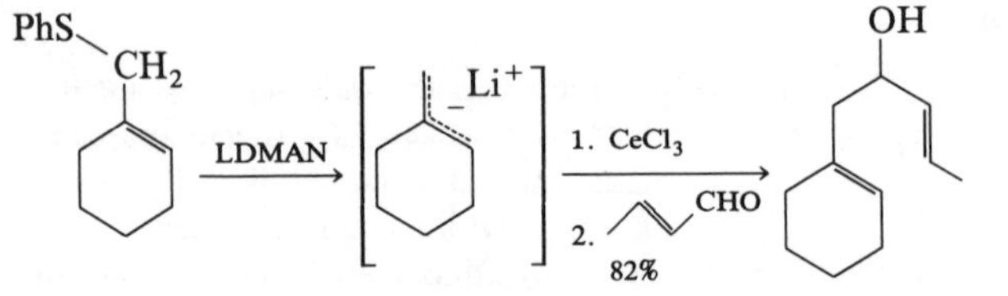

$$(7.8)$$

An especially useful feature of this chemistry is that the cis configurations of allyl anions are usually favored and the capture of these predominantly cis anions with cerium chloride at -78 °C preserves their cis character; upon treatment with aldehydes, they yield cis olefins (Scheme 7.4). By an iterative process, this can lead to cis–cis methylene-interrupted dienes (Scheme 7.4) [12], an important functionality in many natural products. Trienol 6 occurs as its acetate ester in brown algae and has been postulated [17] to be a biosynthetic precursor of some gamete attractants; we have recently been successful in preparing three of these, including the cyclopropyl triene dictyopterine B, from the phosphate ester derived from the now readily available 6 [18].

An application of the reductive lithiation method of producing hydrocarbon allyl anions and of the regiochemical and stereochemical control possible is the economical synthesis of the pheromone of the California red scale [19], a significant citrus pest (Scheme 7.5) [4, 20, 21]. As the inset indicates, the bold parts of the drawing represent both fragments that were inserted as the Biellmann anion 7 and the two starred atoms were inserted in the form of formaldehyde. The latter attacks the allyl anion derived from 8 exclusively at the most-substituted terminus under the influence of Ti(IV) but the allyl anion derived from 9 mainly at the least-substituted terminus under the influence of Ce(III). Although in the latter case 9% of the undesired regioisomer is formed as well, the Z : E ratio (98 : 2) is unusually favorable thanks to the methyl group on the central carbon atom of the allyl system [22]. The overall yield is 23%.

A measure of true stereochemical control can be attained by virtue of the stereochemical equilibration that occurs upon warming the cis-allylcerium from -78 °C, its temperature of generation, to -45 °C. The result is conversion of the very predominantly cis-allylcerium to the trans isomer, which again reacts with carbonyl partners at the least-substituted terminus. One example is shown in eq. (7.9). Fortunately, the ionic allyllithiums prefer the cis configuration, whereas the presumably more covalent cerium allyl π-complexes are more stable in the trans configuration (designated syn by inorganic chemists) so that either cis or trans olefins can be produced (Scheme 7.6).

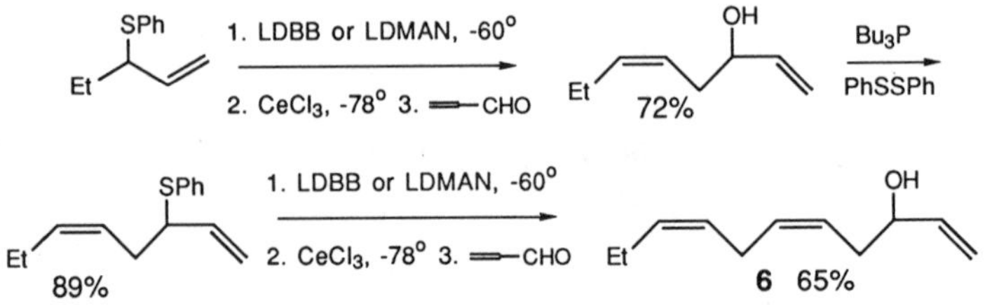

Scheme 7.4

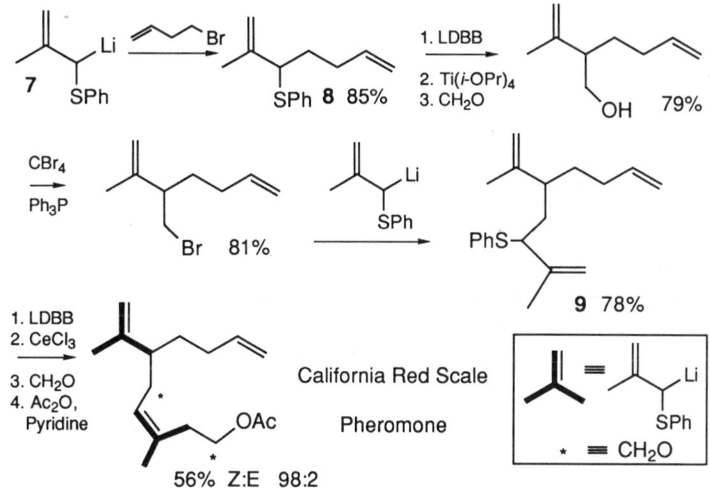

Scheme 7.5

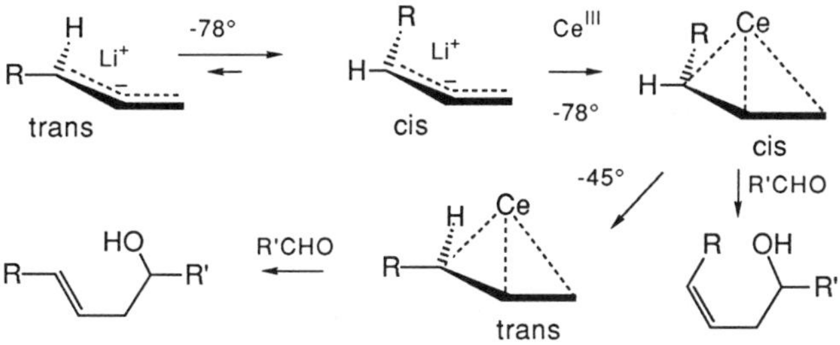

Scheme 7.6

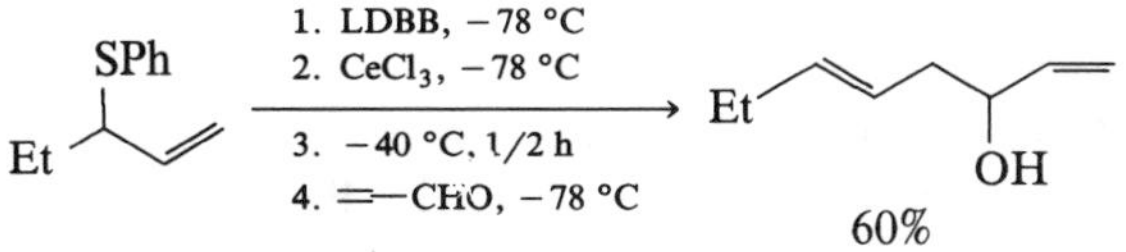

(7.9)

7.2. Reductive Lithiation of Epoxides and Oxetanes

7.2.1. Reductive Lithiation of Epoxides

In 1985, while attempting to generate α-lithioepoxides from α-(phenylthio)epoxides by reductive lithiation, we discovered that epoxides are reductively cleaved in the presence of aromatic radical-anions to yield dianions [eq. (7.10)] [23]. Bartmann [24] has independently discovered this reaction.

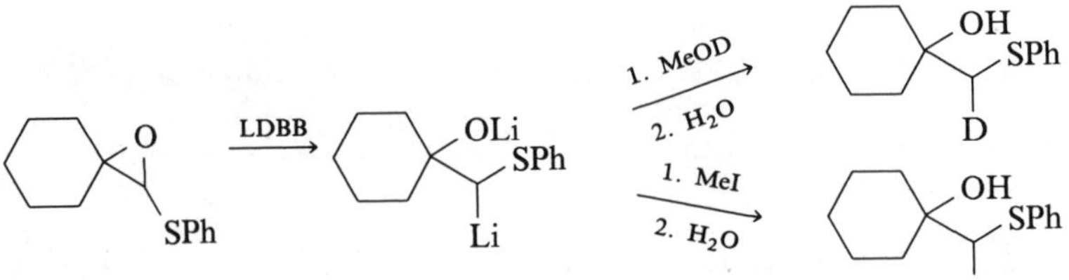

$$(7.10)$$

The direction of opening and the synthetic potential of the method are both of interest. In epoxides substituted with a single alkyl group or with two alkyl groups on the same ring carbon atom and none on the other, cleavage occurs mainly between the oxygen atom and least-substituted ring carbon atom [23,24]. Thus, reduction of **10** produces a good yield of **11**; the regiochemistry is fortunate because the other mode of cleavage results in a dianion **12** that decomposes very rapidly to an olefin and lithium oxide and is thus synthetically useless [eq. (7.11)]. Most 1,2-disubstituted epoxides, upon reductive cleavage, provide similarly unstable dianions, the exception being cyclohexene oxide [23,24]. An example of reductive lithiation of the latter is given in eq. (7.12) [23].

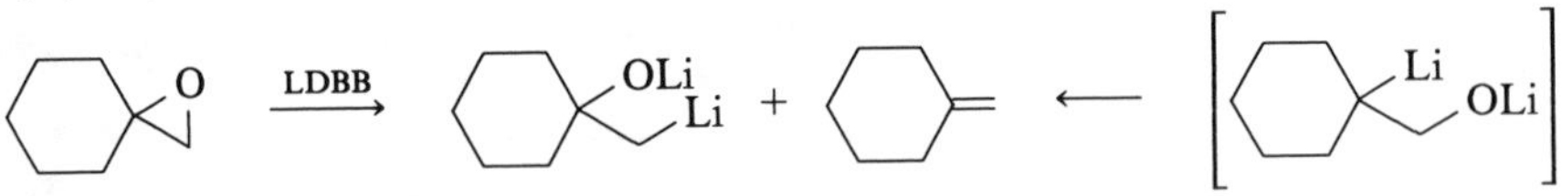

10   **11**   minor product   **12**

$$(7.11)$$

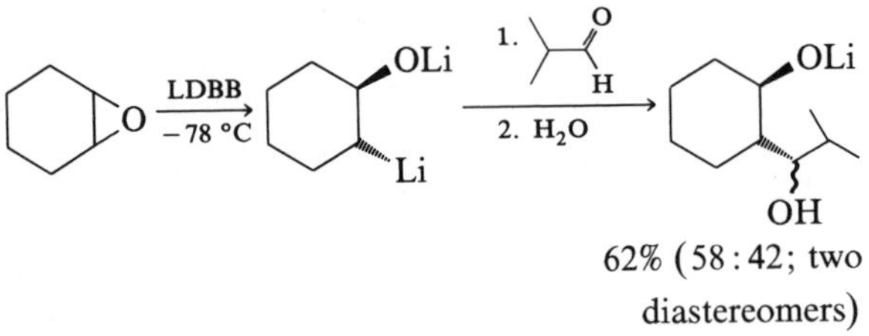

$$(7.12)$$

62% (58:42; two diastereomers)

At first glance the regiochemistry shown in eq. (7.11) is surprising because, assuming the usual mechanism of reductive lithiation, the *least*-substituted radical appears to be generated after cleavage of the radical-anion produced when the epoxide absorbs an electron. However, computations indicated that the putative product **14** of opening of the lithium-epoxide complex **13** is somewhat more stable than the isomer **15** possessing the more branched radical [eq. (7.13)] [25]. The explanation is that the greater stability of the more-branched oxyanion relative to the less-branched one outweighs the lesser stability of the primary radical relative to the tertiary one.

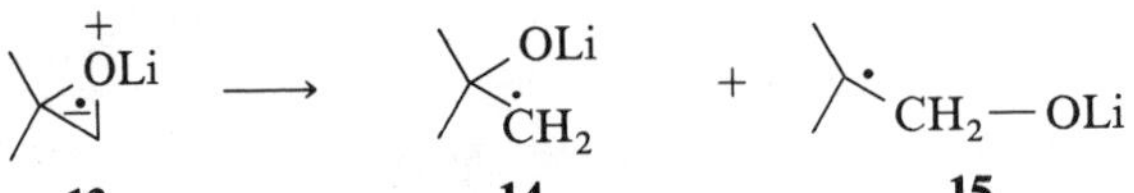

$$(7.13)$$

13   **14**   **15**
             lower energy   higher energy

An illustration of a promising use of the β-lithiooxyanions such as **11** is shown in eq. (7.14). 1,2-Addition to acrolein proceeds smoothly to produce the diol **16**, the secondary hydroxyl group of which can be selectively mesylated and displaced by the oxygen of the free hydroxyl group, with allylic inversion, to yield dihydropyran **17** [23].

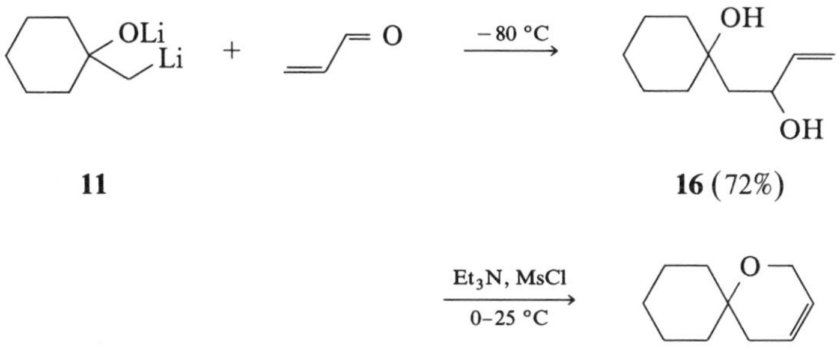

11   **16** (72%)   (7.14)

17 (70%)

Reductive lithiation of ethylene oxide at -95 °C generates the parent β-lithiooxyanion **18** that readily adds to *p*-anisaldehyde (Scheme 7.7) [23]. However, the yield of adduct **19** with 1-acetycyclohexene, although serviceable, is less satisfactory. The major by-product **20** arises from reduction of the enone by **18**, which exhibits such reducing power to one degree or another with aldehydes and ketones; **18** may be regarded as a super Meerwein–Pondorf–Verly reducing agent because both lithium atoms appear capable of bonding to the carbonyl oxygen atom during the hydride transfer [eq. (7.15)]. Diol **19**, like **16**, undergoes ring closure to a dihydropyran but, in this case, because of the great difference in reactivity of the two hydroxyl groups to

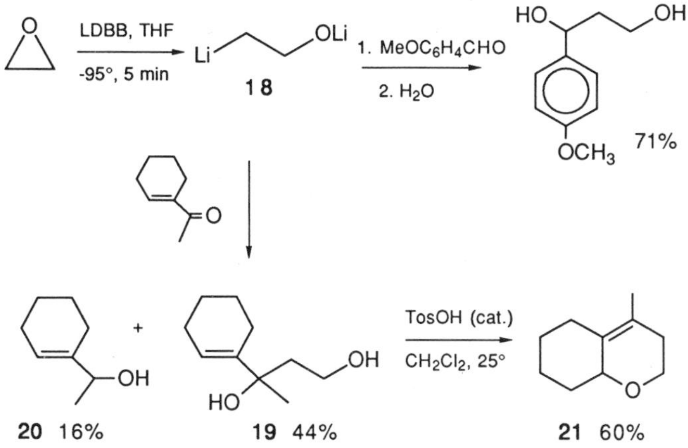

20 16%   **19** 44%   **21** 60%

Scheme 7.7

acid, the ring closure can be acid induced [23].

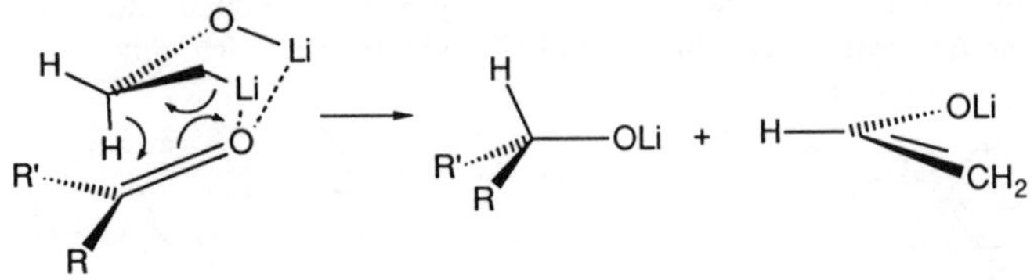

(7.15)

7.2.2. Reductive Lithiation of Oxetanes

Oxetanes have also been found to undergo reductive lithiation [26]. However, whereas epoxides that are unsubstituted on one of the ring carbon atoms can be reductively lithiated with LDBB at -78 °C or below, oxetanes are only reductively cleaved at substantially higher temperatures. The resulting dianions are far more stable than those from epoxides and they do not behave as reducing agents. The reductive lithiation of oxetanes was initiated in order to supply carbanion–oxyanion **23** to a colleague (Dennis Curran) who required it in a synthetic scheme. It is readily produced by the action of LDBB on the inexpensive oxetane **22** [eq. (7.16)].

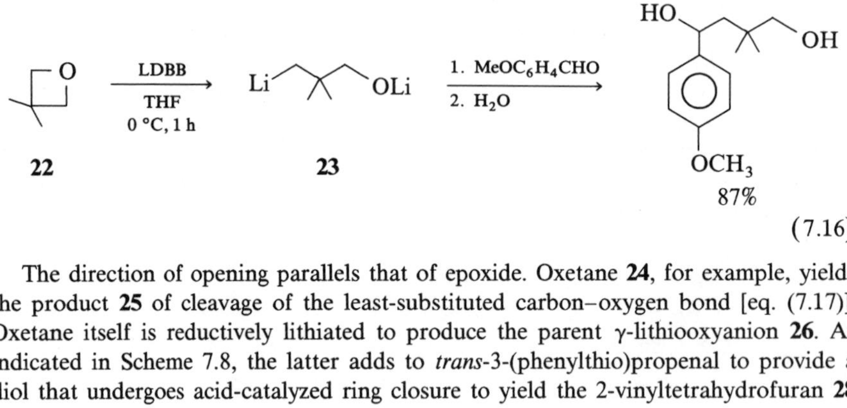

(7.16)

The direction of opening parallels that of epoxide. Oxetane **24**, for example, yields the product **25** of cleavage of the least-substituted carbon–oxygen bond [eq. (7.17)]. Oxetane itself is reductively lithiated to produce the parent γ-lithiooxyanion **26**. As indicated in Scheme 7.8, the latter adds to *trans*-3-(phenylthio)propenal to provide a diol that undergoes acid-catalyzed ring closure to yield the 2-vinyltetrahydrofuran **28** [26].

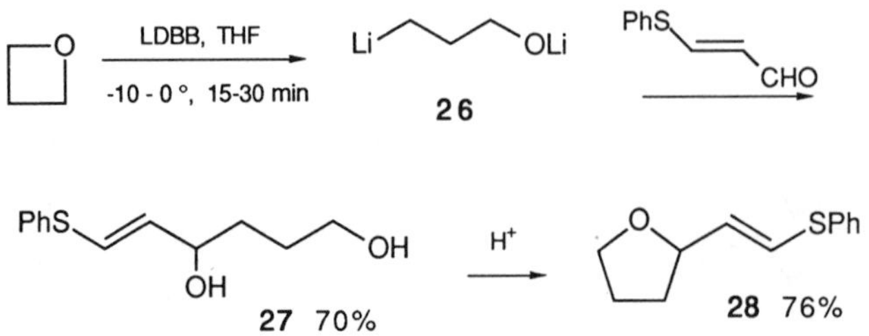

Scheme 7.8

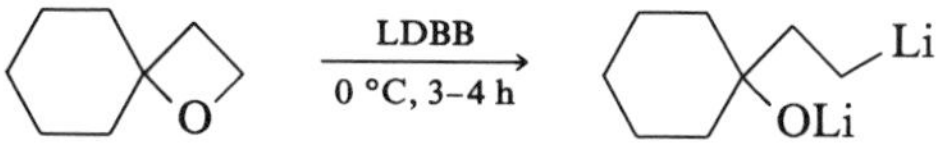

24 **25**

(7.17)

Perhaps the most exciting use of dianions such as **26** is their reactions with lactones followed by acid treatment to provide spiroketals, a type of functionality that is widely distributed in physiologically active natural products [27]. Equation (7.18) illustrates a one-pot synthesis of the pheromone of the Norway spruce beetle [28] using this concept [29]. Six- and seven-membered ring lactones are converted to spiroketals in an analogous fashion (Scheme 7.9).

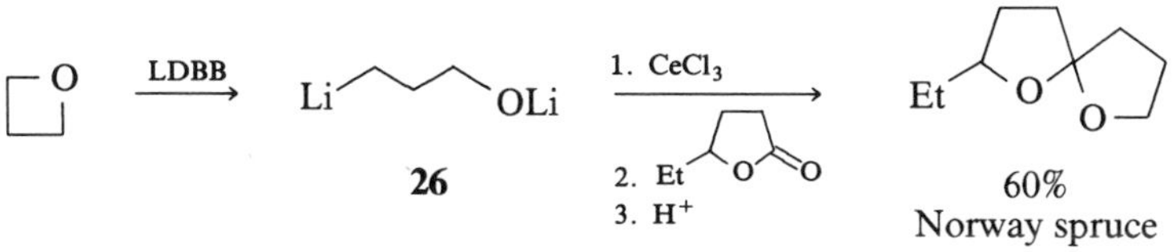

26

60%
Norway spruce
beetle pheromone

(7.18)

Treatment of the γ-lithiooxyanions **23** and **26** with cerium chloride prior to the introduction of the lactone greatly increases the yield in such reactions. The favorable effect of cerium(III) on this reaction was anticipated because the strong oxygen–cerium bond [30] in the intermediate (**29**, M = metal ion) would be expected to be far more resistant to rupture than is the carbon–lithium bond; bond rupture produces a ketone (**30**) that can react with another organometallic molecule to produce a tertiary alcohol **31**, the main unwanted by-product of additions of organolithium compounds to lactones [eq. (7.19)] [31].

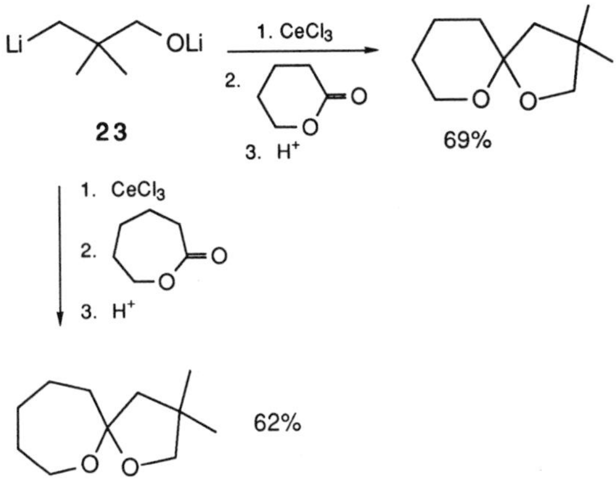

Scheme 7.9

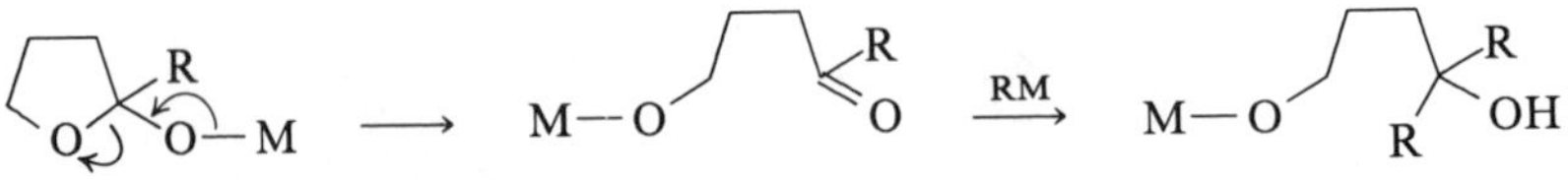

$$(7.19)$$

This technique is strikingly successful even in cases in which the metallooxyanionic intermediate would be particularly prone to ring opening. Examples are the addition of organocerium dianion **32** (from reductive cleavage of oxetane **22**, followed by lithium–cerium exchange) to the coumarin **33** [eq. (7.20)] and even to succinic anhydride [eq. (7.21)] [29]. However, it seems quite unlikely that the metallooxyanionic intermediates formed in the reactions shown in eqs. (7.20) and (7.21) are indeed stable to ring opening because the oxyanions produced upon such openings are very highly stabilized in both cases. Rather, it is probable that they do open but that the carbonyl group is immediately protected by reclosure to a stable cerium salt of a lactol; this process is illustrated in Scheme 7.10 for the addition of **32** to **33**. The efficacy of such internal protection of the carbonyl group is dramatically illustrated by the 50% yield of **34** that is obtained even when the anhydride is added to a THF solution of organometallic **32**, assuring, throughout the addition, an excess of the latter capable of adding to any ketonic intermediate; when the same experiment is repeated using the lithium analogue (**23**) of **32**, only a 13% yield of **34** is produced. The intramolecular attack on the carbonyl group of the intermediate is reminiscent of that observed [32] in the nucleophilic addition of 1,4- and 1,5-di-Grignard reagents to anhydrides, resulting in lactone formation.

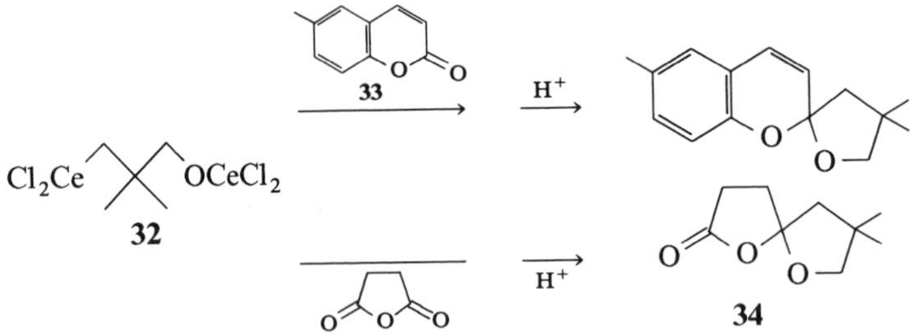

$$76\% \quad (7.20)$$

$$72\% \quad (7.21)$$

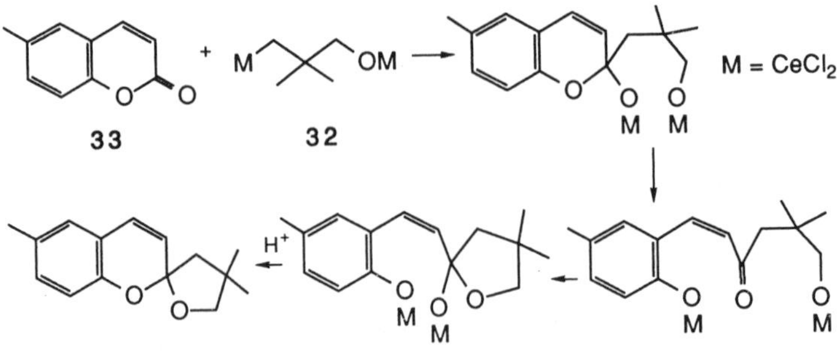

Scheme 7.10

The one-pot preparation of oxaspirolactones such as **34** is noteworthy. There appears to be only one other report of this oxaspirolactone system and the preparation was lengthy [33]. Many uses of such compounds in synthesis can be envisioned.

7.3. Conclusion

This article provides examples of several classes or organic monolithium compounds that can be generated by treatment of phenylthio compounds with aromatic radical-anions. The process is complementary to electrophile exchange as a method of production of carbanions and it proceeds by an intermediate carbon radical. The method is rather general for α-heteroatom organolithium compounds [34] and allyllithiums [12]. In the latter case, methods have been devised to control the regiochemistry and stereochemistry of reactions with aldehydes.

Synthetically useful β- and γ-lithiooxyanions are produced by similar reductive lithiation of epoxides and oxetanes, respectively. The method is more convenient and general than radical-anion–induced reductive lithiation of β- or γ-halooxyanions [7]. When the γ-lithiooxyanions from oxetanes are treated with cerium(III) chloride followed by a five-, six-, or seven-membered lactone and then acid, spiroketals result in a one-pot procedure.

Acknowledgment. I thank my colleagues who performed the work described herein and contributed in numerous other ways to these projects; their names are cited in the references. I also thank the National Science Foundation and the Institute of General Medical Sciences of the National Institutes of Health for generous financial support.

References

1. Durst, T., in Buncel, E., Durst, T. *Comprehensive Carbanion Chemistry. Part B*, Elsevier, New York, 1984, Chapter 5.
2. (a) Cohen, T.; Daniewski, W. M.; Weisenfeld, R. B. *Tetrahedron Lett.*, **1978**, 4665. (b) Cohen, T.; Weisenfeld, R. B. *J. Org. Chem.*, **1979**, *44*, 3601.
3. Screttas, C. G.; Micha-Screttas, M. *J. Org. Chem.*, **1978**, *43*, 1064 and **1979**, *44*, 713.
4. Cohen, T.; Bhupathy, M. *Acc. Chem. Res.*, **1989**, *22*, 152.
5. Cohen, T.; Matz, J. R. *Synth. Commun.*, **1980**, *10*, 311.
6. Freeman, P. K.; Hutchinson, L. L. *J. Org. Chem.*, **1980**, *45*, 1924 and **1983**, *48*, 4705.
7. For example: Barluenga, J.; Flórez, J.; Yus, M. *J. Chem. Soc., Perkin Trans. 1*, **1983**, 3019 and citations therein. Nájera, C.; Yus, M.; Seebach, D. *Helv. Chim. Acta*, **1984**, *67*, 289.
8. Cohen, T.; Lin, M.-T. *J. Am. Chem. Soc.*, **1984**, *106*, 1130.
9. Boeckman, R. K.; Enholm, E. J., Jr.; Demko, D. M.; Charette, A. B. *J. Org. Chem.* **1986**, *51*, 4745.
10. Recently, some α-lithio cyclic ethers have been prepared by tin–lithium exchange. Sawyer, J. S.; Kucerovy, A.; Macdonald, T. L.; McGarvey, G. J. *J. Am. Chem. Soc.*, **1988**, *110*, 842.
11. Cohen, T.; Matz, J. R. *J. Am. Chem. Soc.*, **1980**, *102*, 6900.
12. Guo, B. S.; Doubleday, W.; Cohen, T. *J. Am. Chem. Soc.*, **1987**, *109*, 4710 and citations therein.
13. Fraenkel, G., private communication.
14. Guo, G. S.; Doubleday, W.; Cohen, T., unpublished work.
15. Zimmerman, H. E.; Traxler, M. D. *J. Am. Chem. Soc.* **1957**, *79*, 1920.

16. Imamoto, T.; Sugiura, T.; Takiyama, N. *Tetrahedron Lett.*, **1984**, *25*, 4233.
17. Moore, R. E. *Acc. Chem. Res.*, **1977**, *10*, 40.
18. Abraham, W. D., Ph.D. Thesis, University of Pittsburgh, 1990.
19. Roelofs, W.; Gieselmann, M.; Cardé, A.; Tashiro, H.; Moreno, D. S.; Hendrick, C. A.; Anderson, R. J. *J. Chem. Ecol.*, **1978**, *4*, 211.
20. Cohen, T.; Piccolino, E.; McCullough, D. W., unpublished work.
21. Other syntheses: Oppolzer, W.; Stevenson, T. *Tetrahedron Lett.*, **1986**, *27*, 1139 and citations therein.
22. Schlosser, M. *Proc. Japan Chem. Soc. (Tokyo)*, **1984**, *3*, 1821.
23. Cohen, T.; Jeong, I.-H.; Mudryk, B.; Bhupathy, M.; Awad, M. M. A. *J. Org. Chem.*, **1990**, *55*, 1528.
24. Bartmann, E. *Angew. Chem., Int. Ed. Engl.*, **1986**, *25*, 653.
25. Dorigo, A. E.; Houk, K. N.; Cohen, T. *J. Am. Chem. Soc.*, **1989**, *111*, 8976.
26. Mudryk, B.; Cohen, T. *J. Org. Chem.*, **1989**, *54*, 5657.
27. Boivin, T. L. B. *Tetrahedron*, **1987**, *43*, 3309.
28. First report: Francke, W.; Heeman, V.; Gerken, B.; Renwick, J. A. A.; Vité, J. P. *Naturwissenschaften*, **1977**, *64*, 590. Recent synthesis: Högberg, H.-E.; Hedenström, E.; Isaksson, R.; Wassgren, A.-B. *Acta Chem. Scand., Ser. B*, **1987**, *41*, 694.
29. Mudryk, B.; Shook, C. A.; Cohen, T., unpublished work.
30. Imamoto, T.; Kusumoto, T.; Tawarayama, Y.; Sugiura, Y.; Mita, T.; Hatanaka, Y.; Yokoyama, M. *J. Org. Chem.*, **1984**, *49*, 3904.
31. Cavicchioli, S.; Savoia, D.; Trombini, C.; Umani-Ronchi, A. *J. Org. Chem.*, **1984**, *49*, 1246.
32. Canonne, P.; Bélanger, D.; Lemay, G.; Foscolos, G. B. *J. Org. Chem.*, **1981**, *46*, 3091.
33. Fukuda, H.; Takeda, M.; Sato, Y.; Mitsunobu, O. *Synthesis*, **1979**, 368.
34. Cohen, T.; Sherbine, J. P.; Hutchins, R. R.; Lin, M. T. *Organomet. Synth.*, **1986**, *3*, 361.

8

Asymmetric Reactions of Atropisomeric Sulfur Compounds

Sergio Cossu, Ottorino De Lucchi, Davide Fabbri, Maria Paola Fois, and Paola Maglioli

Dipartimento di Chimica dell'Università
via Vienna 2, I-07100 Sassari
Italy

Giovanna Delogu

Istituto CNR per l'Applicazione delle Tecniche Chimiche Avanzate ai Problemi Agrobiologici
via Vienna 2, I-07100 Sassari
Italy

Bis(phenylsulfonyl)ethylenes, **1** and **2** [1], are becoming popular acetylene equivalents in cycloaddition reactions. They are frequently used to prepare polycyclic dienes through the route outlined in eq. (8.1) [2].

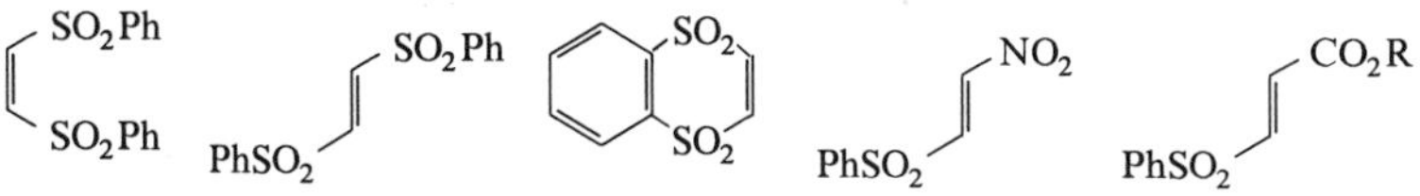

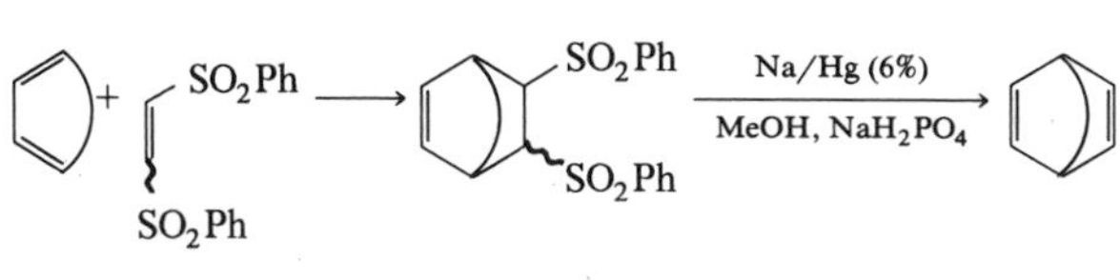

$$(8.1)$$

Principal features of these reagents are the high reactivity, the crystallinity of the intermediates and the simplicity of performance of all reaction steps. Despite the substantial results achieved, this area of investigation is still receiving considerable attention as shown by the recent reports on the even more reactive related reagents **3–5** [3–5]. In a search for new efficient dienophiles, we focused our attention on systems that could function as "asymmetric equivalents of acetylene," that is, systems able to deliver acetylene in a chiral form when reacted with suitable dienes [eq. (8.2)].

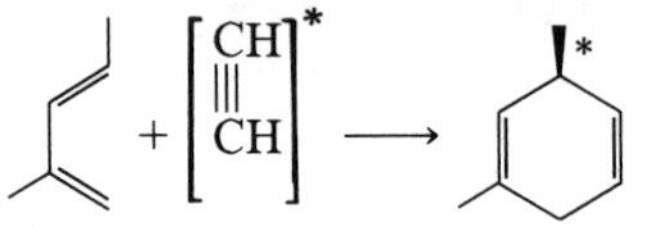

$$(8.2)$$

Several variants were planned, including the chiral vinyl sulfoxides **6** [6], *cis*-bis(phenylsulfonyl)ethylenes containing chiral auxiliaries **7** [7], and, more recently, the chiral atropisomeric sulfur derivative **8**.

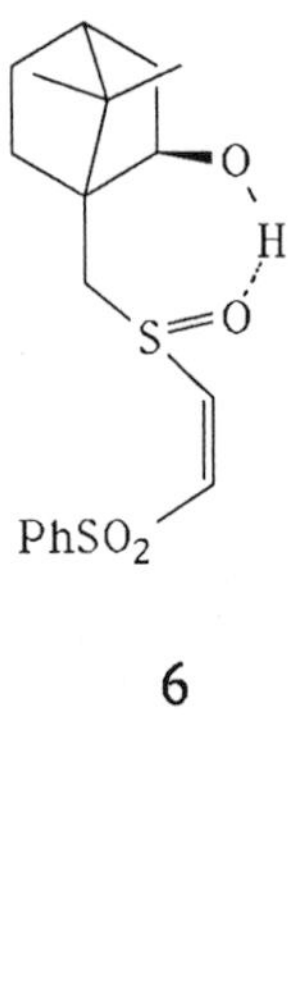

6

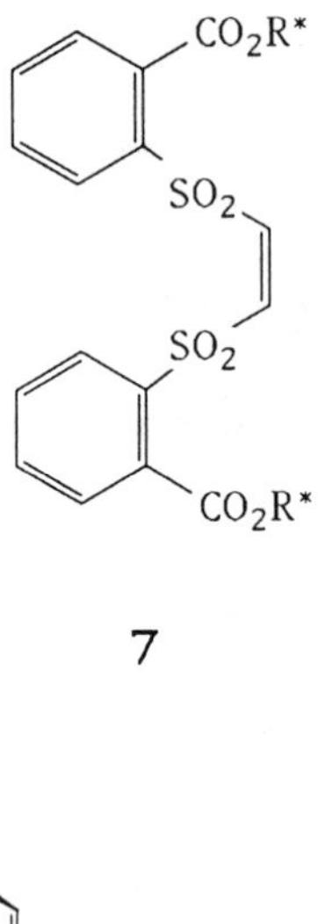

7

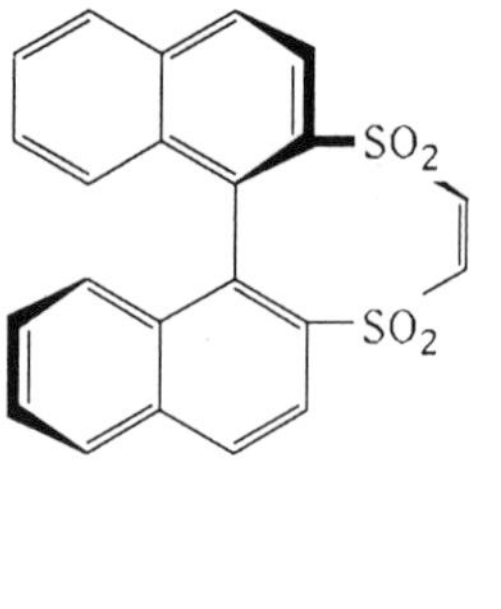

8

The reasoning that led us to conceive of this atropisomeric sulfur variant of **1** is illustrated in the chart in Figure 8.1. Dienophile **8** is, as most binaphthyl compounds, a

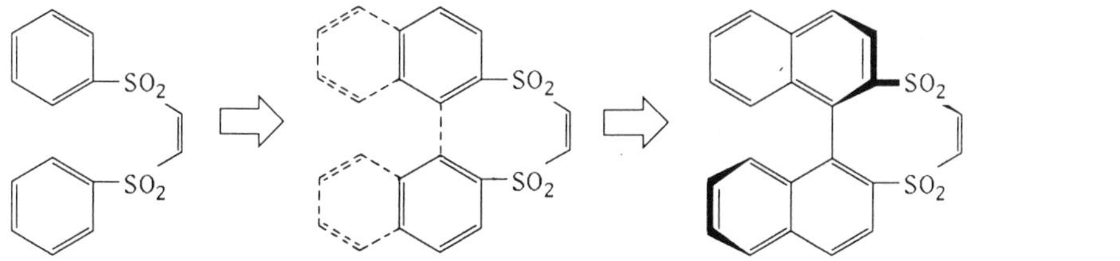

8

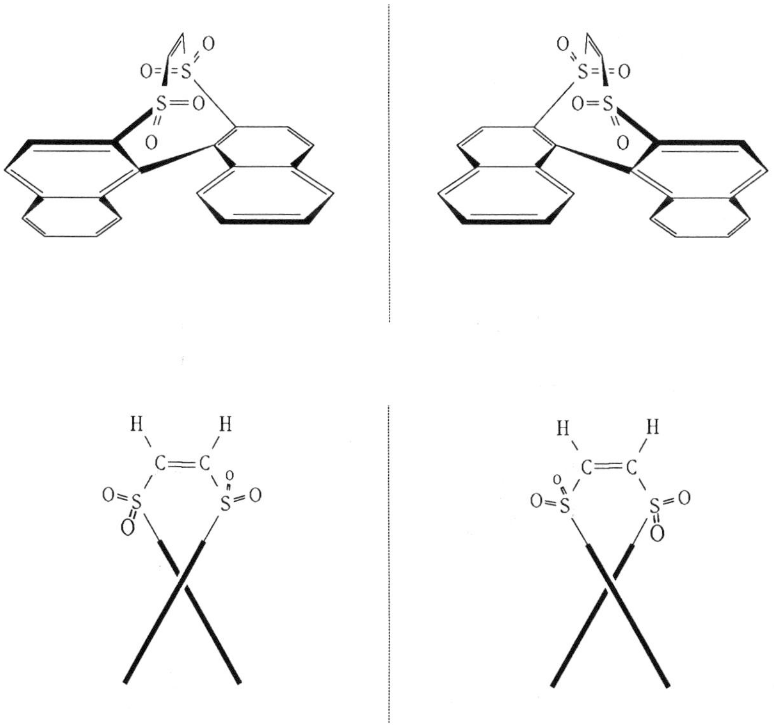

Figure 8.1 ■ Representations of chiral binaphthyl dienophile 8.

chiral molecule with C_2 symmetry. The two enantiomers are represented in Figure 8.1, drawn in two different perspectives. The rigid conformational array of the molecule should be noticed, having two of the four oxygen atoms in the plane containing the double bond and the other two in parallel lines, pointing towards opposite directions. This arrangement is most promising in establishing stereochemical bias in cycloaddition reactions.

Furthermore, a point of interest in **8** is the reasonable amount of strain energy that has been included in the construction of the heterocyclic framework. This feature is particularly evident in the molecular models and derives from the tendency of the double bond to lay planar, against the necessity of the binaphthyl moiety to pucker [eq. (8.3)].

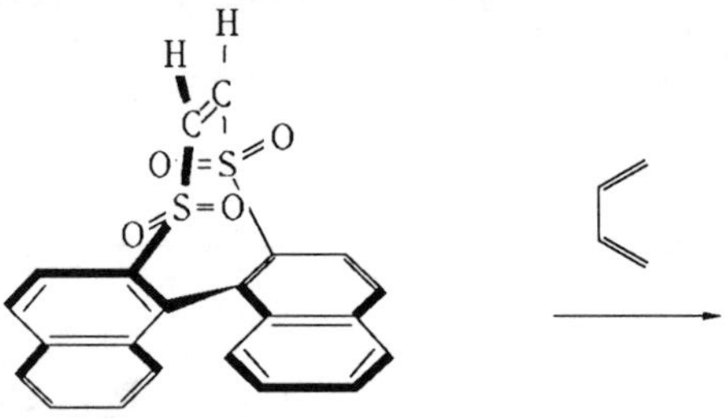

8

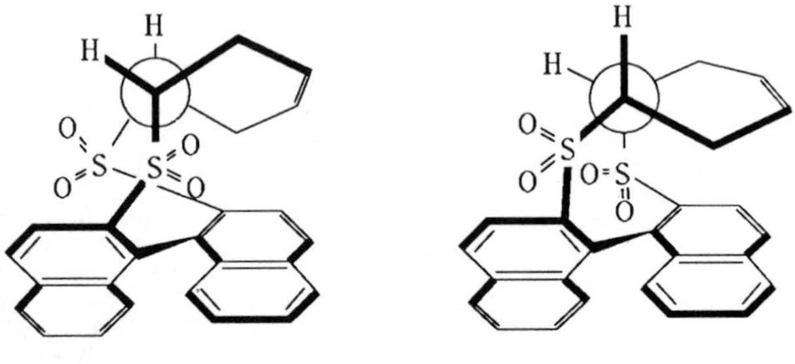

(8.3)

Such strain energy is released in the cycloaddition, that is, on passing from the planar sp^2 bonding of the dienophile to the tetrahedral sp^3 hybridization of the cycloadduct, contributing to the reactivity of the dienophile.

8.1. Synthetic Considerations on the Preparation of the Reagents

The synthesis of **8** (as well as of all other atropisomeric reagents that will be discussed later) requires an efficient preparation of the dithiol **9**. This reagent has been known for a long time (the first preparation of the racemate dates back to 1928 [8a] and the pure optical antipods were reported in 1957 [8b]) and it was originally prepared as shown in eq. (8.4) [8b].

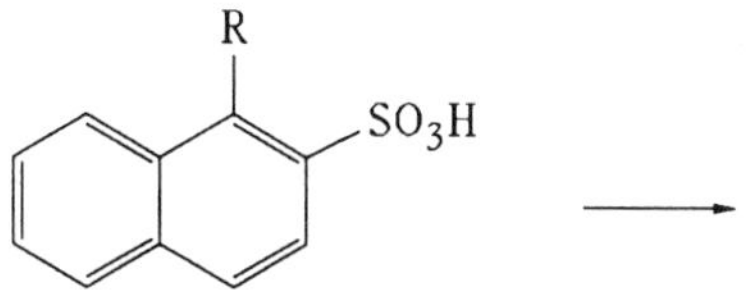

R = NH$_2$
R = N$_2^+$Cl$^-$
R = I

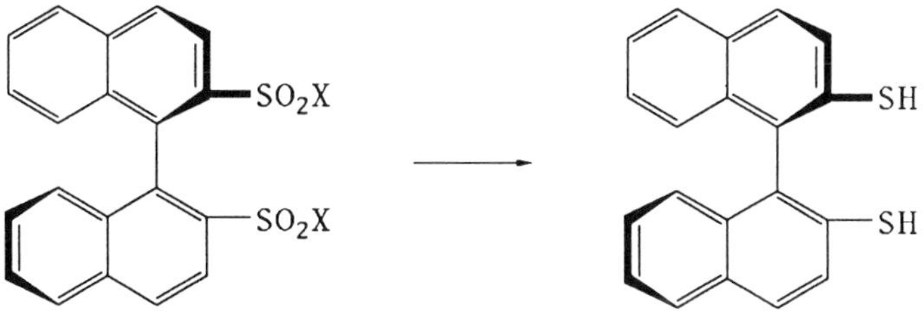

(8.4)

9

X = OH [resolution with (-)-strychnine]
X = Cl

A newer preparation of **9** involves Newman–Kwart rearrangement of the bis(thio-carbamate) obtained from binaphthol and dimethylthiocarbamoyl chloride [8c] [eq. (8.5)]. The latter route constitutes a more viable route to substantial quantities of this thiol.

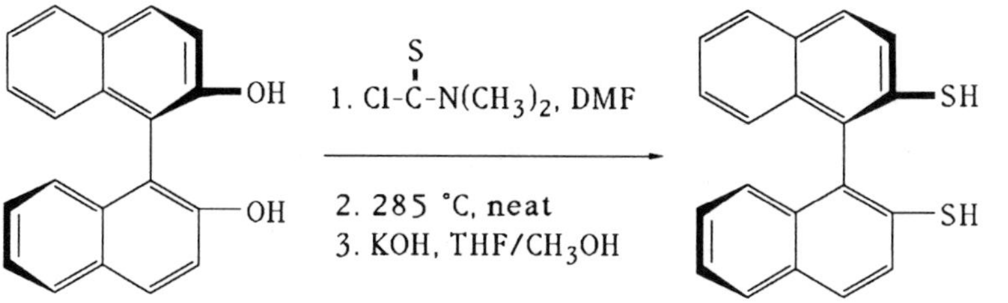

(8.5)

9   60 %

Due to the expected preparation of large amounts of the dithiol **9**, the reaction has been studied in detail, especially with regard to the critical step of the thermal rearrangement. Besides the reported bis(thiocarbamate) **10**, two other products are always formed, irrespective of the reaction conditions, that is, the monorearranged product **11** and the thiophene **12** [eq. (8.6)]. Experiments designed to produce exclusively **10** at the expense of **11** resulted in increased formation of **12**.

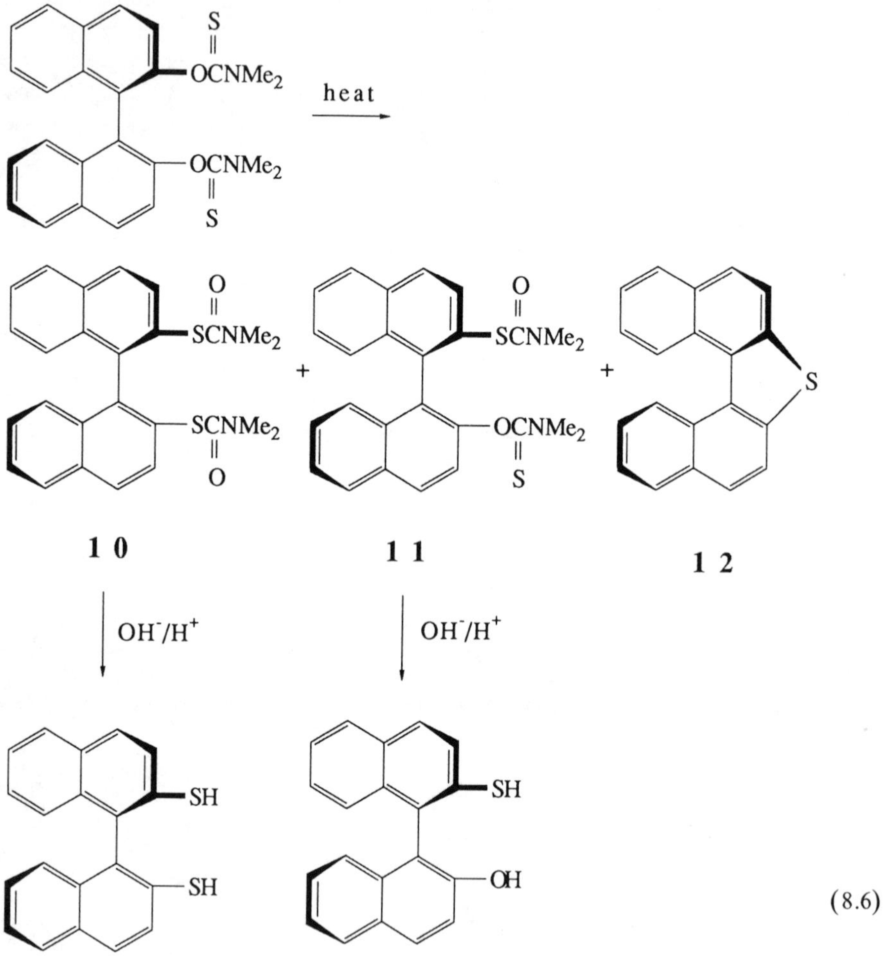

Although in this context we will discuss only the reactions of the dithiol **9**, the mixed ester **11** is the precursor to the equally interesting hydroxythiol **13**, whose synthetic potential is now being examined in our laboratories.

The dinaphthothiophene **12** is not a planar molecule even if constituted by five condensed aromatic rings. The result of an X-ray structural analysis (Figure 8.2) supports this feature [9]. Thiophene **12** is the thermodynamically more favored product of the thermolysis. It is often produced in large amounts, or it can even be obtained as the exclusive product, upon carrying out the thermolysis in a high-boiling solvent such as sulfolane. Because the formation of **12** is detrimental to the otherwise efficient

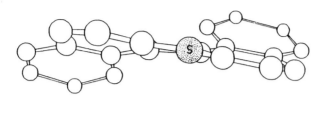

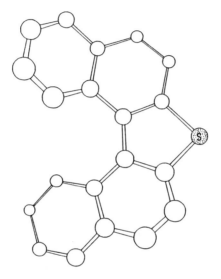

Figure 8.2 ■ Perspective drawings of dinaphtho[2,1-*b*:1′,2′-*d*]thiophene **12**, as determined by X-ray structure analysis [9]. Hydrogen atoms have been omitted.

preparation of **9**, a procedure to reconvert **12** into **9** was sought. Eventually, it was found that treatment of **12** with excess lithium in tetrahydrofuran [eq. (8.7)], followed by quenching with sulfur monochloride (**14**) and reduction by lithium aluminum hydride, efficiently generates **9**.

1. Li, THF
⟶
2. S_2Cl_2 (**14**)
3. $LiAlH_4$, THF

(8.7)

12   **9**

It should be noticed that dinaphthothiophene **12** is readily available through a number of other routes, some of which can be accomplished on large scales [10].

The preparation of **9** from **12**, plus the direct preparation illustrated in eq. (8.6), constitute efficient ways for obtaining **9** in synthetically useful amounts from cheap and readily available starting materials.

8.2. Cycloaddition Reactions of 1,1'-Binaphthalene-2,2'-bis(sulfonyl)-ethylene (Dinaphtho[2,1-e:1',2'-g][1,4]dithiocin,1,1,4,4-tetraoxide)

With **9** in hand, the preparation of dienophile **8** follows standard procedures [eq. (8.8)] [11].

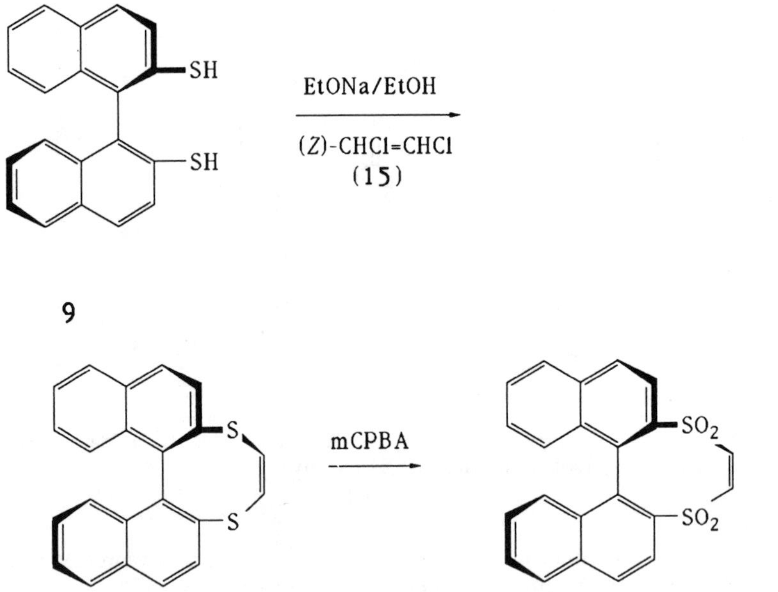

EtONa/EtOH

──────────

(Z)-CHCl=CHCl
(15)

9

mCPBA

╌╌╌╌→

(8.8)

93 %

1 6

72 %

8

The cycloaddition properties of **8** have been tested both in the reaction with symmetric and with nonsymmetric dienes.

8.2.1. Cycloaddition Reactions of 1,1'-Binaphthalene-2,2'-bis(sulfonyl)ethylene with Symmetric Dienes

The Diels–Alder reactions of **8** exhibited the expected high reactivity [see eq. (8.3)]. The reaction is virtually instantaneous with cyclopentadiene, and occurs at room temperature with less-reactive dienes, such as 1,3-cyclohexadiene, cyclooctatetraene, and furan. At variance, the other bis(phenylsulfonyl)-reagents **1** and **2** required higher temperatures or longer reaction time. The stereochemistry of addition in all cases, but cyclopentadiene, is strictly endo. In the reaction with cyclopentadiene, a 9:1 mixture of the endo and exo cycloadducts was produced, regardless of the reaction temperature (from −60 to +100 °C). Because the two isomers did not interconvert even after heating the pure adducts at 160 °C for several hours, no retro Diels–Alder processes operate. This observation suggests that the 9:1 mixture is the kinetic product composition and that the reaction is highly exothermic, as anticipated in eq. (8.3) [12].

In the reaction of **8** with excess cyclopentadiene (> 2 equivalents), it was observed that the crystalline product that separates from the reaction mixture in dichloromethane contained a molecule of cyclopentadiene in about a 1:2 ratio with the cycloadduct. This observation prompted us to prepare the Diels–Alder adducts of **8** with anthracene, with the aim of producing more defined chiral molecular pockets [13]. The reaction with anthracene occurred readily to afford the expected Diels–Alder adduct **17** [eq. (8.9)].

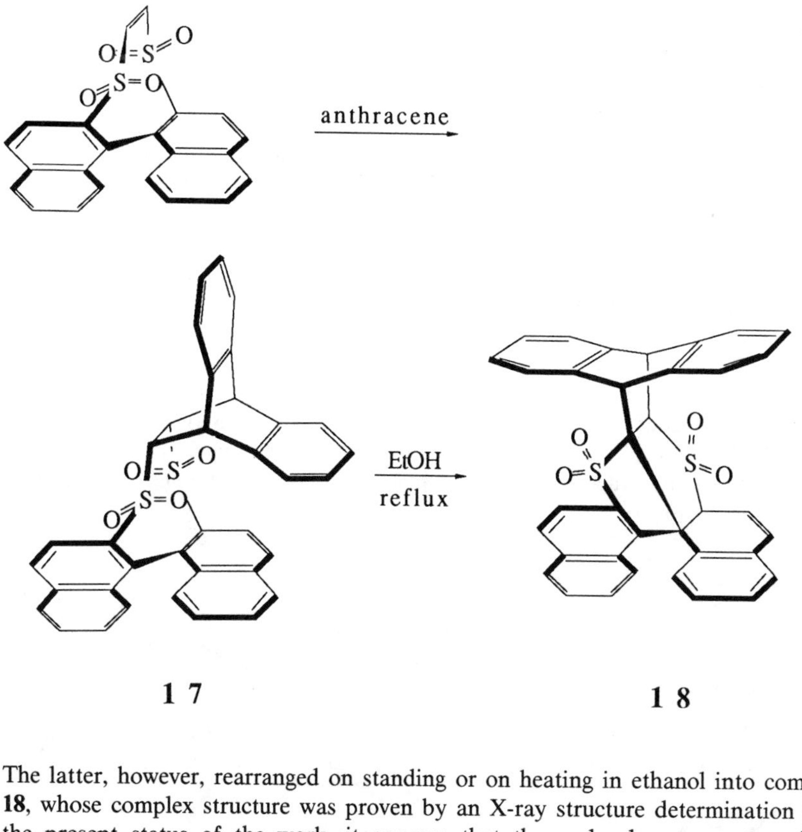

(8.9)

1 7   **1 8**

The latter, however, rearranged on standing or on heating in ethanol into compound **18**, whose complex structure was proven by an X-ray structure determination [9]. At the present status of the work, it appears that the molecular aggregates of these compounds, for example, the one with cyclopentadiene, resemble the inclusion phenomena observed with other anthracene adducts [13].

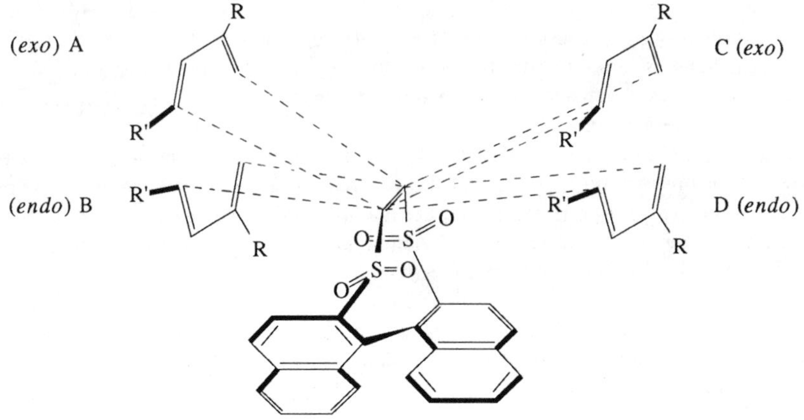

Scheme 8.1

8.2.2. Cycloaddition Reactions of 1,1′-Binaphthalene-2,2′-bis(sulfonyl)ethylene with Nonsymmetric Dienes

The stereochemical issue arises in the cycloaddition with nonsymmetric dienes [eq. (8.2)]. In the cases of 1,3-substituted 1,3-dienes, four possible adducts can be formed, depending upon the approach of the diene to the dienophile, as shown in Scheme 8.1. The adduct derived from the *exo* approach, A (or C), differs from the one derived from the *endo* approach, B (or D), with respect to the stereochemistry of the R′ group whereas between exo approaches A and C (or between endo approaches B and D) the difference is in the regiochemistry of addition.

Upon adding a dichloromethane solution of 1-methoxy-3-trimethylsilyloxy-1,3-butadiene (Danishefsky diene) to a dichloromethane solution of the dienophile **8** in the same solvent, at room temperature, and with a slight excess of the diene, a crystalline precipitate forms after a few seconds. Thin-layer chromatography with several eluants and 200-MHz NMR of the adduct (either of the crystalline material or of the crude material obtained by evaporating the solvent and excess of diene) show that a single adduct is formed [eq. (8.10)].

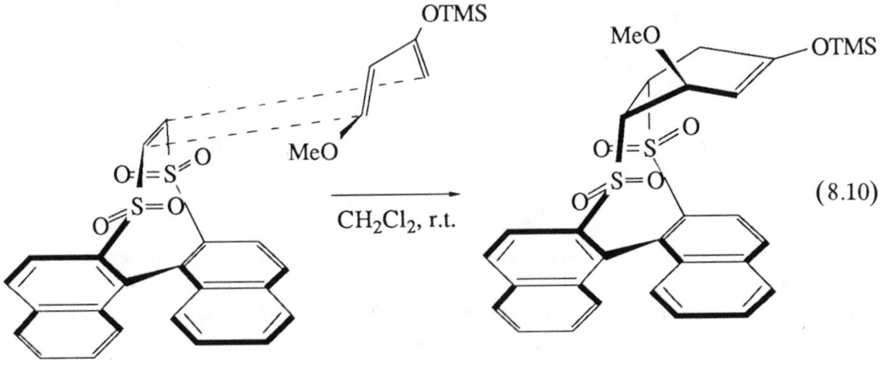

$$(8.10)$$

The illustrated mode of addition and the consequential stereochemistry of the adduct were assigned by combining the results of the ^{1}H NMR spectral data, including Nuclear Overhauser Effect (NOE) measurements, and those of the analysis of the presumed transition states for a concerted Diels–Alder cycloaddition. In fact, of the four possible adducts (Scheme 8.1), only two are in accordance with the NMR data, that is, the adducts derived from approaches A and C. Of the latter two alternatives, the approach of type A is ruled out because in the transition state the methoxy group would interfere with the naphthyl group, whereas in approach C the substituent at C-1 of the diene enters from the open side of the dienophile [Scheme 8.1 and eq. (8.10)].

The result of the stereochemical analysis is different from the reported endo stereochemistry of addition to the Danishefsky diene [14]. We think that in our case the endo approach is inhibited by steric reasons.

In order to understand whether the C-1 or the C-3 substituent directs the stereochemistry of the addition, both 1-substituted and 3-substituted dienes have been reacted with **8**. In the case of 1-substituted dienes [eq. (8.11)] a single product was always observed, irrespective of the nature of the substituent.

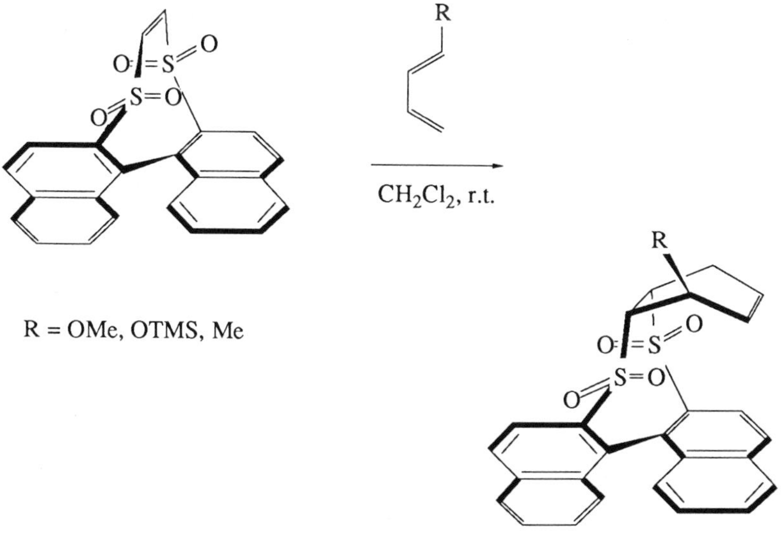

$$R = OMe, OTMS, Me \qquad\qquad CH_2Cl_2, \text{ r.t.} \qquad\qquad (8.11)$$

It is noteworthy that in the adduct of piperylene (R = Me), virtually complete diastereoselection has been achieved with a single methyl group. The stereochemistry of addition in this case has been rigorously defined by X-ray structure determination (Figure 8.3) [9]. X-ray analysis confirms the exo mode of addition and corroborates the stereochemical assignment of the adduct to the Danishefsky diene.

Other 1-substituted dienes [eq. (8.11)] are believed to produce similar products, as can be expected by the similarity of the NMR spectra.

1-Substituted cyclic dienes, such as 2-methoxyfuran and 1-methoxy-1,3-cyclohexadiene, similarly afford a single stereoadduct, to which the structures reported in eq. (8.12) have been assigned on the basis of similar reasoning.

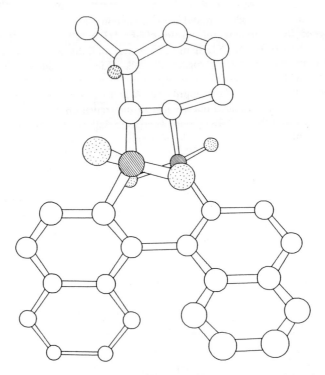

Figure 8.3 ■ Perspective drawing of the adduct of **8** to piperylene, as determined by X-ray structure analysis [9]. All hydrogen atoms have been omitted, except for the hydrogen α to the methyl group.

$$\text{(8.12)}$$

R = OMe, OTMS X = O
R = OMe X = (CH$_2$)$_2$

Of the two 2-substituted dienes that have been reacted with **8** [eq. (8.13)], only 2-trimethylsilyloxy-1,3-butadiene gave a single adduct, whereas isoprene (R = Me) produced about an 8 : 2 mixture of diastereoisomers.

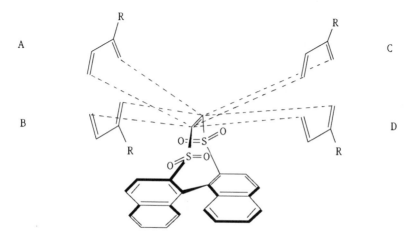

(8.13)

R = OTMS, Me

In the latter case, at variance with the cycloadditions reported so far, the reaction required higher temperatures and longer time (48 h at reflux in chloroform). In the assignment of the configuration, it should be noted that both exo and endo approaches result in the same adduct (Scheme 8.2) that is, the stereochemical bias is only dependent on the face of the dienophile in which the cycloaddition occurs, so there is only regiochemical concern. Of the two possible conformations in accordance with the NMR spectrum, corresponding to the approaches C/D and A/B [eq. (8.14)], the former is preferred because it is the same as for the cycloadducts so far discussed and because it appears to be more stable from molecular models and simple molecular mechanics calculations.

Scheme 8.2

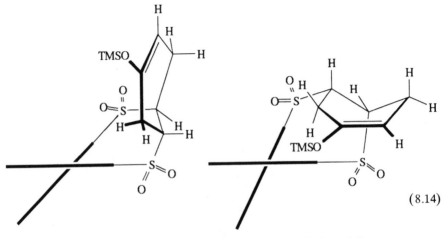

approach C and D

approach A and B

(8.14)

Such assignment is further supported by the similarity of the NMR data and of the NOE experiments with those of the previous systems. However, an X-ray structure determination is in progress to secure the structure assignment.

The possibility for the system to work as an equivalent of acetylene has been verified for the model compound **19** [eq. (8.15)].

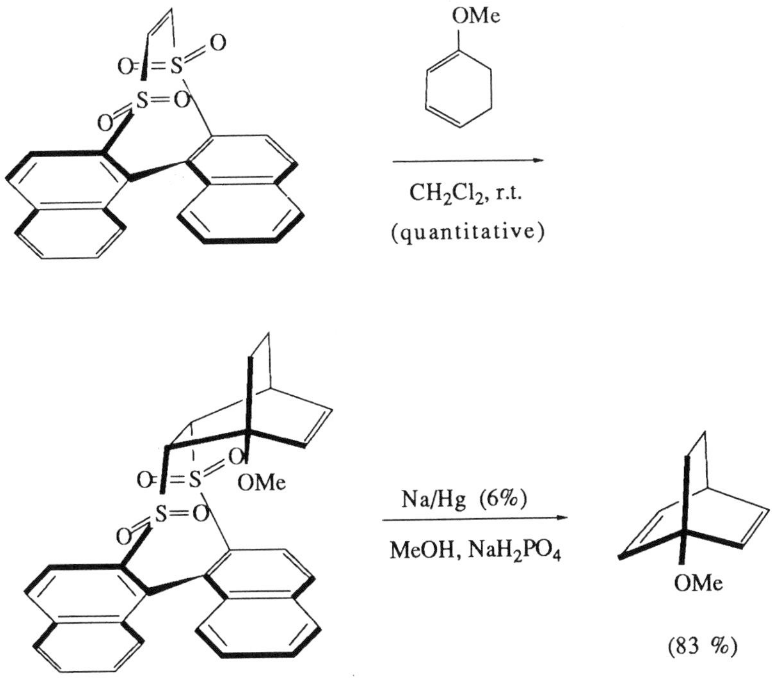

1 9

Although no chiral material can be formed in this case, it is demonstrated that the adducts readily desulfonylate under sodium amalgam reduction, and preliminary experiments suggest that the chiral auxiliary can be recovered and recycled.

8.3. Generation and Reactions of the Carbanions of Dinaphtho[2,1-*d*:1′,2′-*f*][1,3]dithiepine and its *S*-Oxides

Condensation of the dithiol **9** with molecules different from (*Z*)-1,2-dichloroethylene of eq. (8.8), may serve to prepare other compounds of synthetic potential, as, for example, the atropisomeric chiral molecule of C_2 symmetry shown in eq. (8.16).

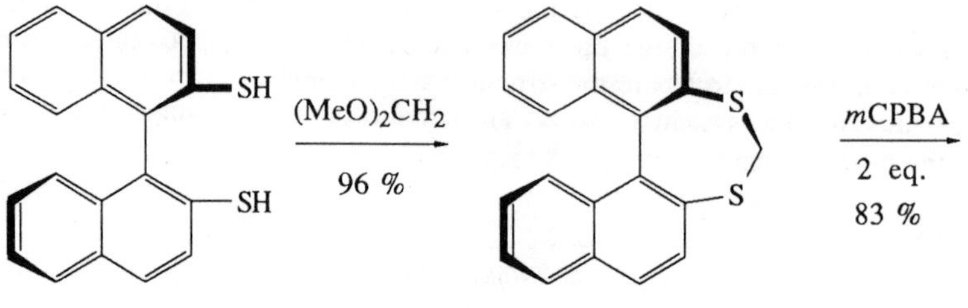

(8.16)

9   **2 0**

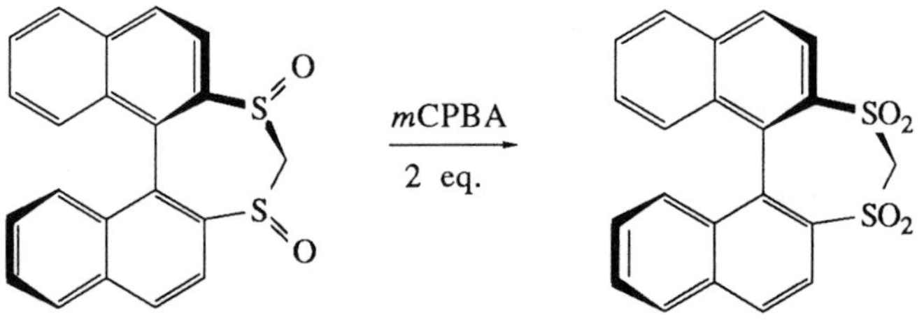

2 1   **2 2**

The bis-sulfoxide **21** and bis-sulfone **22** are prepared from **20** via oxidation with 2 or 4 equivalents of *m*-chloroperbenzoic acid. Reagents **20–22** are atropisomeric chiral variants of the vast class of dithiomethylene compounds [15] and their *S*-oxides [16], which are well-known carbonyl anion equivalents and reagents of established utility.

The reaction of the anions of these molecules with electrophiles, such as deuterated water, methyl iodide, and so on, is of little significance because no new stereocenters are formed (the carbon that is substituted is in the symmetry axis). On the other hand, in the reaction with benzaldehyde a new stereocenter results. Hence, in principle, the reaction may give two diastereoisomers. Indeed, in the reaction of the anion of **20**, generated by *n*-butyllithium in tetrahydrofuran, a mixture of the two possible diastereoisomers was formed [eq. (8.17)], whereas no reaction was observed in the case of the oxidized substrates **21** and **22**.

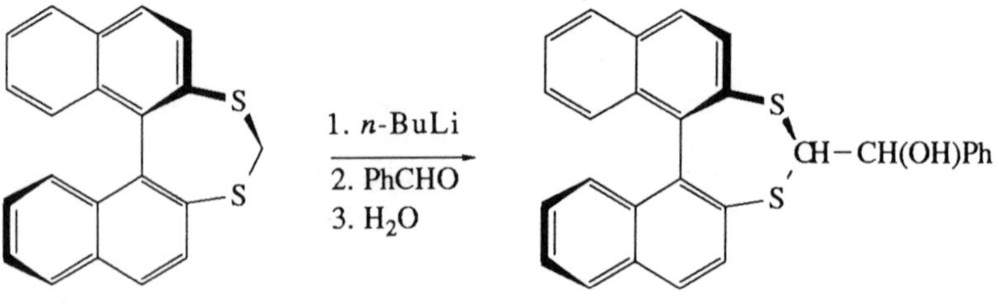

1. *n*-BuLi
2. PhCHO
3. H$_2$O

(8.17)

2 0   two diastereoisomers

Because in the latter cases the carbanion was indeed formed, as demonstrated by independent experiments (quenching with methyl iodide or deuterated water), it is presumed that the anion is too stable to prevent the retro-aldol reaction.

It is important to notice that the bis-sulfoxide **21** is the only diastereoisomer that is formed in the bis-oxidation. No other diastereomeric sulfoxides have been noticed.

Consequently, the monosulfoxide from the oxidation with one mole of peracid is also a single diastereomer.

Reaction of the anion of the monosulfoxide with methyl iodide affords a single diastereoisomer [eq. (8.18)].

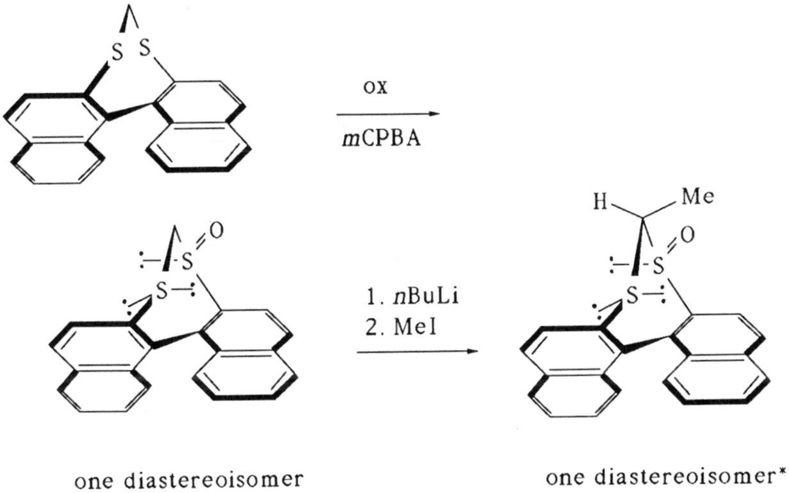

one diastereoisomer   one diastereoisomer*

(8.18)

No attack of the electrophile from the other side could be detected. In a very similar way, addition of benzaldehyde to the same anion also affords a single diastereoisomer [eq. (8.19)].

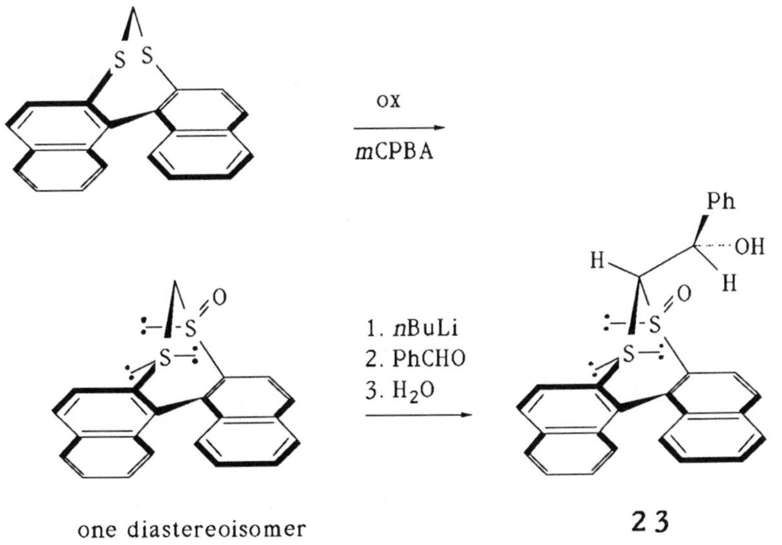

one diastereoisomer   **23**

(8.19)

one diastereoisomer

*Stereochemistry not yet defined.

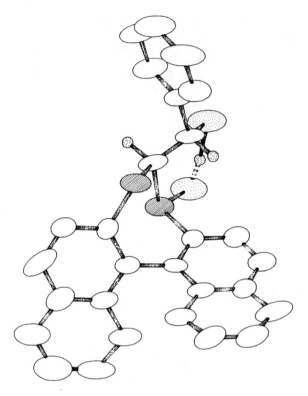

Figure 8.4 ■ Perspective drawing of compound **23**, as determined by X-ray structure analysis [9]. Hydrogen atoms have been omitted, except those of stereochemical interest.

In this case, overall, three contiguous chiral centers have been constructed stereospecifically from the single asymmetric unit of the binaphthyl residue.

The stereochemistry of the sulfoxide oxygen was suggested by analysis of molecular models and confirmed by X-ray structure determination of the benzaldehyde adduct **23** (Figure 8.4) [9]. The configuration of the two newly formed carbon centers corresponds to the one expected on the basis of a transition state of the type **24** (shown in two different perspectives) that has been postulated for acyclic dithioacetal oxides [17, 18].

2 4   **2 4'**

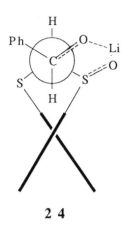

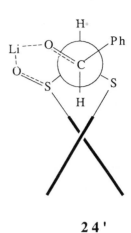

2 4   **2 4'**

Of the two possible conformations possessing an equatorial phenyl group, **24** is favored over **24'** because in the latter the aldehyde hydrogen would approach the naphthyl residue too closely.

It has to be pointed out that the related acyclic formaldehyde ditolylthioacetal S-oxide [eq. (8.20)] is reported to afford in the reaction with benzaldehyde a mixture of three out of the four possible diastereoisomers in a 5 : 3.5 : 1.5 ratio [18].

$$O=S* \overset{CH_2}{\diagdown} S \quad \xrightarrow[\substack{2.\ PhCHO \\ 3.\ H_2O}]{1.\ nBuLi,\ THF} \quad O=S* \overset{CHCH(OH)Ph}{\diagdown} S \qquad (8.20)$$

diastereomeric ratio: 50 : 35 : 15 : 0

Of the two other possible nonsymmetric combinations of oxides, that is, the sulfone–sulfide **25** and the sulfone-sulfoxide **26**, the former afforded with benzaldehyde variable mixtures of the four diastereoisomers, whereas the latter did not react.

2 5   **2 6**

8.4. Conclusions and Perspectives

Although the preliminary experiments so far examined have been performed on racemic substrates, the reagents are available in optically active form by known methodologies [8b] or, more efficiently, with our method of enantioselective sulfide oxidation based on modified Sharpless reagents [19]. The synthetic strategy presented here is now going to be applied to other chiral frameworks derived from largely available, optically active natural products possessing C_2 symmetry, such as tartaric acid and sugar derivatives.

The investigation on efficient and readily obtainable chiral variants of the several dithioacetals and oxides that are known in the literature should permit the synthesis of important classes of compounds, such as α-aminoacids, 2-arylpropionates, alkoxyarylacetates, and so on, in an optically active form [20].

References

1. De Lucchi, O.; Lucchini, V.; Pasquato, L.; Modena, G. *J. Org. Chem.*, **1984**, *49*, 596.
2. De Lucchi, O.; Modena, G. *Tetrahedron*, **1984**, *40*, 2585. De Lucchi, O.; Pasquato, L. *Tetrahedron*, **1988**, *44*, 6755.
3. Wenkert, E.; Broka, C. A. *Finn. Chem. Lett.*, **1984**, 126. Nakayama, J.; Nakamura, Y.; Hoshino, M. *Heterocycles*, **1985**, *23*, 1119.
4. Ono, N.; Kamimura, A.; Kaji, A. *Tetrahedron Lett.*, **1986**, *27*, 1595.
5. Okada, K.; Okamoto, K.; Oda, M. *J. Am. Chem. Soc.*, **1988**, *110*, 8736.
6. De Lucchi, O.; Lucchini, V.; Marchioro, C.; Valle, G.; Modena, G. *J. Org. Chem.*, **1986**, *51*, 1457. De Lucchi, O.; Buso, M.; Modena, G. *Tetrahedron Lett.*, **1987**, *28*, 107.
7. This approach was suggested by the observation that the two sulfonyl groups exert a strong influence on each other if they are assembled in a cis relationship (De Lucchi, O.; Pasquato, L.; Modena, G.; Valle, G. *Zeit. Krist.*, **1985**, *170*, 267). This fact may constitute an important feature for transmitting the asymmetric information in the Diels–Alder reaction. Delogu, G.; De Lucchi, O.; Fois, M. P., unpublished results.
8. (a) Barber, H. J.; Smiles, S. *J. Chem. Soc.*, **1928**, 1141. (b) Armarego, W. L. F.; Turner, E. E. *J. Chem. Soc.*, **1957**, 13. (c) Cram, D. M.; Helgeson, R. C.; Koga, K.; Kyba, E. P.; Madan, K.; Sousa, L. R.; Siegel, M. G.; Moreau, P.; Gokel, G. W.; Timko, J. M.; Sogah, G. D. Y. *J. Org. Chem.*, **1978**, *43*, 2758.
9. The X-ray structural analyses were performed by G. Valle (Centro Studi Biopolimeri del C.N.R., via Marzolo 1, I-35131 Padova, Italy).
10. Gogte, V. N.; Palkar, V. S.; Tilak, B. D. *Tetrahedron Lett.*, **1960**, 30. Tedjamulia, M. L.; Tominaga, Y.; Castle, R. N.; Lee, M. L. *J. Heterocycl. Chem.*, **1983**, *20*, 1143.
11. The intermediate unoxidized bis-sulfide **16** is one of the few examples of 1,4-dithiocin, an aromatic 10-π-electrons system: Eggelte, H. J.; Bickelhaupt, F.; Loopstra, B. O. *Tetrahedron*, **1978**, *34*, 3631. Schroth, W.; Billig, F.; Zschunke, A. *Z. Chem.*, **1969**, *9*, 184.
12. The double bond that forms in the Diels–Alder reaction is in an asymmetric environment. In view of recent results in the addition of electrophiles to sulfonyl adducts, this olefin is well set up for asymmetric additions. See: Cadogan, J. I. G.; Cameron, D. K.; Gosney, I.; Highcock, R. M.; Newlands, S. F. *J. Chem. Soc., Chem. Commun.*, **1986**, 766. Cossu, S.; De Lucchi, O.; Dilillo, F., unpublished results.
13. Czugler, M.; Weber, E.; Ahrendt, J. *J. Chem. Soc., Chem. Commun.*, **1984**, 1632. Weber, E.; Csöregh, I.; Ahrendt, J.; Finge, S.; Czugler, M. *J. Org. Chem.*, **1988**, *53*, 5831.

14. Danishefsky, S.; Yan, C.-F.; Singh, R. K.; Gammill, R. B.; McCurry, P. M., Jr.; Fritsch, N.; Clardy, J. *J. Am. Chem. Soc.*, **1979**, *101*, 7001.
15. Gröbel, B.-T.; Seebach, D. *Synthesis*, **1977**, 357.
16. Colombo, L.; Gennari, C.; Resnati, G.; Scolastico, C. *Synthesis*, **1981**, 74. Colombo, L.; Gennari, C.; Resnati, G.; Scolastico, C. *J. Chem. Soc., Perkin Trans. 1*, **1981**, 1278. Guanti, G.; Narisano, E.; Pero, F.; Banfi, L.; Scolastico, C. *J. Chem. Soc., Perkin Trans. 1*, **1984**, 189. Ogura, K.; Fujita, M.; Inaba, T.; Takahashi, K.; Iida, H. *Tetrahedron Lett.*, **1983**, *24*, 503.
17. This work strictly correlates with a recent study on applications of optically active vinyl sulfoxides obtained via a self-induced diastereoselective oxidation mediated by chiral auxiliaries containing an hydroxy group [6].
18. Colombo, L.; Gennari, C.; Scolastico, C.; Guanti, G.; Narisano, E. *J. Chem. Soc., Perkin Trans. 1*, **1981**, 1278.
19. Di Furia, F.; Licini, G.; Modena, G.; De Lucchi, O. *Tetrahedron Lett.*, **1989**, *30*, 2575.
20. Aminoacids: Ogura, K.; Tsuchihashi, G. *J. Am. Chem. Soc.*, **1974**, *96*, 1960. Hirama, M.; Hioki, H.; Ito, S. *Tetrahedron Lett.*, **1988**, *29*, 3125. 2-Arylpropionates: Ogura, K.; Mitamura, S.; Kishi, K.; Tsuchihashi, G. *Synthesis*, **1979**, 880. Alkoxyarylacetates: Ogura, K.; Watanabe, J.; Takahashi, K.; Iida, H. *J. Org. Chem.*, **1982**, *47*, 5404.

Reactions of Sulfoxides with Organometallic Reagents: Is Sulfurane an Intermediate?

Naomichi Furukawa

Department of Chemistry
University of Tsukuba
Tsukuba, Ibaraki 305
Japan

9.1. Introduction

Recently, heteroatoms belonging to the third to fifth rows of the periodic table in organic compounds have attracted considerable attention. One of the prominent features of heteroatoms such as Si, P, S, and so on, is that they can expand valency beyond the normal octet to decets or even dodecets, producing numerous so-called hypervalent (HV) compounds. These substances are interesting not only for their unusual structures as compared to those based on first-row elements of the periodic table, but also for their important roles as intermediates in nucleophilic or electrophilic substitution reactions.

In organic sulfur chemistry, hypervalent compounds called σ-sulfuranes of pentacoordinate species or persulfuranes of hexacoordinate species are formed either by expanding the valency of the central sulfur atom from octet to decet or dodecet. According to the Martin nomenclature, these σ-sulfuranes are classified as (10-S-5) and (12-S-6), respectively [1].

Normally, σ-sulfuranes have four ligands and one lone pair of electrons, forming a trigonal bipyramidal structure (TBP structure) in which lone-pair electrons always occupy an equatorial position and among the four ligands, the two most electronegative ligands take the apical positions (a) whereas the two others occupy the remaining two equatorial positions (e). On the other hand, persulfuranes (12-S-6) have a square bipyramidal structure (SBP structure) as shown in Scheme 9.1.

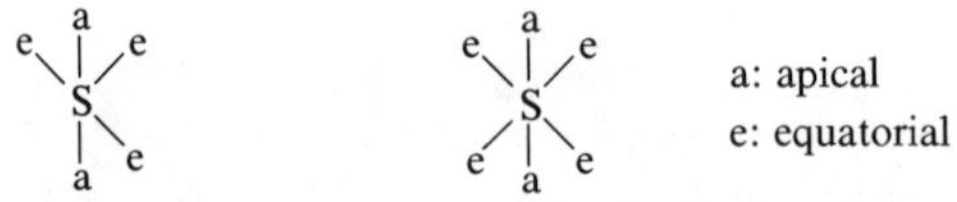

Sulfurane (10-S-5)
TBP structure

Persulfurane (12-S-6)
SBP structure

a: apical
e: equatorial

Scheme 9.1 ■ Hypervalent sulfur compounds.

In the TBP structure, the two atoms in the apical ligands and the central sulfur atom form a three-center–four-electron bond. An approximate molecular orbital (MO) model of a three-center–four-electron bond presented by Musher [2] is shown in Scheme 9.2. Among the four bonding electrons, two electrons fill the lowest bonding MO (Ψ_1) and the other two electrons are the nonbonding MO (Ψ_2). The third MO is antibonding. These molecular orbitals are closely related to those of the π-allylic system.

The three-center–four-electron bonds have the following characteristics:

1. More electronegative ligands tend to occupy the apical positions.
2. Facile ligand interchange takes place without bond breaking via pseudorotation or turnstile rotation.
3. Lone-pair electrons always occupy an equatorial position.
4. Five-membered rings span at the apical and the equatorial positions.
5. In general, nucleophilic substitution on the sulfur atom takes place from the apical position and the leaving group is eliminated from the apical position (a–a process) so that the stereochemistry should be inversion.

However, recent investigations reveal that the reactions take place not only via an a–a process but also via an a → e (or an e → a) and even via an e–e process. Thus, many

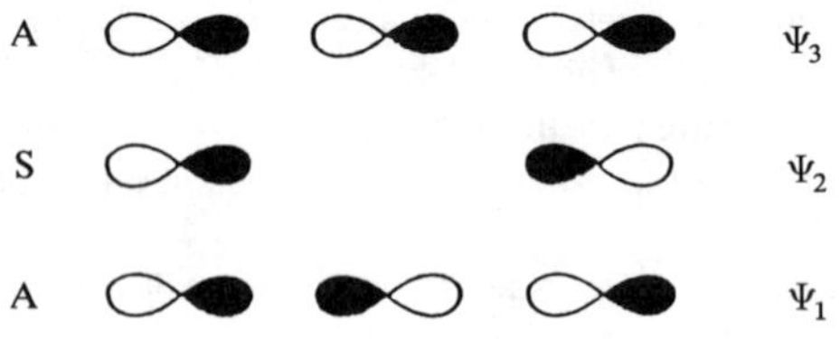

A Ψ_3

S Ψ_2

A Ψ_1

O—M—O

$\overset{+}{Me_4}NI_3^-$

-0.541 $+0.083$ -0.541 CNDO/2

$(I \text{———} I \text{———} I)^-$

-0.50 $+0.05$ -0.50 nqr

(0.50)   (0.50)

Scheme 9.2

stereochemical directions have been reported for the reactions [3]. The most essential feature of hypervalent species is that the central atom is valence-shell expanded, for example, the sulfur atom in a σ-sulfurane assumes the decet. Therefore, hypervalent species are relatively unstable and the central atom tends to resume the normal valency by extruding a ligand bearing a pair of electrons or a pair of ligands affording a stable compound recovering the octet in the sulfur atom.

Although SF_4, SF_6, and some other stable σ-sulfuranes have been known for more than 100 years, considerable attention has focused on the chemistry of these new types of molecules since Martin and Arhart [4] and Kapovits and Kalman [5] reported their pioneering work on synthesis of stable sulfuranes in 1971. Since that time a number of stable sulfuranes and persulfuranes have been prepared and their structures and reactions investigated, as highlighted in reviews by Martin and co-workers [6].

It has long been argued whether sulfuranes are actually formed as intermediates in the nucleophilic substitution reactions of organosulfur compounds or as a simple transition state analogous to that for an S_N2 reaction on carbon [7]. Thus, two distinct processes are conceivable in the reactions: one is a simple one-step process and the other is constructed from at least two steps involving sulfurane as an intermediate.

Tricoordinated sulfur compounds, such as sulfonium salts, sulfoxides, and related derivatives, undergo nucleophilic substitutions upon treatment with nucleophiles and hence are useful targets for investigation of the mechanism of the reactions on the sulfur atom. The reactions of these compounds have been thought to proceed via initial formation of a σ-sulfurane from which either ligand exchange or a ligand-coupling reaction takes place to afford the products, although only a few σ-sulfuranes have been detected in the reactions by spectroscopic methods [8].

In order to shed more light on the problem, namely, whether sulfuranes are actually involved as intermediates, to establish diagnostic methods for the formation of sulfuranes as intermediates, and also to extend the nucleophilic reactions of organosulfur compounds for organic synthesis, we have systematically studied the reactions of sulfoxides with organometallic regents, such as organolithium or Grignard reagents. This chapter summarizes the results and presents evidence for σ-sulfuranes as unstable intermediates in the reactions of organosulfur compounds.

9.2. Stable σ-Sulfuranes

Recently, many stable organic hypervalent compounds of Si, P, S, Se, and so on, have been synthesized [6]. In order to synthesize stable hypervalent compounds, we have to design the attached ligands on the basis of the essential concepts to stabilize the hypervalent bonds as follows:

1. strong electronegative group or elements such as F and O should be used as apical ligands;
2. five-membered ring annelation such as o,o-benzoannelation should be used to fix both the apical and equatorial ligands;
3. electropositive π-donating carbon bonds are required as equatorial ligands.

Several stable σ-sulfuranes are illustrated in Scheme 9.3.

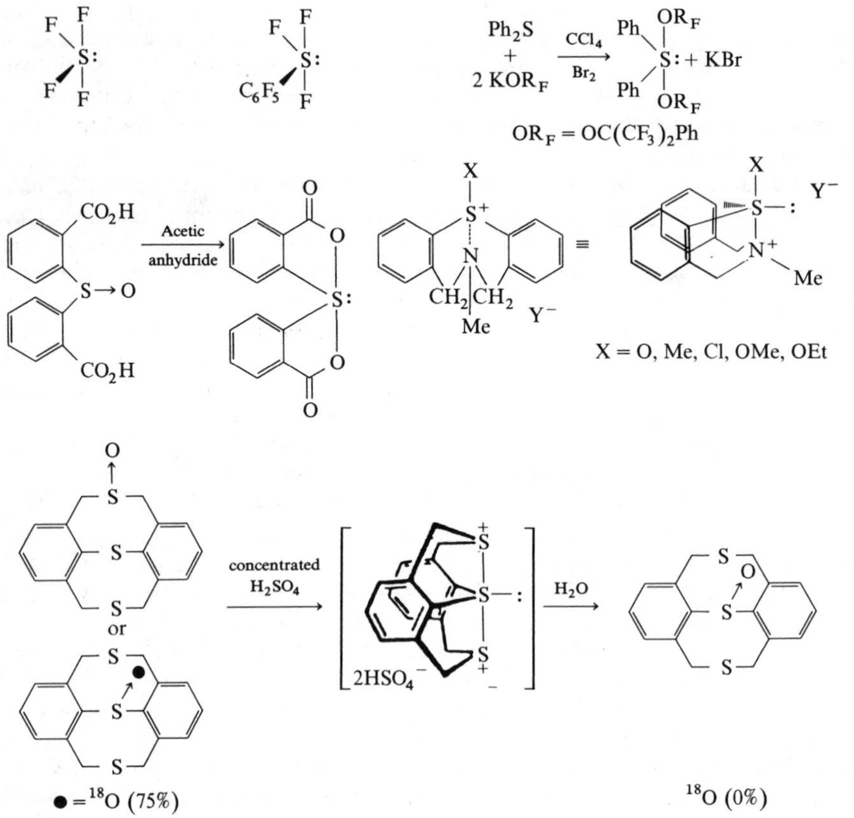

Scheme 9.3

9.3. σ-Sulfuranes as Reactive Intermediates or Transition States

For many years we have been studying nucleophilic substitution reactions of organic sulfur compounds, that is, acid-catalyzed oxygen exchange reactions [9] and nucleophilic substitution reactions using sulfilimines [10]. Here I would like to discuss the reactions of sulfoxides and selenoxides with organometallic reagents together with mechanisms and their synthetic application.

9.3.1. Reactions of Sulfonium Salts with RLi

Since Wittig and Fritz found that triarylsulfonium salts react with organolithium reagents to give numerous compounds including ligand coupling biaryls [11], many sulfonium salts were treated with organolithium reagents under various reaction conditions. The reactions are assumed to proceed via initial formation of a σ-sulfurane as an unstable intermediate which on decomposition affording biaryl as a coupling product and diaryl sulfide [12].

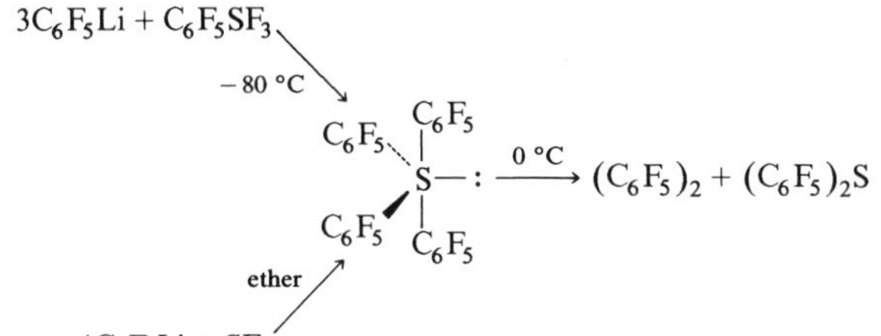

$$3C_6F_5Li + C_6F_5SF_3$$

$$-80\ °C$$

$$\underset{\overset{|}{C_6F_5}}{\overset{C_6F_5}{\underset{C_6F_5}{\overset{C_6F_5}{S-:}}}} \xrightarrow{0\ °C} (C_6F_5)_2 + (C_6F_5)_2S$$

$$4C_6F_5Li + SF_4 \qquad \textbf{A}$$

ether

Scheme 9.4

$$Ph_3\overset{+}{S}:X^- + PhLi \longrightarrow [Ph_4S:] \longrightarrow Ph-Ph + PhSPh + LiX$$

Formation of sulfurane as an intermediate as shown in the above reaction was confirmed on the basis of stereochemical results and ^{14}C-tracer experiments. Many other examples have been reported by Trost, Mislow, Hori, Andersen, and others [12–16]. Furthermore, Sheppard [17] reported the formation of tetraaryl sulfurane in the reaction of pentafluoro derivatives by ^{19}F NMR as the first clear-cut example of a distinct sulfurane intermediate as shown in Scheme 9.4.

^{19}F NMR spectra indicate that the sulfurane A has three ^{19}F NMR peaks corresponding to o-, m-, and p-fluorine atoms at below $-20\ °C$, which shift to the NMR absorption peaks corresponding to the two products upon warming. At present, this is the only known σ-sulfurane having four carbon ligands.

Similar treatment of triphenylselenium salts with PhLi also gave biphenyl and diphenyl selenide as products, suggesting that tetraphenyl selenurane should be formed as an intermediate, although it is not isolated or detected [18]. Furthermore, tetrachlorotellurane or dichlorodiphenyl tellurane reacts with PhLi to give tetraphenyl tellurane as a stable crystalline compound, which has been purified and identified by conventional spectroscopic analysis [19]. Upon heating, this stable tellurane is converted to biphenyl and diphenyl telluride. Therefore, these results reveal that both the stable and unstable tetraphenyl hypervalent compounds of group VI elements give biphenyl and diphenyl sulfide, selenide, and telluride. The reactions are summarized in Scheme 9.5. The formation of these products becomes diagnostic for the formation of σ-sulfurane and other hypervalent compounds as intermediates.

$$Ph_3S^+X^- + PhLi \longrightarrow \underset{\text{unstable}}{[Ph_4S:]} \longrightarrow Ph-Ph + PhSPh$$

$$Ph_3Se^+X^- + PhLi \longrightarrow \underset{\text{unstable}}{[Ph_4Se:]} \longrightarrow Ph-Ph + PhSePh$$

$$TeCl_4 + 4PhLi$$

$$Ph_2TeCl_2 + 2PhLi \longrightarrow \underset{\substack{\text{stable}\\ \text{(isolated mp 105 °C)}}}{Ph_4Te:} \longrightarrow Ph-Ph + PhTePh$$

Scheme 9.5 ▪ Hypervalent (HV) compounds.

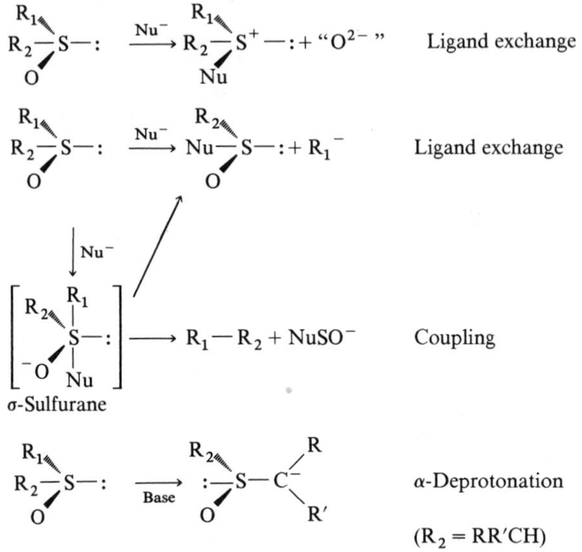

Scheme 9.6

9.3.2. Reaction of Sulfoxides with Organolithium and Grignard Reagents

In contrast to the reactions of sulfonium salts, sulfoxides react with organolithium and Grignard reagents by four different reaction pathways. One pathway involves nucleophilic substitution with Walden inversion at the sulfinyl sulfur atom ("ligand exchange"). A second pathway, recently discovered in our laboratory, involves concomitant ligand exchange and disproportionation. A third one is ligand-coupling reactions, also discovered in our laboratory, and a fourth reaction is proton abstraction to form α-sulfinyl carbanions. The four reactions are illustrated in Scheme 9.6.

These reactions, except α-proton abstraction, are presumed to proceed via a σ-sulfurane as an intermediate or a transition state. Because we have accumulated many results on the substitution reactions of sulfoxides with organolithium and Grignard reagents, we would like to summarize here the results and discuss aspects of the mechanism of the substitution reactions of tricoordinated sulfur as well as selenium atoms.

9.3.2a. Ligand-Exchange Reactions. Ligand-exchange reactions have been intensively studied using sulfoxides and other tricoordinated sulfur compounds. These reactions proceed with inversion of configuration on the sulfur atom and generally are observed in the oxygen exchange reactions of sulfoxides, nucleophilic substitution reactions of sulfilimines and alkaline hydrolysis of oxysulfonium salts [20].

Previously, sulfoxides were found to undergo facile ligand-exchange reactions upon treatment with organolithium or Grignard reagents in ether solvent. In the reactions using alkyl aryl sulfoxides or haloalkyl aryl sulfoxides, the most electronegative ligand is usually replaced by organometallic reagents to give dialkyl sulfoxides with complete inversion of the sulfur center [21] or alkyl aryl sulfoxides with replacement of the haloalkyl moiety [22]. Therefore, these reactions are believed to proceed via an S_N2-type process in which a σ-sulfurane is a transition state. Similar reactions of sulfoxides involving the t-Bu group as either a ligand or a reagent were found to give racemized products. The reactions have been proposed to proceed via an electron transfer reaction [21]. Recently, these ligand-exchange reactions of sulfoxides have been utilized to provide several new organolithium or Grignard reagents that are difficult to generate by conventional procedures. Hence the exchange reaction is useful for organic synthesis [23]. Diaryl sulfoxides have been known to react with aryl Grignard reagents to afford the corresponding sulfonium salts in which the oxygen ligand serves as a leaving group [24]. Andersen and his co-workers have reported that diaryl sulfoxides react with aryllithium reagents at room temperature initially affording triaryl sulfonium salts that generate either tetraaryl sulfuranes by the attack of ArLi or benzyne by proton abstraction. The sulfurane formed undergoes further coupling reaction to give biaryls and diaryl sulfides analogous to the reactions with sulfonium salts [25].

9.3.2b. Ligand-Exchange and Disproportionation Reactions.

Interestingly, when we tried o-lithiation of aryl phenyl sulfoxides using the corresponding aryl phenyl sulfoxides, that is, phenyl p-tolyl sulfoxide, with n-BuLi or lithium diisopropylamide (LDA) at low temperature (-70 to -110 °C), we found that optically pure phenyl p-tolyl and β-naphthyl p-tolyl sulfoxides undergo facile racemization on treatment with LDA at low temperature. However, when we used excess $(i$-Pr$)_2$NH in the reactions, o-lithiation was found to occur without racemization at the sulfinyl sulfur atom [26]. Then we carried out the reactions of phenyl p-tolyl sulfoxide (2) with n-BuLi and we found that less than one equivalent of n-BuLi reacts with sulfoxide (2) to give a mixture of three sulfoxides, that is, 2, diphenyl (1), and di-p-tolyl (3) in close to the $2:1:1$ statistical ratio together with n-butyl p-tolyl sulfoxide as a ligand-exchange product. Similarly, other organolithium reagents react with sulfoxide 2 to afford the disproportionation products. Even a 0.1 molar equivalent of n-BuLi also gave a mixture of three sulfoxides soon after mixing with the sulfoxide 2 at -85 °C. Other diaryl sulfoxides also gave a mixture of three disproportionated sulfoxides in a nearly statistical $1:2:1$ ratio. An equivalent mixture of phenyl and p-tolyl sulfoxides also gave similar disproportionated sulfoxides in a $1:2:1$ ratio. The results are summarized in Scheme 9.7 and Table 9.1 [27].

As shown in Table 9.1, not only simple p-substituted sulfoxides but also bulky mesityl phenyl sulfoxide undergo facile disproportionation with racemization of the recovered sulfoxide. Obviously, the reaction should be initiated by attack of organolithium reagent on the sulfinyl sulfur atom to eliminate one aryl group as a ligand-exchange product. This ArLi once formed may again attack the starting sulfoxide to form an ate complex (σ-sulfurane) from which another ligand is ejected. These exchange processes may be repeated rapidly even at $-85 \sim -110$ °C to give a mixture of disproportionation products. Furthermore, interestingly, the initially substituted n-butyl p-tolyl sulfoxide was also found to be completely racemized. These result suggest that the reaction is not a simple S_N2-like substitution observed in the reactions of alkyl aryl sulfoxides. These facile reactions of diaryl sulfoxides with organolithium reagents at low temperature have never been known before and contrast markedly with the results obtained by Andersen and co-workers [25]. We assume that the differences between our results and those of Andersen are due to the difference in the temperature and concentrations of the lithium reagents employed in the reactions. Because our

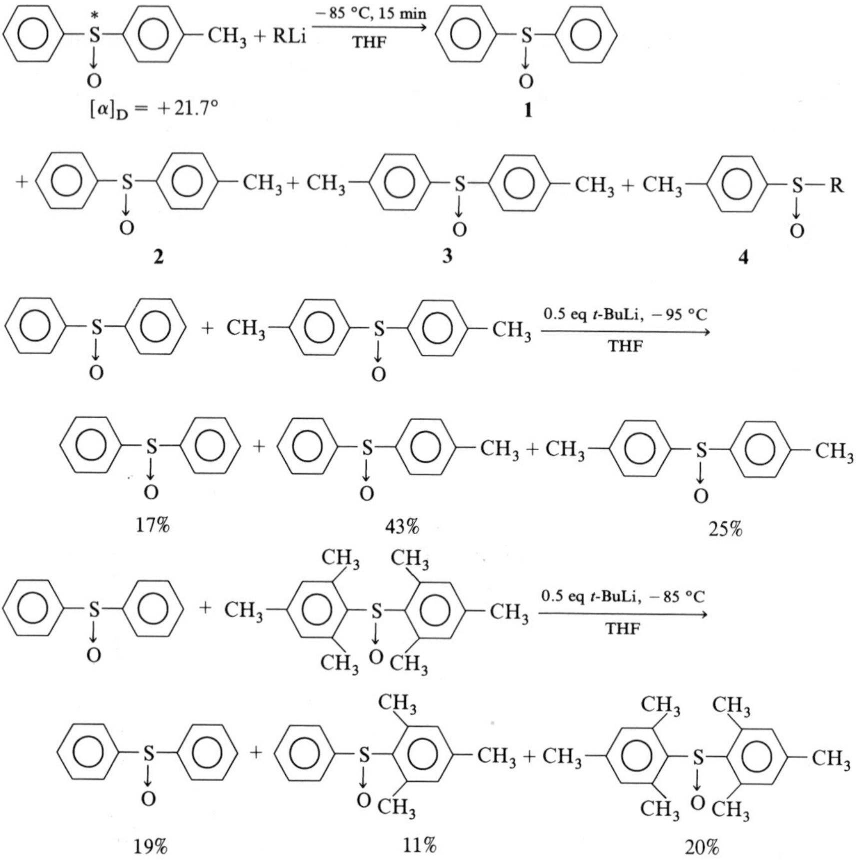

Scheme 9.7

Table 9.1 ■ Products from the Reaction of Optically Pure Phenyl p-Tolyl Sulfoxide with Organdithium Reagents

R	Molar equivalents	Yield (%)				$[\alpha]_D$	
		1	2	3	4	2	4
n-Bu	2.0	0	0	0	20[a]	—	0
t-Bu	4.0	0	0	0	29	—	0
t-Bu	2.0	0	0	0	30[b]	—	0
t-Bu	1.0	10	29	21	24	0	
t-Bu	0.5	14	35	22	15	0	
t-Bu	0.2	17	41	21	7	0	
Ph	0.2	24	48	22	0	0	—
LDA	1.0	0	quant.	0	0	20.3	—

[a] $(n\text{-Bu})_2S \rightarrow O$, 55%.
[b] $t\text{-Bu}-S-S-t\text{-Bu}$, 21%.
$\qquad\qquad\downarrow$
$\qquad\qquad O$

experiments were carried out at around $-80 \sim 0$ °C, our hypothesis is that the initial step takes place by an approach of R—Li coordinating to the sulfinyl oxygen atom from the back side of the more electronegative aryl group to form an unstable σ-sulfurane that may undergo rapid pseudorotation or turnstile rotation and then give ligand-exchange products, namely, racemic n-butyl p-tolyl sulfoxide and PhLi. On the other hand, at the higher temperature employed by Andersen, in addition to [or instead of (?)] initial σ-sulfurane formation, RLi may approach from the back side of the OLi group and displace OLi, because it is probable that the OLi group should be a better leaving group than the phenyl group. In the sulfinyl compounds the C—S bond energy (60–70 kcal/mol) is less than that of the S—O bond (85–90 kcal/mol) so that it is natural that the C—S bond should be broken more readily than the S—O bond at low temperature. At higher temperature both S—O and C—S bond cleavage should be possible (see analogous results below with selenoxides). It is well known that under acidic conditions, such as acid-catalyzed oxygen exchange reactions or oxysulfonium salts, oxygen serves as a leaving group rather then the carbon residue. Under Andersen's conditions it is possible that the leaving ability of oxygen may be enhanced through an increase in the positive charge at the sulfur atom. Thus, we propose the following probable mechanism for these facile ligand-exchange and disproportionation reactions (Scheme 9.8).

Like sulfoxides, diaryl selenoxides also undergo facile ligand-exchange and disproportionation reactions. However, selenoxides seem to react more readily than sulfoxides. For example, phenyl p-tolyl selenoxide reacts with 0.5 equiv of t-BuLi (or PhLi) to give both a mixture of three selenoxides and three reduced selenides, in nearly $1:2:1$ ratio, with the concomitant formation of biaryls as shown below. The ratio of these ligand-exchange and reduction products also depends largely on the temperature and mole ratio of RLi reagents. The higher the temperature, the more the reduction

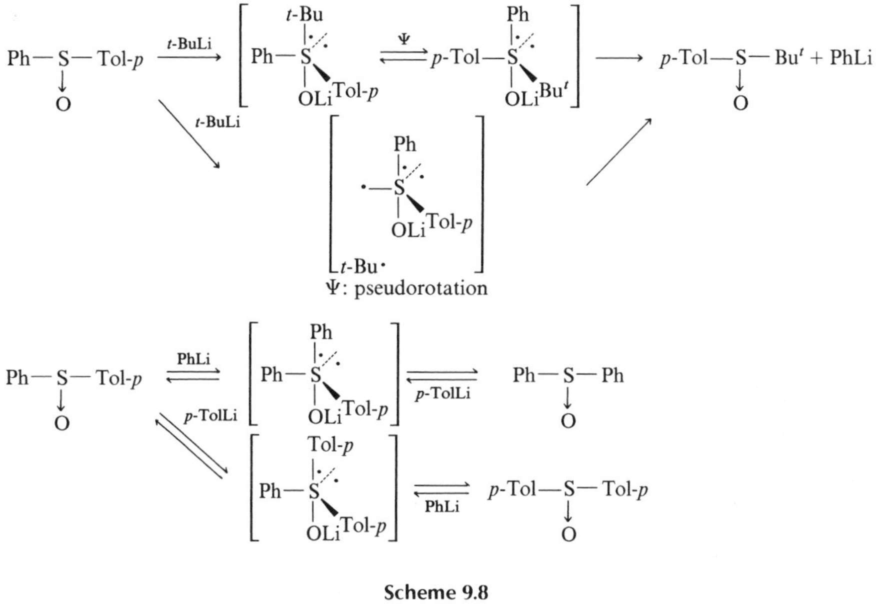

Scheme 9.8

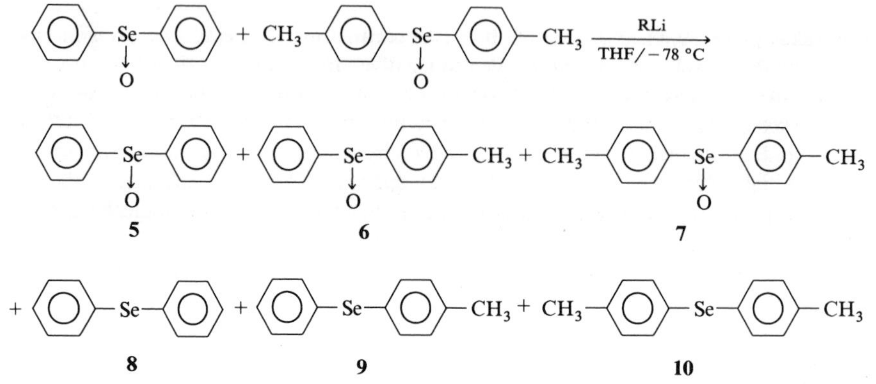

Scheme 9.9

products and biaryls were obtained. The selenides did not react with t-BuLi (or PhLi) under the reaction conditions or even in refluxing THF. Therefore, the disproportionation should take place rapidly to give initially a mixture of disproportionated selenoxides [28] (see Scheme 9.9 and Table 9.2).

In these reactions, we postulated the formation of sulfurane or selenurane as an unstable intermediate on the basis of product analysis. In order to confirm the formation of these hypervalent compounds directly, we tried to detect the hypervalent compounds by observations of the ^{1}H and ^{13}C NMR spectra of diphenyl sulfoxide or diphenyl selenoxide with PhLi at -80 °C. However, the spectra observed were consistent with those of a mixture of the sulfoxide (or selenoxide) and PhLi. Furthermore, we followed the ^{77}Se NMR spectra of diphenyl selenoxide with PhLi at -100 °C. The ^{77}Se NMR spectra showed absorption peaks corresponding to diphenyl

Table 9.2 ▪ Products from the Reaction of Phenyl p-Tolyl Selenoxide
with Organolithium Reagents

RM	Molar equivalents	Temperature (°C)	Yield (%)					
			5	6	7	8	9	10
n-BuLi	0.2	−78	17	31	22	2	2	1
n-BuLi	0.5	−78	5	15	11	4	8	3
n-BuLi	1.0	−78	4	13	9	5	10	5
n-BuLi	2.0	−78	—	—	—	8	21	11
t-BuLi	0.2	−78	12	58	10	—	1	—
t-BuLi	0.5	−78	10	20	16	6	15	9
t-BuLi	1.0	−78	3	6	4	1	1	1
t-BuLi	2.0	−78	—	—	—	1	3	3
t-BuLi	0.5	−20	8	14	8	4	10	6
PhLi	0.2	−78	16	27	13	2	2	1
PhLi	1.0	−78	6	7	2	2	2	1

selenoxide and selenide, respectively. The results indicate that the disproportionation reaction actually takes place but even triaryl selenurane oxide is highly reactive and rapidly decomposes to the products. σ-Sulfurane or σ-selenurane intermediates cannot be detected by the present NMR techniques with the preceding reaction conditions and substrates [29].

9.3.2c. Ligand-Coupling Reactions.

Ligand coupling is a well-known reaction of hypervalent species. In hypervalent species, apical positions are occupied by electronegative ligands using p orbitals, whereas equatorial positions, which use sp^2 hybridized orbitals, are presumed to be taken up by electron-donating substituents or π-donating groups. In general, ligand coupling is considered to take place between two electronegative apicophilic ligands [6]. However, the mechanism for the coupling, namely, whether two apical ligands undergo concerted extrusion at their apical positions or whether one ligand occupies the apical position and another one shifts to the equatorial position by pseudorotation or turnstile rotation and then these two ligands come close together for coupling has not been resolved. However, on consideration of stereoelectronic requirements for ligand-coupling reactions, the latter mechanism must be a better explanation, because the stereochemistry of the carbon atoms attached at the coupling center involves retention of configuration indicating that the reaction should proceed via concerted front-side attack of one ligand on the other via a three-membered transition state or intermediate without bond-breaking. Thus, if there is any cohesive interaction between the two ligands, they would be extruded from the central valence-shell expanded atom in a concerted manner, affording a ligand-coupling product in which the original configuration of the two coupled ligands is retained completely, as shown in the reaction of triphenyl sulfonium salt with PhLi (Scheme 9.10) and as will be discussed more fully below with examples containing chiral centers at the migrating carbon atoms.

As shown previously, many sulfonium salts undergo ligand-coupling reactions in a concerted manner upon treatment with organolithium reagents. However, in the

Ligand
coupling

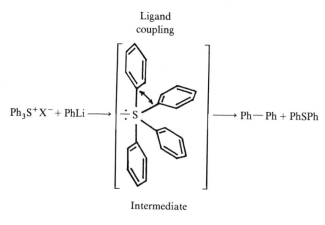

$$Ph_3S^+X^- + PhLi \longrightarrow \qquad \longrightarrow Ph{-}Ph + PhSPh$$

Intermediate

Scheme 9.10

reactions of sulfoxides with organolithium and Grignard reagents, ligand coupling was unknown until we recently found that benzyl pyridyl sulfoxide undergoes readily and exclusively the ligand-coupling reaction upon treatment with Grignard or organo-lithium reagents [30]. We have subsequently found many examples of ligand-coupling reactions of sulfoxides bearing at least one strongly electron-withdrawing heteroaryl group, such as 2-pyridyl and 2-quinolyl. As a counter ligand for the coupling process, not only benzyl but also allylic sec- and tert-alkyl and phenyl groups can couple with the pyridyl and quinolyl groups upon treatment with an appropriate Grignard reagent at room temperature or even upon cooling or heating. An aromatic ligand bearing an electron-withdrawing group such as (p-phenylsulfonyl)phenyl also can replace a het-eroaryl group to achieve a smooth coupling [31]. Therefore, ligand coupling becomes a general phenomenon in the reactions of sulfoxides bearing heteroaryls with Grignard or organolithium reagents. Grignard reagents are better to use than organolithium reagents, because organolithium reagents work as both thiophilic reagents and bases toward heterocyclic compounds to give a complex mixture of intractable products. As a typical example, benzyl 2-pyridyl sulfoxide reacts with PhMgBr in THF at room temperature to give 2-benzylpyridine in a quantitative yield whereas phenyl 2-pyridyl sulfoxide also affords 2-benzylpyridine in nearly an identical yield upon treatment with $PhCH_2MgCl$. These results clearly indicate that the coupling reactions proceed via a common intermediate, σ-sulfurane, which can undergo pseudorotation or turnstile rotation. The remaining organic sulfur species is PhSOMgX, which can be converted to methyl phenyl sulfoxide or can be quenched with water to give PhS(O)SPh (via PhSOH) and its disproportionation products, as shown in Scheme 9.11.

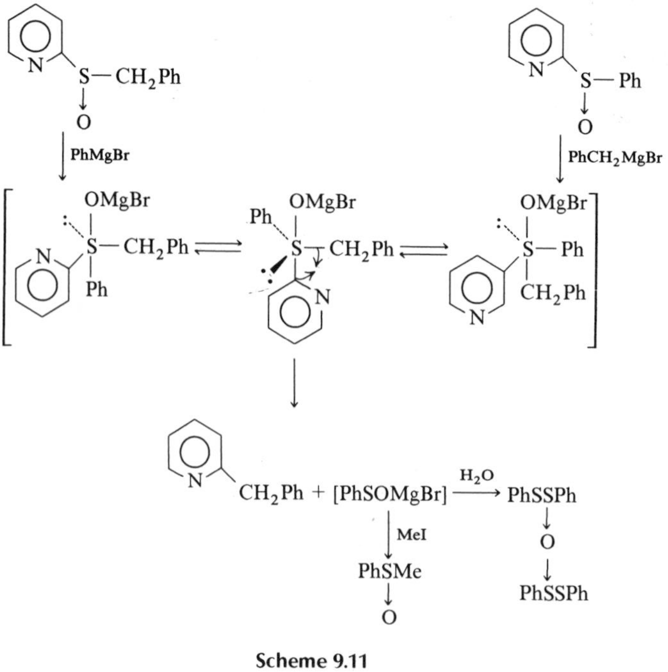

Scheme 9.11

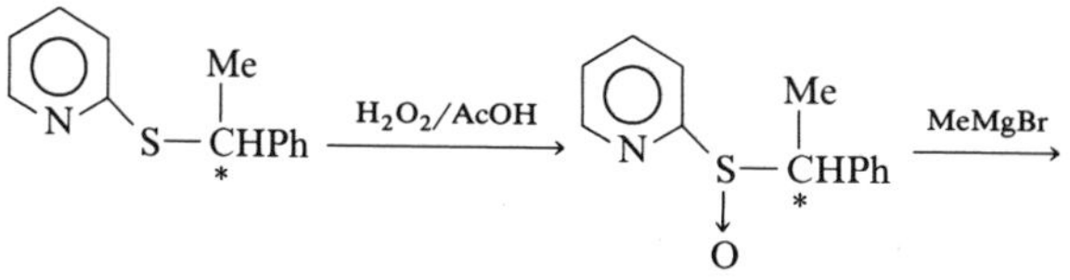

(S)–(–) $[\alpha]_D^{25°}$ = –375° (c = 1.156, benzene) $S_S C_S$ (36)
(R)–(+)   +375° (c = 0.742, benzene) $S_R C_S$ (37)

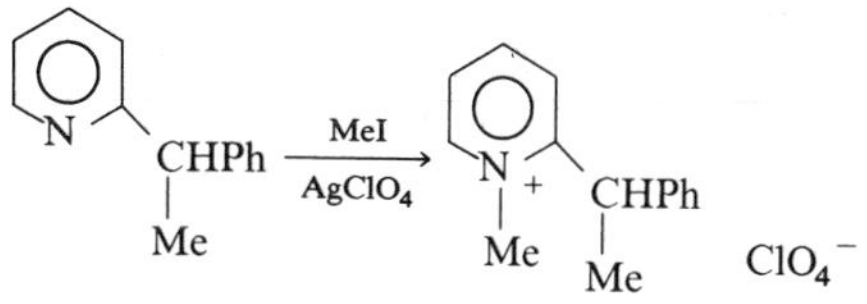

(S)–(+) $[\alpha]_D^{25°}$ +63°   (R)–(–) $[\alpha]_D^{25°}$ = –55°
(R)–(–)   –65°   (c = 1.146, $CHCl_3$)
(c = 1.131, benzene)
97% ee

Scheme 9.12

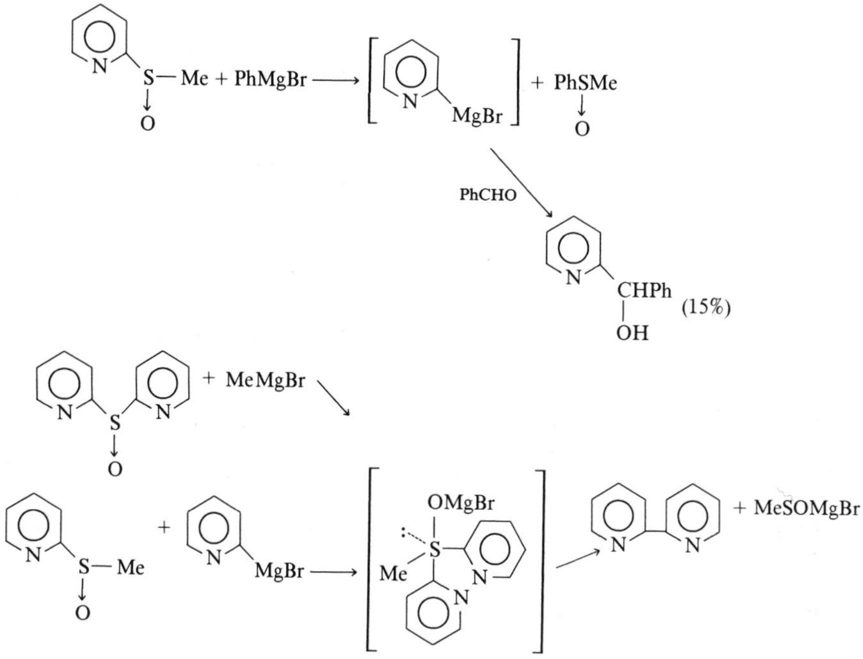

Scheme 9.13

A crossover experiment using an equimolar combination of two different sulfoxides was carried out. Thus, a mixture of benzyl 2-pyridyl sulfoxide and (4-methylphenyl)-methyl 6-picolyl sulfoxide were treated with PhMgBr, giving exclusively 2-benzyl-pyridine and 2-(4-methylphenyl)-6-methylpyridine in good yields. None of the crossover products was obtained, establishing that the coupling reaction proceeds intramolecu-larly. Therefore, the formation of σ-sulfurane is a rational explanation for the reaction mechanism. Because the coupling was found to proceed nearly quantitatively, a stereochemical study of the coupling reaction was carried out using optically active 1-phenylethyl 2-pyridyl sulfoxide [32]. The results are shown in Scheme 9.12.

The result clearly demonstrates that the configuration of the migrating carbon center in the sulfoxide is completely retained after the coupling reaction takes place. The configuration of 2-(1-phenylethyl)pyridine was confirmed by X-ray crystallo-graphic analysis after converting it to the crystalline N-methylpyridinium perchlorate [32].

Table 9.3 ▪ Coupling Reactions of Pyridyl Sulfoxides with Grignard Reagents[a]

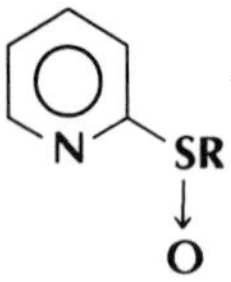

R	R'	Product (%)
CH_2Ph	CH_3	2-Benzylpyridine (83)
CH_2Ph	C_2H_5	2-Benzylpyridine (90)
CH_2Ph	Ph	2-Benzylpyridine (98)
CH_2Ph	n-BuLi	2-Benzylpyridine (46)
CH_2Ph	α-Pyridyl	2-Benzylpyridine (26)
CH_3	CH_3	2,2-Bipyridine (73)
CH_3	Ph	2,2-Bipyridine (79) $PhSCH_3$ (36)
C_2H_5	C_2H_5	2-Ethylpyridine (55)
$CH(CH_3)_2$	C_2H_5	2-Ethylpyridine (identified) 2-Isopropylpyridine (identified)
$C(CH_3)_3$	C_2H_5	2-Ethylpyridine (39) 2-t-Butylpyridine (24)
$CH_2CH{=}CH_2$	Ph	2-Allylpyridine (61)
Ph	C_2H_5	2,2'-Bipyridine (42)
Ph	$PhCH_2$	2-Benzylpyridine (71)
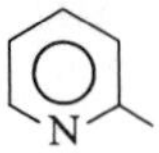	C_2H_5	2,2'-Bipyridine (63)
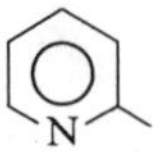	$PhCH_2$	2-Benzylpyridine (40)

[a]An equimolar or half-molar equivalent of Grignard reagent was used.

Table 9.4 ■ Preparation of Bipyridines

X	R	Yield (%)
H	CH_3	30
6-SCH_3	CH_3	61
4-CH_3	C_2H_5	12
5-CH_3	C_2H_5	10
6-CH_3	C_2H_5	57
5-Cl	CH_3	40
6-Cl	CH_3	55
5-Br	CH_3	40
6-Br	CH_3	50
3,5-Cl_2	CH_3	52
6-C_2H_5	CH_3	30
6-piperidyl	CH_3	22

Recently, several other examples of the stereochemical experiments using not only optical isomers but also geometrical isomers, such as vinylic and crotyl sulfoxides, have been reported by Oae and co-workers [33] and Gibbs and Okamura [34]. The stereochemical investigations of the ligand-coupling reactions reveal that they proceed with complete retention of configuration at the coupling center. Therefore, one can presume that the coupling reactions of sulfoxides having electron-withdrawing ligands proceed via the formation of σ-sulfurane and hence, one can conclude that the ligand-coupling reactions are one of the diagnostic methods for formation of hypervalent compounds as intermediates.

We have found other exchange and ligand-coupling reactions of sulfoxides as shown in Scheme 9.13 [35].

Alkyl 2-pyridyl sulfoxides give predominantly 2,2′-bipyridyls that may arise from the cross-coupling reactions involving the initial formation of 2-PyMgBr, by Grignard exchange, that reacts with starting sulfoxide soon after generation to provide σ-sulfurane. The results of alkyl 2-pyridyl sulfoxides are shown in Table 9.3. This cross-coupling reaction can only be applied to 2-pyridyl sulfoxides and used for preparation of symmetrical 2,2′-bipyridyl derivatives. Several results are shown in Table 9.4.

Phenyl 2-pyridyl sulfoxide gives different types of products by changing the temperature and Grignard reagents. At lower temperature (e.g., −78 °C), with ArMgBr this sulfoxide gives only the ligand-exchange product. With p-TolMgBr, for example, at low temperature a mixture of phenyl and p-tolyl pyridyl sulfoxides was obtained. At room temperature p-TolMgBr gives ligand-coupling products, namely, a mixture of 2-phenyl and 2-p-tolylpyridine and 2,2′-bipyridyl. The result suggests that both the coupling and cross-coupling reactions take place simultaneously. When one uses EtMgBr at room

Table 9.5

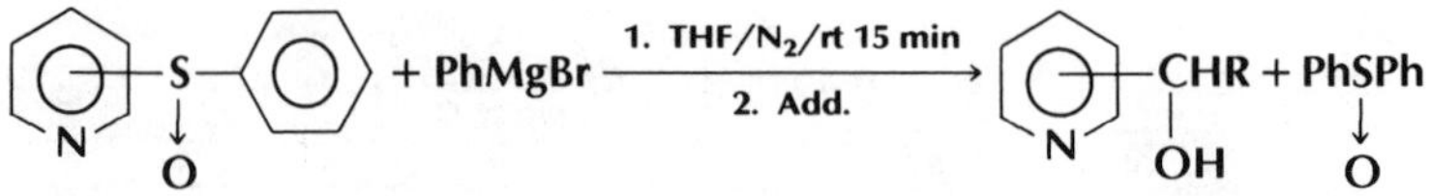

$$\text{2-PySOPh structure} + \text{PhMgBr} \xrightarrow[\text{2. Add.}]{\text{1. THF/N}_2\text{/rt 15 min}} \text{Py}-\text{CHR} \;(\text{with OH}) + \text{PhSPh}\;(\to \text{O})$$

Sulfoxide[a]	Add.	Alcohol	Yield (%)
2-PySOPh	PhCHO	Py—Py	31
		Py—Ph	17
		Py—CH(OH)—Ph	13
2-QuSOTol-p	H₂O	Qu—Qu	30
		Qu—Ph	40
3-PySOPh	PhCHO	Py—CH(OH)—Ph	88
	α-napht.CHO	Py—CH(OH)—napht.-α	80
4-PySOPh	PhCHO	Py—CH(OH)—Ph	64
	α-napht.CHO	Py—CH(OH)—napht.-α	63
	 MeO—C₆H₄—CHO	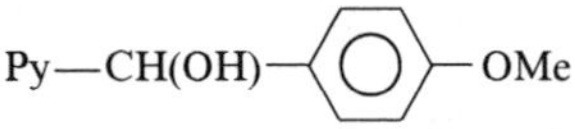 Py—CH(OH)—C₆H₄—OMe	73
	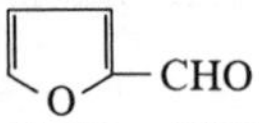 furan-2-CHO	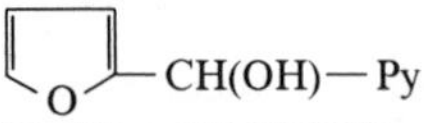 furan-2-CH(OH)—Py	81
	PhCH=CHCHO	PhCH=CHCH(OH)—Py	60
3-PySOPh	PhCOMe	PyPhMeCOH	47
	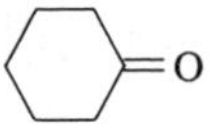 cyclohexanone	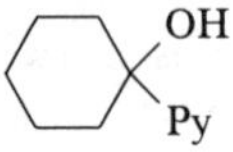 1-(Py)cyclohexan-1-ol	54
	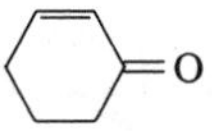 cyclohex-2-enone	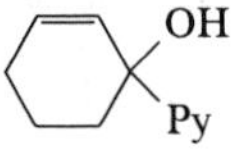 1-(Py)cyclohex-2-en-1-ol	61
	PhCH=CHCOPh	PyPhCH—CH₂COPh	64
	(PhCO)₂O	PyCOPh	75
4-PySOPh	PhCOMe (60 °C)	PyPhMeOH	26
	PhCOPh (60 °C)	PyPh₂COH	7
	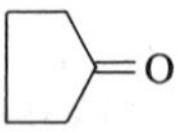 cyclopentanone	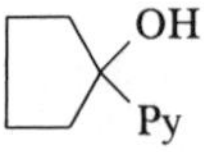 1-(Py)cyclopentan-1-ol	55
	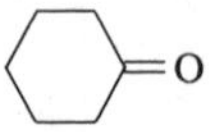 cyclohexanone	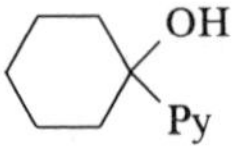 1-(Py)cyclohexan-1-ol	38
	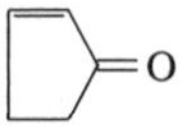 cyclopent-2-enone	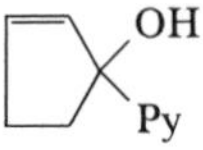 1-(Py)cyclopent-2-en-1-ol	64
	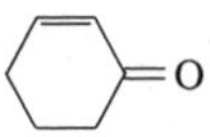 cyclohex-2-enone	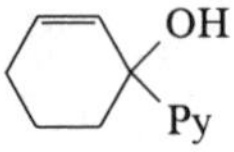 1-(Py)cyclohex-2-en-1-ol	66
	PhCH=CHCOPh	PyPhCH—CH₂COPh	60
	(PhCO)₂O	PyCOPh	70
	PhCOCl	PyCOPh	24
	PhCO₂Et (60 °C)	No reaction	

Table 9.5 ■ Continued

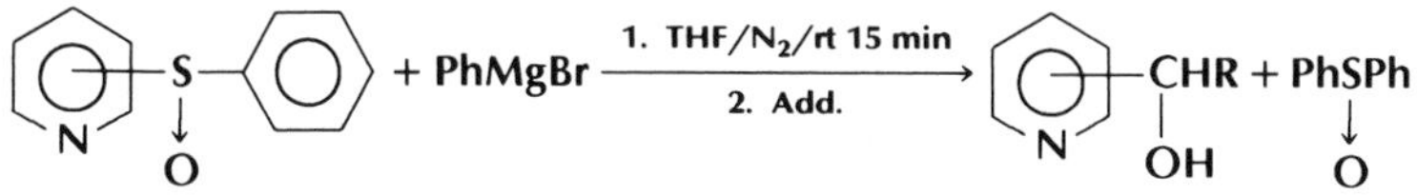

Sulfoxide[a]	Add.	Alcohol	Yield (%)
4-QuSOTol-p	PhCHO	4-Qu—CH(OH)—Ph	80
	PhCH=CHCHO	PhCH—CHCH(OH)—Qu-4	53
	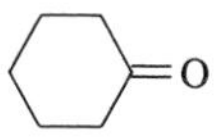	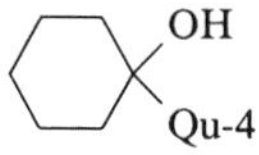	16
	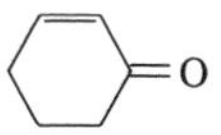	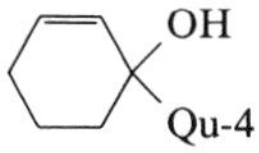	20
	PhCH=CHCOPh	4-QuPhCH—CH₂COPh	72

[a]Qu = quinoline, Py = pyridine.

temperature, the products are again both 2-phenylpyridine and 2,2′-bipyridyl [36]. These different reaction modes of phenyl 2-pyridyl sulfoxide with Grignard reagents can be explained in terms of the formation of Mg metal complex by the 2-pyridyl nitrogen atom and the sulfinyl oxygen atom, which forces the 2-pyridyl and the sulfinyl oxygen atom to take the two equatorial positions and thus, the p-tolyl group can approach only from the back side of the electronegative phenyl group to form a

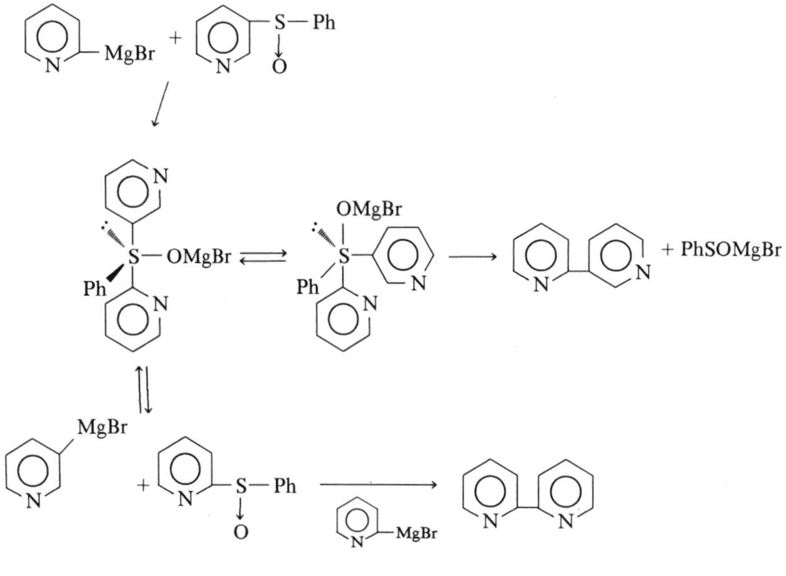

Scheme 9.14

Table 9.6

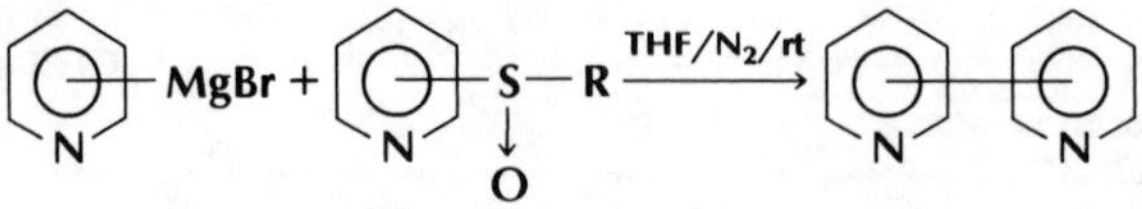

PyMgBr	Sulfoxide	Product	Yield (%)
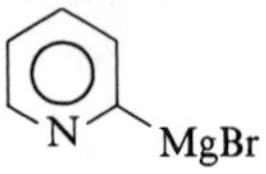	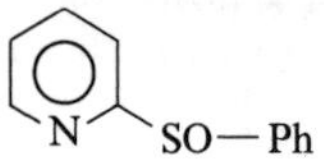	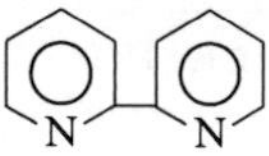	75
	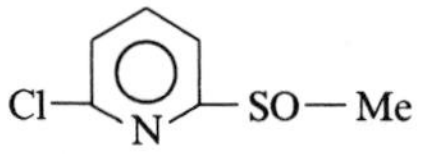	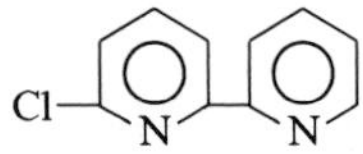	43
	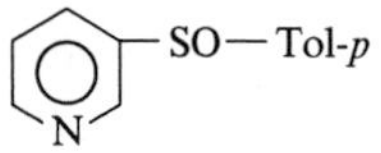	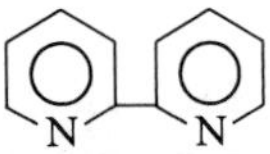	23
		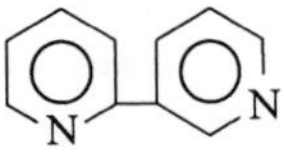	37
	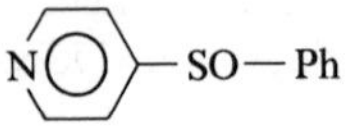	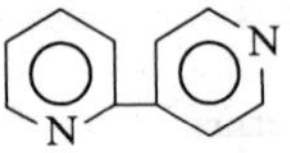	58
	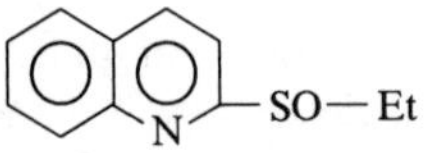	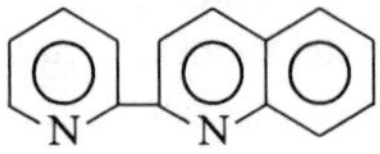	62
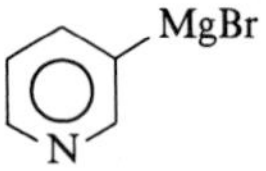	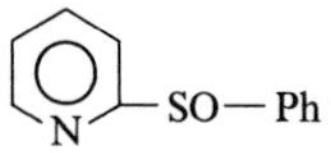	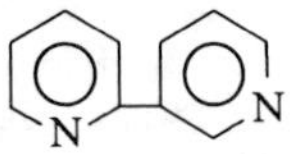	63
	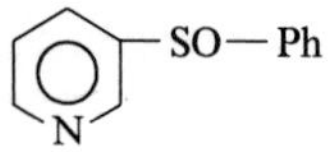	No reaction	
	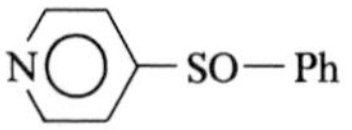	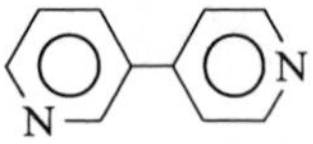	34
		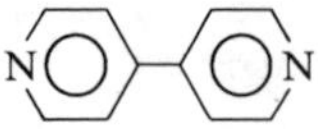	14
	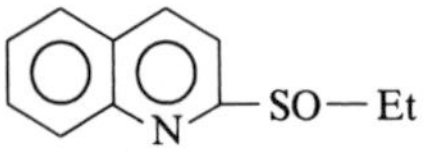	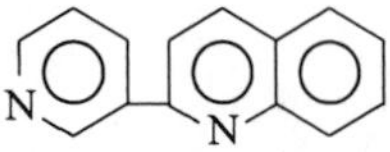	52
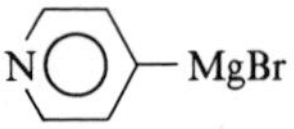	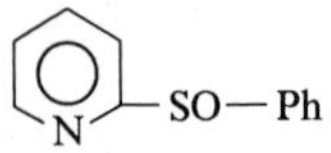	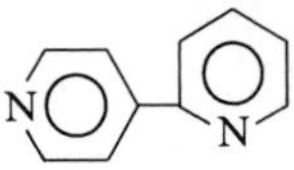	63
	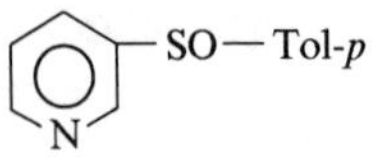	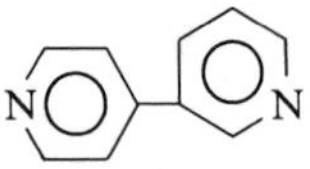	13

Table 9.6 ■ Continued

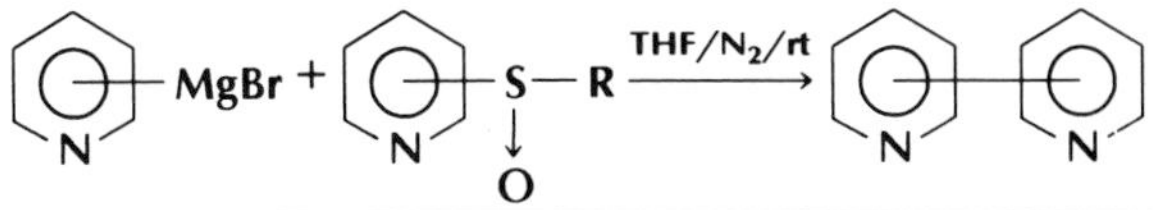

PyMgBr	Sulfoxide	Product	Yield (%)
		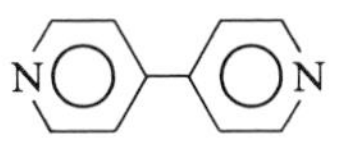 N◯—◯N	25
	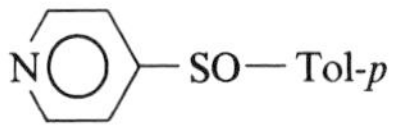 N◯—SO—Tol-p	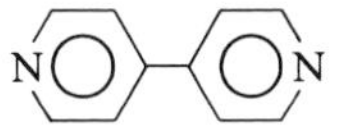 N◯—◯N	50
	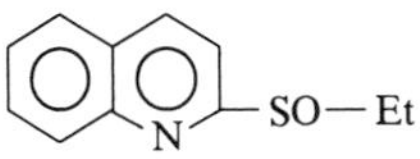 ◯◯ₙ—SO—Et	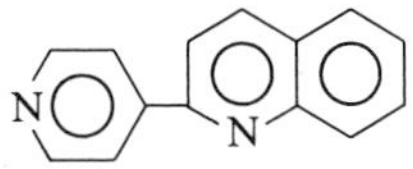 N◯—◯◯ₙ	56
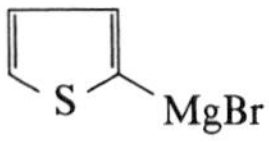 S◯—MgBr	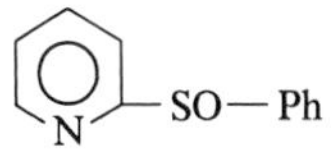 ◯ₙ—SO—Ph	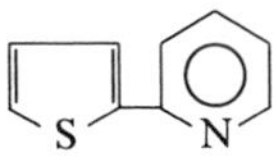 S◯—◯ₙ	66
	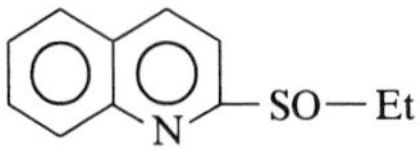 ◯◯ₙ—SO—Et	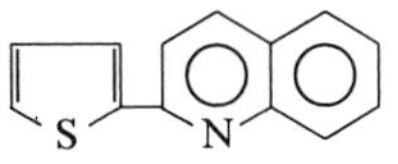 S◯—◯◯ₙ	68

σ-sulfurane. At higher temperature, the coordination of MgBr by the ligands should be diminished and hence, the pyridyl ring must be placed at the equatorial position when the p-tolyl group approaches the central sulfur atom.

In contrast to 2-pyridyl sulfoxide, both 3- and 4-pyridyl p-tolyl (or phenyl) sulfoxides were found exclusively to undergo the ligand-exchange reactions to eliminate stereospecifically 3- and 4-pyridyl Grignard reagents and phenyl p-tolyl sulfoxide. The optical activity of phenyl p-tolyl sulfoxide indicated that it had completely inverted. The resulting 3- and 4-PyMgBr can be trapped with several electrophiles, such as aldehydes and ketones [37]. Some examples are shown in Table 9.5.

These results indicate that chelation of Mg by both the pyridyl nitrogen and the sulfinyl oxygen atoms plays an important role in determining the stability of the intermediate σ-sulfurane involving pyridyl rings.

As an extension of the ligand-coupling reactions, we tried the cross-coupling reactions using various phenyl pyridyl sulfoxides with pyridyl Grignard reagents, as shown in Scheme 9.14. Various cross-coupling biheteroaryls thus obtained are summarized in Table 9.6 [38]. These results suggest that the reactivity of pyridyl sulfoxides and PyMgBr is in the following order, namely, 2-pyridyl ≥ 4-pyridyl ≫ 3-pyridyl. This cross-coupling reaction is a convenient procedure for providing numerous unsymmetrical biheteroarylic derivatives.

Furthermore, these ligand coupling and exchange reactions can be used for synthetic purposes for preparation of unsymmetrically substituted bipyridyls and

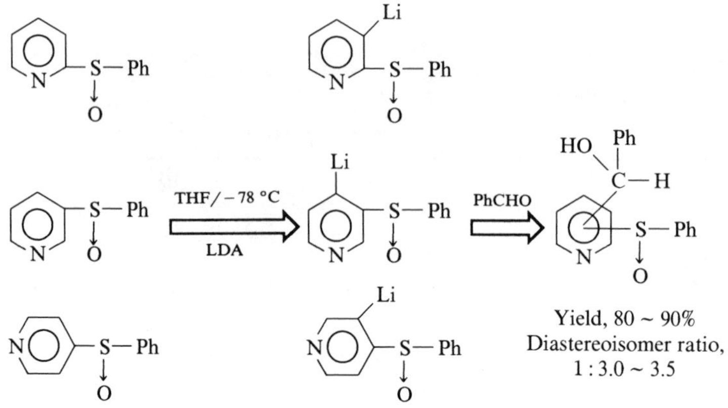

Scheme 9.15

phenylpyridine derivatives. In order to introduce substituents in the pyridine ring, phenyl pyridyl sulfoxides were treated with LDA to give the regiospecific lithiation in the pyridine ring of the sulfoxides. Upon treatment with electrophiles, the lithio derivatives give various o-substituted pyridyl phenyl sulfoxides. From these compounds the sulfinyl group can be removed or cross ligand-coupling reactions effected to afford numerous pyridine derivatives bearing functional groups at appropriate positions or bipyridyl derivatives [39]. Several examples are summarized in Scheme 9.15.

Selenoxides bearing a pyridine ring also undergo similar ligand-coupling reactions to afford products via the formation of σ-selenuranes as intermediates [40].

9.4. Conclusion

Four reaction modes are observed in the reaction of organometallic reagents with sulfoxides and selenoxides:

1. simple ligand exchange;
2. ligand exchange and disproportionation;
3. ligand coupling;
4. α-deprotonation with rearrangement (not mentioned here) [41].

These remarkable variations in the reaction modes depend on the substituent attached to the sulfinyl sulfur atom or to the selenenyl selenium atom. In general, the more electronegative substituents attached at the sulfur, the more the reactions tend to give the ligand-coupling products or rapid ligand exchange and cross-coupling products, because electron-withdrawing ligands are known to stabilize hypervalent bonds. If we presume that ligand coupling is diagnostic for formation of hypervalent species, then the present results are consistent with this concept, namely, that there is a direct relationship between the structure of sulfoxides and the stability of σ-sulfurane. Scheme 9.16 summarizes the reaction mechanisms and the relation between the mode of reactions and the nature of ligands attached to the sulfur or selenium atom.

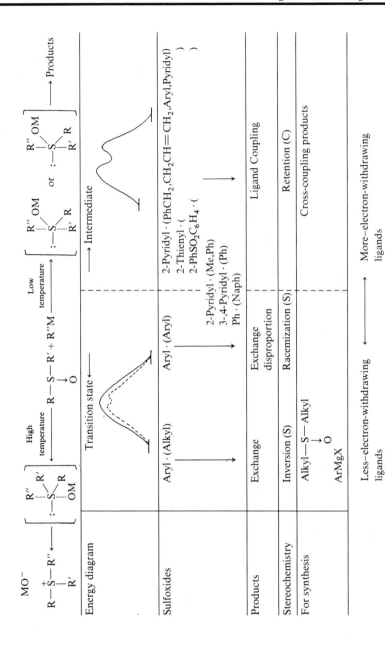

Scheme 9.16

Thus, to the question, "Is σ-sulfurane an intermediate in the reactions of sulfoxides with organometallic reagents?," the answer is "Yes" but σ-sulfuranes or σ-selenuranes are quite unstable and become a transition state (or a frozen transition state) when substituted by more electropositive ligands. Although we did not succeed in directly detecting σ-sulfuranes or σ-selenuranes by spectroscopic methods, we could get evidence for the existence of these species as reactive intermediates in these reactions. We also found that substitution reactions of sulfoxides could be used for preparation of organolithium or Grignard reagents that were difficult to prepare by common methods.

References

1. Perkins, W. C.; Martin, J. C.; Arduengo, A. J.; Lau, W.; Alegria A.; Kochi, J. K. *J. Am. Chem. Soc.*, **1980**, *102*, 7753.
2. Musher, J. I. *Angew. Chem., Int. Ed. Engl.*, **1969**, *8*, 54.
3. (a) Oae, S. *Organic Sulfur Chemistry*, Vol. 1 (in Japanese), Kagakudojin, Kyoto, 1982, p. 206. (b) Rayner, D. R.; von Schriltz, D. M.; Day, J.; Cram, D. J. *J. Am. Chem. Soc.*, **1968**, *90*, 2721. (c) Oae, S.; Yokoyama, M.; Kise, M.; Furukawa, N. *Tetrahedron Lett.*, **1968**, 4131.
4. Martin, J. C.; Arhart, R. J. *J. Am. Chem. Soc.*, **1971**, *93*, 2339.
5. Kapovits, I.; Kalman, A. *J. Chem. Soc., Chem. Commun.*, **1971**, 649.
6. Hayes, R. A.; Martin, J. C.; Bernardi, F.; Csizmadia, I. G. *Organic Sulfur Chemistry, Theoretical and Experimental Advances*, Mangini, A., ed., Elsevier, Amsterdam, 1985, Chapter 8, p. 408.
7. Kice, J. K. *Adv. Phys. Org. Chem.*, **1980**, *17*, 65.
8. (a) Johnson, C. R.; Rigau, J. J. *J. Am. Chem. Soc.*, **1969**, *91*, 5398. (b) Calo, V.; Scorrano, G.; Modena, G. *J. Org. Chem.*, **1969**, *24*, 2020. (c) Owsley, D. C.; Helmkamp, G. K.; Rettig, F. *J. Am. Chem. Soc.*, **1969**, *91*, 5239.
9. Oae, S.; Kise, M. *Tetrahedron Lett.*, **1967**, 1409.
10. Oae, S.; Furukawa, N. *Sulfilimines and Related Derivatives*, American Chemical Society ACS Monograph 179, 1983, p. 145.
11. Wittig, G.; Fritz, H. *Liebigs Ann. Chem.*, **1952**, *577*, 39.
12. (a) Trost, B. M.; Ziman, S. D. *J. Am. Chem. Soc.*, **1971**, *93*, 3825. (b) Trost, B. M.; Atkins, R. C.; Hoffman, L. *J. Am. Chem. Soc.*, **1973**, *95*, 1285. (c) Trost, B. M.; Arndt, H. C. *J. Am. Chem. Soc.*, **1973**, *95*, 5288.
13. Trost, B. M.; Schinski, W. L.; Mantz, I. D. *J. Am. Chem. Soc.*, **1969**, *91*, 4320. (b) LaRochelle, R. W.; Trost, B. M. *J. Am. Chem. Soc.*, **1971**, *93*, 6077.
14. Harrington, D.; Weston, J.; Jacobus, J.; Mislow, K. *J. Chem. Soc., Chem. Commun.*, **1972**, 1079.
15. Hori, H.; Kataoka, T.; Shimizu, H.; Miyagaki, M. *Chem. Pharm. Bull.*, **1974**, *22*, 1711, 2044, 2020.
16. Andersen, K. K.; Yeager, S. A.; Peynircioglu, N. B. *Tetrahedron Lett.*, **1970**, 2485.
17. Sheppard, W. A. *J. Am. Chem. Soc.*, **1971**, *93*, 5597.
18. Kwart, H.; King, K. G. *Chemistry of Silicon, Phosphorus and Sulfur*, Springer-Verlag, New York, 1977, p. 11.
19. Wittig, G.; Fritz, H. *Liebigs Ann. Chem.*, **1952**, *577*, 39. (b) Hellwinkel, D.; Fahrbach, G. *Liebigs Ann. Chem.*, **1968**, *712*, 1. (c) Glover, S. A. *J. Chem. Soc., Perkin Trans. 1*, **1980**, 1338.
20. Johnson, C. R.; McCants, D. *J. Am. Chem. Soc.*, **1965**, *87*, 5404.
21. (a) Lockard, J. P.; Schroeck, C. W.; Johnson, C. R. *Synthesis*, **1973**, 485. (b) Durst, T.; Lebell, M. J.; Van Den Elzen, R.; Tin, K.-C. *Can. J. Chem.*, **1974**, *52*, 761.

22. Hojo, M.; Masuda, R.; Sakai, T.; Fujimori, K.; Tsutsumi, S. *Synthesis*, **1977**, 789.
23. (a) Satoh, T.; Oohara, T.; Ueda, Y.; Yamakawa, K. *Tetrahedron Lett.*, **1988**, *29*, 313, 4039. (b) Theobald, P. G.; Okamura, W. H. *Tetrahedron Lett.*, **1987**, *28*, 6565. (c) Furukawa, N.; Shibutani, T.; Fujihara, H. *Tetrahedron Lett.*, **1987**, *28*, 2727.
24. (a) Wildi, B. S.; Taylor, S. W.; Potratz, H. A. *J. Am. Chem. Soc.*, **1951**, *73*, 1965. (b) Chung, S.-K.; Sakamoto, S. J. *Org. Chem.*, **1981**, *46*, 4590. (c) Bonini, B. F.; Maccognani, G.; Mazzanti, G.; Piccinelli, P. *Tetrahedron Lett.*, **1979**, 3987. (d) Carpino, L. A.; Chen, H.-W. *J. Am. Chem. Soc.*, **1971**, *93*, 785.
25. (a) Andersen, K. K.; Caret, R. L.; Karup-Nielsen, I. *J. Am. Chem. Soc.*, **1974**, *96*, 8026. (b) Ackerman, B. K.; Andersen, K. K.; Karup-Nielsen, I.; Peynircioglu, N. B.; Yeager, Y. A. *J. Org. Chem.*, **1974**, *39*, 964.
26. Furukawa, N.; Matsumura, K.; Ogawa, S.; Fujihara, H., unpublished results.
27. Furukawa, N.; Ogawa, S., unpublished results.
28. Furukawa, N.; Ogawa, S.; Satoh, S., unpublished results.
29. Ogawa, S.; Furukawa, N., unpublished results.
30. (a) Oae, S.; Kawai, T.; Furukawa, N. *Tetrahedron Lett.*, **1984**, *25*, 69. (b) Nef, G.; Eder, U.; Seeger, S. *Tetrahedron Lett.*, **1980**, *21*, 903. (c) Oae, S. *Phosphorus and Sulfur*, **1986**, *27*, 13.
31. Kawai, T.; Kodera, Y.; Furukawa, N.; Oae, S.; Ishida, M.; Takeda, T.; Wakabayashi, S. *Phosphorus and Sulfur*, **1987**, *34*, 139.
32. Oae, S.; Kawai, T.; Furukawa, N.; Iwasaki, F. *J. Chem. Soc., Perkin Trans. 2*, **1987**, 405.
33. Oae, S.; Takeda, T.; Wakabayashi, S. *Tetrahedron Lett.*, **1988**, *29*, 4441, 4445.
34. Gibbs, R. A.; Okamura, W. H. *J. Am. Chem. Soc.*, **1988**, *110*, 4062.
35. Kawai, T.; Furukawa, N. F.; Oae, S. *Tetrahedron Lett.*, **1984**, *25*, 2549.
36. Furukawa, N.; Shibutani, T., unpublished results.
37. Furukawa, N.; Shibutani, T.; Fujihara, H. *Tetrahedron Lett.*, **1986**, *27*, 3899.
38. Furukawa, N.; Shibutani, T.; Fujihara, H. *Tetrahedron Lett.*, **1987**, *28*, 5845.
39. Furukawa, N.; Shibutani, T., unpublished results.
40. Furukawa, N.; Ogawa, S.; Satoh, S., unpublished results.
41. (a) Oda, R.; Yamamoto, K. *J. Org. Chem.*, **1961**, *26*, 4679. (b) Manya, P.; Sekera, A.; Rumpf, P. *Tetrahedron*, **1970**, *26*, 467.

Facile Syntheses of Unsaturated Aldehydes, Ketones, Amides, and Related Natural Products via Arsonium Salts

Yao-Zeng Huang, Li-Lan Shi, Jian-Hua Yang, Wen-Juan Xiao, Sao-Wei Li, Wei-Bo Wang

Shanghai Institute of Organic Chemistry
Academia Sinica
345 Lingling Lu, Shanghai 200032,
China

10.1. Introduction

In this paper are described the facile syntheses of unsaturated aldehydes, ketones, and amides that involve the reaction of a variety of aldehydes with arsonium salts in the presence of K_2CO_3 under phase transfer condition. The general formulation is:

$$R(CH=CH)_nCHO + Ph_3\overset{+}{As}CH_2(CH=CH)_mX\ Br^- \xrightarrow[-Ph_3AsO]{K_2CO_3,\ mixed\ solvent,\ rt}$$

$$R(CH=CH)_{n+m+1}X$$

$$n = 0,1;\ m = 0,1,\ and\ X = CHO,\ COCH_3,\ CONR_1R_2$$

When $m = 0$, all the reactions are highly stereoselective in the E configuration; when $m = 1$, the reactions are stereoselective in the E,E configuration in the case of $X = CONR_1R_2$ and in favor of the E,E configuration in the case of $X = CHO$ and $COCH_3$. The E,Z configuration can be isomerized to that of E,E by treating with a catalytic amount of iodine. Using our method, some biologically active natural products have been synthesized. The simplicity of our method, the mildness of the reaction conditions, the high stereoselectivity, and the excellent yields enable this method to be a practical approach to the syntheses of polyenals, polyenones, polyenamides, and related natural products.

In our previous publications we reported that arsonium ylides bearing an electron-withdrawing substituent in the alkylidene moiety are more reactive than the corresponding phosphonium ylides. The greater reactivity is attributed to the fact that the overlap of p orbitals of carbon with d orbitals of arsenic is less effective than with d orbitals of phosphorus. Consequently, the arsonium ylides are more nucleophilic than the corresponding phosphonium ylides [1]. Direct evidence for the superiority in behavior of arsonium ylides over phosphonium ylides comes from a comparison of the structure of benzoylmethylenetriphenyl-arsorane with that of benzoylmethylenetriphenyl-phosphorane [2], as determined by X-ray crystallography.

Recently, we found that it is unnecessary to isolate the ylides themselves because a variety of aldehydes reacted with arsonium salts equally well in the presence of potassium carbonate under phase transfer conditions to achieve olefination.

$$R(CH{=}CH)_nCHO + Ph_3\overset{+}{A}sCH_2(CH{=}CH)_m X\ \overset{-}{B}r \xrightarrow[-Ph_3AsO]{K_2CO_3,\ \text{mixed solvent}}$$

$$R(CH{=}CH)_{n+m+1}X,$$

where $n = 0,1$; $m = 0,1$, and $X = CHO, COCH_3, CONR_1R_2$

In most cases, the reactions were highly stereoselective in the E configuration or in favor of the E configuration. Using our present method, it is convenient to synthesize enals, polyenals, enones, polyenones, enamides, polyenamides, and so on, which are important classes of biologically active natural products. In comparison with common methods for olefination (by Wittig) and phosphoryl reagents that require rather drastic conditions, the simplicity of our method, the mildness of the reaction conditions (room temperature and weak basicity), the high stereoselectivity and the excellent yields enable this reaction to be a practical approach to the syntheses of polyenals, polyenones, polyenamides, and related natural products via arsonium salts. Furthermore, the advantages of this method lie in the avoidance of the isolation of unstable ylides in certain cases and the readiness of reconverting triphenylarsenic oxide into triphenylarsine.

10.2. A Facile Route for the Synthesis of Enals and Its Application to Insect Sex Attractants

Reactions that convert carbonyl compounds to α,β-unsaturated aldehydes are highly useful operations, especially for the synthesis of biologically active natural products. When formylmethylenetriphenylphosphorane is used, rather drastic conditions are required [3,4]. Alternatively, when the formyl-group masked Wittig reagent or phosphoryl reagent is used, acid hydrolysis of products is required and this may be incompatible with complex synthesis [5,6]. Another method for formylolefination involves treating the carbonyl compounds with cis-2-ethoxyvinyllithium, which is difficult to prepare [7].

$$\underset{\textbf{1}}{RCHO} + \underset{\textbf{2}}{Ph_3\overset{+}{A}sCH_2CHO\ \overset{-}{B}r} \xrightarrow[THF/Et_2O,\ rt]{K_2CO_3(s),\ trace\ H_2O} \underset{\textbf{3}}{RCH{=}CHCHO + Ph_3AsO}$$

Table 10.1 ▪ Formylolefination of Aldehydes

Substrate	Reaction time (h)	Product[a] (E isomer > 97%)	Yield (%)
1 $p\text{-}O_2NC_6H_4CHO$	4	$p\text{-}O_2NC_6H_4CH=CHCHO$	96
2 $p\text{-}ClC_6H_4CHO$	20	$p\text{-}ClC_6H_4CH=CHCHO$	93
3 C_6H_5CHO	20	$C_6H_5CH=CHCHO$	87
4	16		86[b]
5	9		98
6 $n\text{-}C_5H_{11}CHO$	17	$n\text{-}C_5H_{11}CH=CHCHO$	81
7 $n\text{-}C_8H_{17}CHO$	24	$n\text{-}C_8H_{17}CH=CHCHO$	89
8	3		91
9	13		85

[a]All the products were confirmed by ^{1}H NMR, IR, and mass spectroscopy E isomers were determined by gas chromatography and ^{1}H NMR.

[b] ⬡—CHO : **2** = 1 : 2.4 (mole ratio).

We found that treatment of various aldehydes with formylmethyltriphenylarsonium bromide (**2**) in the presence of potassium carbonate in THF–Et$_2$O (trace H$_2$O) at room temperature gave α,β-unsaturated aldehydes in 81–98% yields, of which the E isomer amounted to 97% [8]. The results are shown in Table 10.1.

A substantial number of conjugated dienes function as sex attractants for insects. Among the four geometrical isomers, a great number of dienes with the (Z,E) configuration have already been identified as the components of sex pheromones, for example, (3Z,5E)-3,5-tetradecadien-1-ol acetate (**4**) for the Carpenterworm moth (*Prionoxystus robiniae*) [9], (5Z,7E)-5,7-dodecadien-1-ol (**5**) for *Dendrolimus spectabilis* [10], (5Z,7E)-5,7-dodecadiene-1-ol acetate (**6**) for *Dendrolimus punctatus* [11], and (5E,7Z)-5,7-dodecadienal (**7**) for *Malacosoma californicum* [12]. Recently, **4** and **7** have been synthesized by nonstereoselective Wittig reaction [9,13,14], whereas **6** was obtained with only 85% stereoselectivity and in 31% overall yield [15].

4

5

6

7

By means of our formylolefination method, we succeeded in preparing the requisite key intermediates, α,β-unsaturated aldehydes **8a**, **8b**, and **8c** in excellent yields (75–90%), of which the E configuration amounted to 98% [16].

$$Ph_3\overset{+}{As}CH_2CHO\ Br^- + RCHO \xrightarrow[-Ph_3AsO]{(i)} RCH=CHCHO$$

8a: $R = CH_3(CH_2)_7$
8b: $R = CH_3(CH_2)_3$
8c: $R = THPO(CH_2)_4$

(i) Et_2O–THF$(7:3)$–trace H_2O/K_2CO_3, 20 °C, 18–24 h

The syntheses of **4, 5, 6**, and **7** are shown by Scheme 10.1.

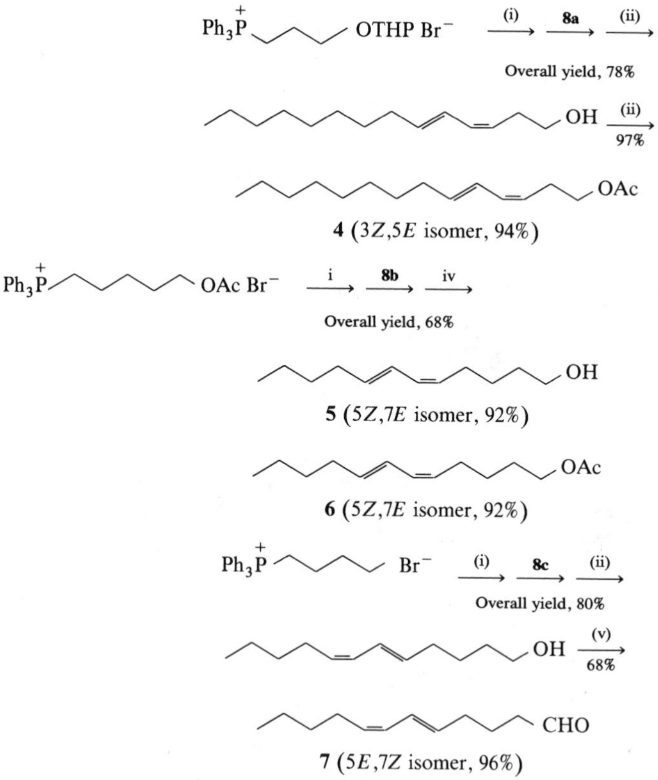

$$Ph_3\overset{+}{P}\diagdown\diagup\diagdown\diagup\ OTHP\ Br^- \xrightarrow{(i)} \overset{8a}{\longrightarrow} \xrightarrow{(ii)}$$

Overall yield, 78%

$$\diagup\diagdown\diagup\diagdown\diagup\diagdown\diagup\diagdown=\diagdown\diagup OH \xrightarrow[97\%]{(ii)}$$

$$\diagup\diagdown\diagup\diagdown\diagup\diagdown\diagup\diagdown=\diagdown\diagup OAc$$

4 (3Z,5E isomer, 94%)

$$Ph_3\overset{+}{P}\diagup\diagdown\diagup\diagdown\diagup\ OAc\ Br^- \xrightarrow{i} \overset{8b}{\longrightarrow} \xrightarrow{iv}$$

Overall yield, 68%

$$\diagup\diagdown\diagup=\diagdown\diagup\diagdown\diagup OH$$

5 (5Z,7E isomer, 92%)

$$\diagup\diagdown\diagup=\diagdown\diagup\diagdown\diagup OAc$$

6 (5Z,7E isomer, 92%)

$$Ph_3\overset{+}{P}\diagdown\diagup\diagdown\diagup\ Br^- \xrightarrow{(i)} \overset{8c}{\longrightarrow} \xrightarrow{(ii)}$$

Overall yield, 80%

$$\diagup\diagdown\diagup=\diagup\diagdown\diagup\diagdown\diagup OH \xrightarrow[68\%]{(v)}$$

$$\diagup\diagdown\diagup=\diagup\diagdown\diagup\diagdown CHO$$

7 (5E,7Z isomer, 96%)

(i) THF–HMPA$(1:1)/n$-BuLi, -30 °C, 2 h
(ii) MeOH/H$^+$, rt, 20 h
(iii) Ac_2O/Py, 5–10 °C, 8 h
(iv) EtOH/OH$^-$, 24 h
(v) PCC/CH_2Cl_2, 15°C, 2 h

Scheme 10.1

10.3. A Facile Route for the Synthesis of Dienals and Its Application to the Synthesis of Leukotriene A_4 Methyl Ester

2,4-Dienals are key intermediates for the synthesis of some important biologically active natural products. To our knowledge, few methods are available for direct formation of 2,4-dienal from aldehydes by one-step reactions. We found a facile route for the synthesis of 2,4-dienal by means of arsonium salts. Thus, treatment of a variety of aldehydes with 3-formylallyltriphenylarsonium bromide (**10**) and potassium carbonate in a mixed solvent at room temperature gave a mixture of $2E,4E$ and $2E,4Z$ dienals in 81–98% yields, in favor of the former. The latter compounds could be isomerized to the former compounds by treatment with a catalytic amount of iodine under daylight [17].

$$RCHO + Ph_3 \overset{+}{As} \diagup\!\!\!\diagdown\!\!\!\diagup CHO \ \overset{-}{Br} \xrightarrow[\text{THF–Et}_2O\ (1:9),\ rt,\ -Ph_3AsO]{K_2CO_3(s),\ \text{trace } H_2O}$$

$$\textbf{9} \qquad\qquad \textbf{10}$$

$$R\diagup\!\!\!\diagdown\!\!\!\diagup\!\!\!\diagdown CHO \xrightarrow{I_2,\ h\nu} R\diagup\!\!\!\diagdown\!\!\!\diagup\!\!\!\diagdown CHO$$

$$\textbf{11} \qquad\qquad\qquad \textbf{12}$$

The results are shown in Table 10.2.

By application of this method, we succeeded in synthesizing a key intermediate E,E-conjugated dienaldehyde (**14**) from aldehyde (**13**) in excellent yield and achieved the stereoselective synthesis of leukotriene A_4 (LTA_4) methyl ester, in cooperation with Yulin Wu and his co-workers [18]. (See Scheme 10.2.)

Table 10.2 ■ Formylenylolefination of Aldehyde via Arsonium Salt 10[a]

	Aldehyde R	Reaction time (h)	Product	Yield[b] (%)	$(2E,4E):$ $(2E,4Z)$[c]
a	p-$O_2NC_6H_4$—	12	p-$O_2NC_6H_4(CH=CH)_2CHO$ (3a,3a′)	98	4:1
b	C_6H_5—	12	$C_6H_5(CH=_2CHO$ (3b,3b′)	85	3.5:1
c	(furanyl)	19	(furanyl)$(CH=CH)_2CHO$ (3c,3c′)	81	1.4:1
d	(dioxolane)	8	(dioxolane)$(CH=CH)_2CHO$ (3d,3d′)	81	1:1
e	C_5H_{11}—	20	$C_5H_{11}(CH=CH)_2CHO$ (3e,3e′)	79	4:1
f	C_9H_{19}—	23	$C_9H_{19}(CH=CH)_2CHO$ (3f,3f′)	85	4:1

[a]All reactions were carried out at room temperature.
[b]Isolated yields.
[c]Estimated by ^1H NMR.

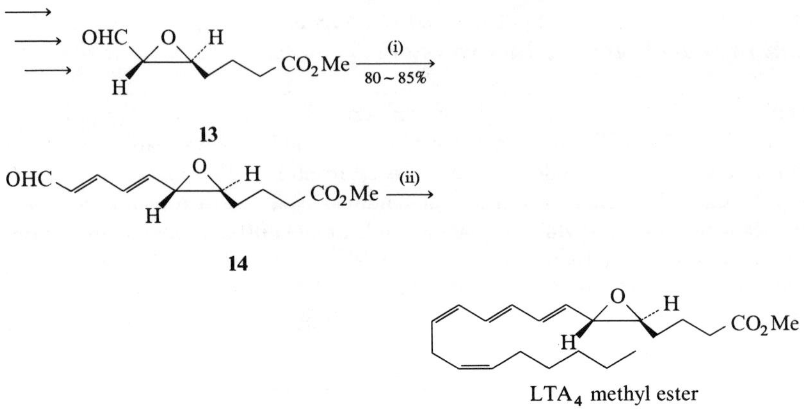

13

14

LTA$_4$ methyl ester

(i) $Ph_3As^+CH_2CH=CHCHO\ Br^-$, K_2CO_3, Et_2O/THF (trace H_2O), I_2

(ii) $Ph_3\overset{+}{P}CH_2CH_2CH=CHC_5H_{11}\ I^-$, BuLi, THF/HMPA

Scheme 10.2

10.4. A Facile Route for the Synthesis of α-Enones and Its Application to the Synthesis of Key Intermediates of Brassinosteroid and Prostagladin PGF$_{2α}$

An efficient and highly stereoselective synthesis of (E)-α-enones by the reaction of aldehydes **16** with two new arsonium salts **15** and potassium carbonate (trace of water) at room temperature has been achieved [19]. The results are shown in Table 10.3.

Table 10.3 ▪ (E)-1-Alkenyl Ketones 17 Prepared

17	R$_1$	R$_2$	Solvent[a]	Reaction time (h)	Yield (%)
a	i-C$_3$H$_7$	C$_6$H$_5$	I	18	96
b	i-C$_3$H$_7$	n-C$_9$H$_{19}$	I	18	91
c	i-C$_3$H$_7$	4-NO$_2$C$_6$H$_4$	I	16	90
d	i-C$_3$H$_7$	2-pyridyl	I	6	96
e	i-C$_3$H$_7$	4-CH$_3$OC$_6$H$_4$	I	18	79
f	n-C$_5$H$_{11}$	C$_6$H$_5$	I	25	98
			II	5	99
g	n-C$_5$H$_{11}$	n-C$_9$H$_{19}$	I	21	92
			II	6	95
h	n-C$_5$H$_{11}$	2-pyridyl	II	3	92
i	n-C$_5$H$_{11}$	4-ClC$_6$H$_4$	II	3	87
i	n-C$_5$H$_{11}$	C$_6$H$_5$CH=CH	II	6	71

[a]Solvent I = CH$_3$CN, solvent II = CH$_2$Cl$_2$/THF (3:1)

$$Ph_3\overset{+}{As}CH_2COR_1 \;\overset{-}{Br} + R_2CHO \xrightarrow[\substack{71-99\%}]{\substack{CH_2Cl_2/THF/H_2O(trace)/K_2CO_3 \\ or\; MeCN/H_2O(trace)/K_2CO_3 \\ 25\,^\circ C,\; 3-25\; h}} R_2CH=CHCOR_1$$

15   **16**   **17**

$$R_1 = i\text{-}C_3H_7,\; n\text{-}C_5H_{11}$$

Using this method, we have prepared the key intermediates **20** and **23** for the synthesis of brassinosteroid and prostaglandin $PGF_{2\alpha}$ starting from **18** and **21**, respectively.

The aldehyde **19**, prepared by ozonolysis of the Δ^{22}-steroid **18**, reacted with the arsonium salt **15** ($R_1 = i$-Pr) in acetonitrile under the previously mentioned conditions (12 h) to give the key intermediate **20** in 78% yield. (See Scheme 10.3.) The ^{1}H NMR spectrum of **20** showed that the 3- and 6-acetoxy groups remained intact during the reaction; the signal of the 20-methyl group appeared at $\delta = 1.09(d)$ and thus proved that epimerization at C-20 had not occurred.

The construction of the 1-alkenyl side chain of $PGF_{2\alpha}$ and its analogues has frequently been accomplished by olefination of the Corey aldehyde **22** with phosphonic esters or phosphoranes.

The reaction conditions must be strictly controlled and in large-scale preparations the dienone **24** is often formed by elimination of acetic acid. Furthermore, the phosphonic ester must be prepared from the α-lithio derivative of the dimethyl phosphonate. An alternative approach to the $PGF_{2\alpha}$ side chain from lactone-aldehyde **22** involving the BF_3-mediated reaction of a sulfone with an aldehyde has been reported, but the transformation of lactone-aldehyde **22** to the sulfone requires six steps. When the reaction of lactone-aldehyde **22** with reagent **15** ($R_1 = n$-C_5H_{11}) is

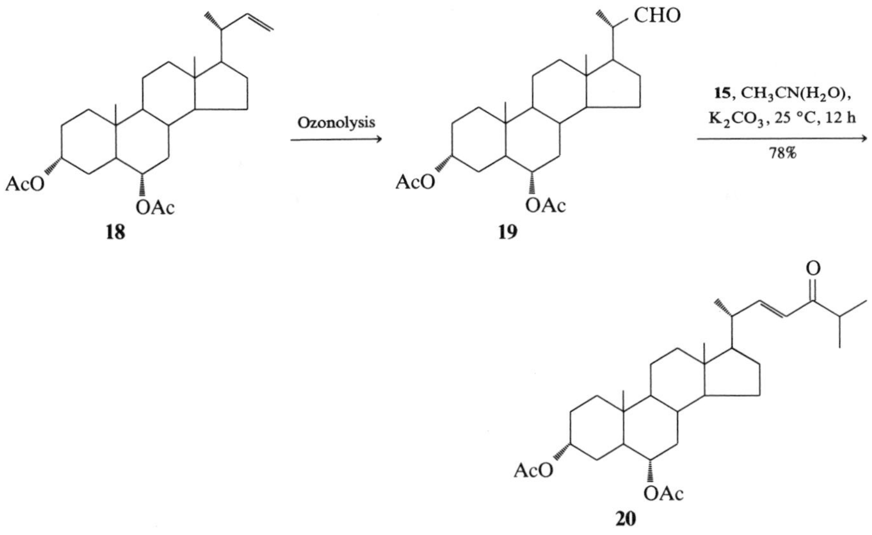

18   Ozonolysis   **19**   **15**, CH$_3$CN(H$_2$O), K$_2$CO$_3$, 25 °C, 12 h   78%

20

Scheme 10.3

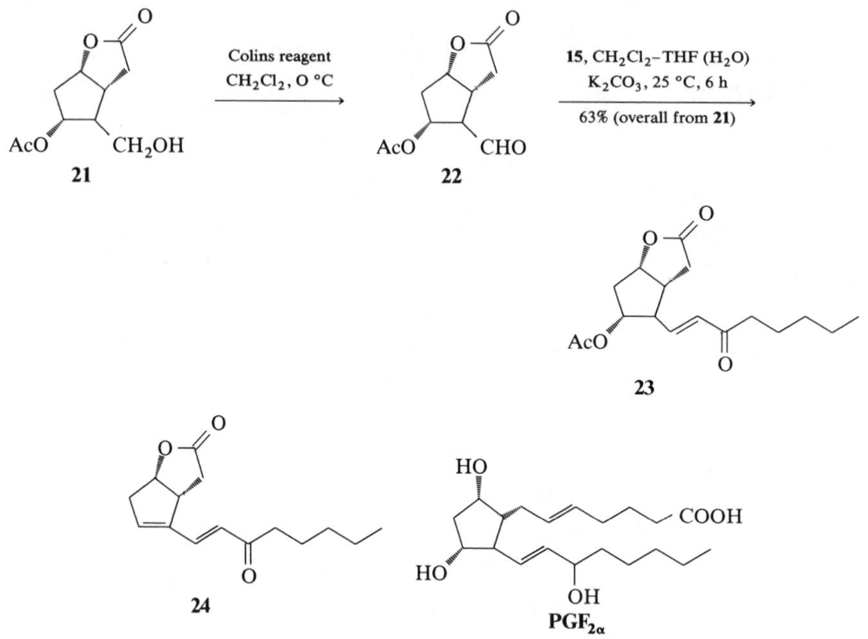

Scheme 10.4

carried out in acetonitrile, the elimination product **24** is formed predominantly. However, with the CH_2Cl_2/THF solvent system in which the salt **15** ($R_1 = n\text{-}C_5H_{11}$) is more soluble, the elimination is completely prevented. Thus the unstable aldehyde **22** reacts smoothly with **15** ($R_1 = n\text{-}C_5H_{11}$) (6 h) to give $PGF_{2\alpha}$ intermediate **23** in 63% overall yield from **21**. (See Scheme 10.4.)

In conclusion, (E)-α-enones can be prepared stereoselectively and in high yields from aldehydes under mild conditions. This method appears to be especially suitable for unstable substrate aldehydes in the synthesis of natural products.

10.5. A Facile Route for the Syntheses of Polyenones and Its Application to the Synthesis of Navenone A

The synthesis of ω-substituted conjugated polyenones (**25**) such as navenones, the trail-breaking alarm pheromones of the sea slug [20], has attracted the attention of synthetic organic chemists.

$$R(CH{=}CH)_nCOCH_3$$

25

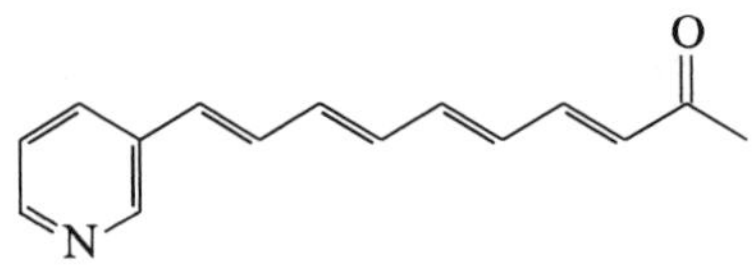

Navenone A

The Wittig or Wittig–Horner reaction is the most general approach toward the synthesis of olefins. However, it has been shown that the triphenylphosphine reacted with 3-bromo-3-penten-2-one to give a mixture of bromides that, in turn, on treatment with potassium *t*-butoxide reacted with benzaldehyde to give a mixture of desired (3*E*,5*E*)-6-phenyl-3,5-hexadien-2-one (**26**) and undesired (*Z*)-3-benzylidene-4-penten-2-one (**27**) in very low yields [21].

$$PhCH{=}CHCH{=}CHCOCH_3 \qquad CH_2{=}CH{-}\underset{\underset{\displaystyle CHPh}{\|}}{C}{-}COCH_3$$

$$\textbf{26} \qquad\qquad\qquad \textbf{27}$$

The abnormal Arbuzov reaction of γ-halogeno-α,β-unsaturated ketone prevents the preparation of the corresponding phosphonate (**28**) [21]. The ketone group masked phosphorane (**29**) or phosphonate (**30**) reacted with aldehydes to give a mixture of E and Z olefination products. Hydrolysis of the resulting dioxolane products requires several days and the overall yields were very low.

$$\underset{\textbf{28}}{(CH_3O)_2\overset{\overset{\displaystyle O}{\|}}{P}CH_2CH{=}CHCOCH_3} \qquad\qquad \underset{\textbf{29}}{Ph_3P{=}CHCH{=}CH{-}\underset{}{C}{-}CH_3}$$

$$\underset{\textbf{30}}{(CH_3O)_2\overset{\overset{\displaystyle O}{\|}}{P}CH_2CH{=}CH{-}\underset{}{C}{-}CH_3}$$

Furthermore, the starting material, (*E*)-5-bromo-3-penten-2-one ethylene ketal, was rather difficult to prepare [22].

In contrast to triphenylphosphine, triphenylarsine smoothly reacted with the same bromoketone to give a pure arsonium bromide (**31**). It reacted with a variety of aromatic, heteroaromatic, and aliphatic aldehydes in the presence of potassium carbonate under phase transfer conditions at low temperature to afford conveniently polyenones in good yield [23]. The 3*E*,5*Z* isomers could be isomerized to 3*E*,5*E* isomers with catalytic iodine in daylight.

$$R(CH{=}CH)_n CHO + Ph_3\overset{+}{A}sCH_2CH{=}CHCOCH_3\ \bar{Br} \xrightarrow[\text{0–3 °C, } -Ph_3AsO]{\text{K}_2CO_3(s),\ Et_2O\ (trace\ H_2O)}$$

$$\textbf{31}$$

$$R(CH{=}CH)_{n+2}COCH_3, \qquad n = 0,1$$

$$\textbf{32}$$

The results are shown in Table 10.4.

The trail-breaking alarm pheromones of the sea slug, navenone A, has been synthesized by a seven-step sequence in 1.1% overall yield [20].

In our hands, the synthesis of navenone A was achieved by a two-step reaction via arsonium salts, as shown in Scheme 10.5.

Table 10.4 ▪ Synthesis of Conjugated Polyenones

RCHO	Temperature/time (°C/h)	Yield (%)	(3E,5E):(3E,5Z)
1 C_6H_5CHO	(0–3)/36	81	77:23
2 $p\text{-}O_2NC_6H_4CHO$	(0–3)/8	91	74:26
3 $p\text{-}ClC_6H_4CHO$	25/16	87	84:16
4 $C_6H_5CH{=}CHCHO$	(0–3)/28	84	85:15
5 $n\text{-}C_5H_{11}CHO$	(0–3)/20	56	78:22
6 $n\text{-}C_8H_{17}CHO$	(0–3)/6.5	67	79:21
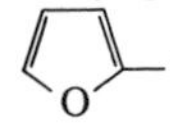 7 (furyl)—CHO	(0–3)/19	71	74:21

Reaction conditions: The solvent used was Et_2O (trace H_2O), except $Et_2O/THF(1:1)$–trace H_2O in entry 3. All compounds were characterized by elemental analysis (except for the known compounds in entries 1, 2, and 4), mass spectroscopy, IR, and ^{1}H NMR. The ratio of the isomers was estimated by gas chromatography or thin-layer cromatography scanning and ^{1}H NMR.

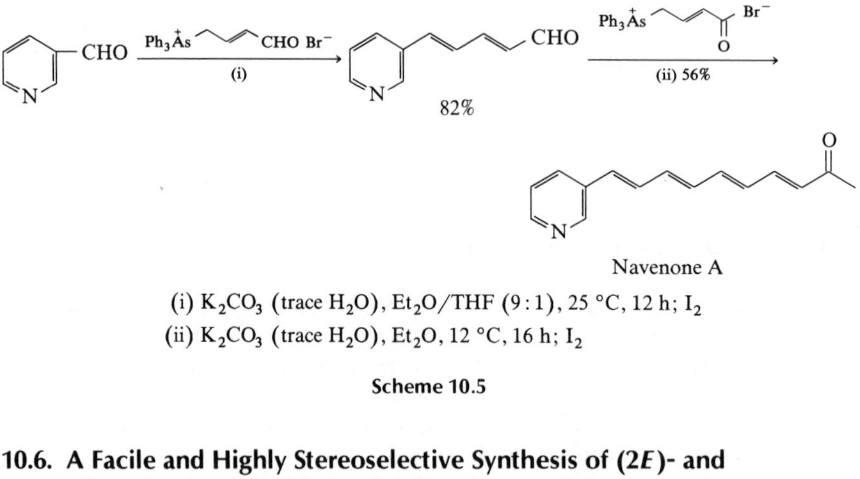

Navenone A

(i) K_2CO_3 (trace H_2O), Et_2O/THF (9:1), 25 °C, 12 h; I_2
(ii) K_2CO_3 (trace H_2O), Et_2O, 12 °C, 16 h; I_2

Scheme 10.5

10.6. A Facile and Highly Stereoselective Synthesis of (2E)- and (2E,4E)-Unsaturated Amides and Related Natural Products via Arsonium Salts

(2E)-Unsaturated and (2E,4E)-unsaturated amides belong to an important class of natural products that show both physiological and insecticidal activities [24]. Several methods have appeared in the literature for synthesizing these products, involving Wittig reaction, Knoevenagel condensation, and the double elimination method. For example, the synthesis of 7-thiophene-heptatrienoic acid piperidide isolated from *Otanthus maritima* has been achieved through a multiple-step reaction in 3.5% overall yield [25], whereas synthesis of trichonine was accomplished in 27.5% overall yield [26]. A method for the synthesis of (2E,4E)-dienamides through a double elimination reaction of acetoxy sulfones employing t-BuOK as a base [27] was reported. All the previously mentioned methods and procedures involve multiple-step reactions or suffer from the difficulty of getting the requisite intermediates.

We have found that $(2E)$- and $(2E,4E)$-unsaturated amides are easily accessible from various aldehydes with high stereoselectivity by means of organoarsonium bromides 34 [28]. The outline of our method is illustrated in the following scheme and the results are summarized in Table 10.5.

$$R(CH{=}CH)_n CHO \ + \ Ph_3 \overset{+}{A}sCH_2 \overset{\overset{\displaystyle O}{\parallel}}{C}NR_1R_2 \ \overset{-}{Br} \ \xrightarrow[K_2CO_3(s),\ 25\ °C,\ -Ph_3AsO]{THF\ or\ CH_3CN/trace\ H_2O}$$

33	**34**
33a: $n = 0$	**34a**: $R_1R_2 = -(CH_2)_4-$
33b: $n = 1$	**34b**: $R_1R_2 = -(CH_2)_5-$

$$R(CH{=}CH)_{n+1}\overset{\overset{\displaystyle O}{\parallel}}{C}NR_1R_2$$

35
35a: $R_1R_2 = -(CH_2)_4-$
35b: $R_1R_2 = -(CH_2)_5-$

The reagents 34a and 34b were easily prepared. Thus triphenylarsine reacted with readily available 2-bromoacetyl pyrrolidine and 2-bromoacetyl piperidine in benzene under reflux for 5–6 h in 85% and 96% yields, respectively. The respective ylides from the resultant two reagents reacted with saturated aldehydes 33a to afford $(2E)$-unsaturated amides, or reacted with (E)-α,β-unsaturated aldehydes 33b to afford $(2E,4E)$-dienamides smoothly in excellent yields.

We have applied this method to the synthesis of tetradeca-$2E,4E$-dienoic acid pyrrolidide (36), naturally occurring in *Achillea arten* [29], and 7-(thiophene)-$2E,4E,6E$-heptatrienoic acid piperidide (37), isolated from *Otanthus maritima* [25]. Scheme 10.6 outlines our approach.

Similarly, the trichonine has been synthesized as shown in Scheme 10.7 [30].

10.7. A General and Highly Stereoselective Approach to Unsaturated Isobutylamides via Arsonium Salt; New Synthesis of Pellitorine

Lipid amides such as $(2E)$- and $(2E,4E)$-unsaturated isobutylamides occur widely in higher plants. Such amides are comparatively unstable and difficult to access, occurring only in small amounts in plants [31]. Representative examples are spilanthol (39) and pellitorine (40), which are naturally occurring insecticides isolated from *Spilanthes oleranceae* [32] and *Anacyclus pyrethrum* roots [33], respectively.

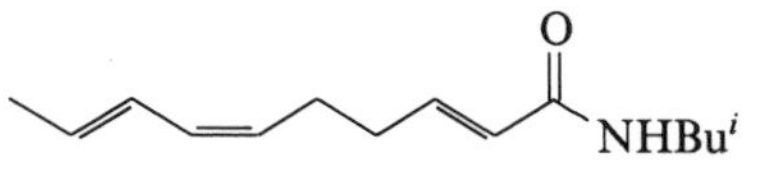

39 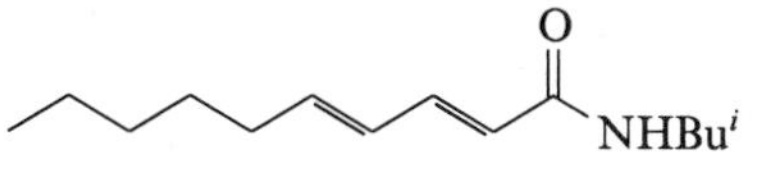   **40**

Table 10.5 ■ The Reaction of 33 with 34a and 34b[a]

Reactant	Product[b]	Reaction time (h)	Yield[c] (%)
1 p-O$_2$NC$_6$H$_4$CHO	p-O$_2$NC$_6$H$_4$CH=CHCON (pyrrolidine)	9	81
	p-O$_2$NC$_6$H$_5$CH=CHCON (piperidine)	8	98
2 C$_6$H$_5$CHO	C$_6$H$_5$CH=CHCON (pyrrolidine)	10	95
	C$_6$H$_5$CH=CHCON (piperidine)	6	98
3 (furyl)—CHO	(furyl)CH=CHCON (pyrrolidine)	4	98
	(furyl)—CH=CHCON (piperidine)	4	97
4 (pyridyl)—CHO	(pyridyl)—CH=CHCON (pyrrolidine)	9	95
	(pyridyl)CH=CHCON (piperidine)	9	96
5 n-C$_6$H$_{13}$CHO	n-C$_6$H$_{13}$CH=CHCON (pyrrolidine)	10	96
	n-C$_6$H$_{13}$CH=CHCON (piperidine)	8	99
6 C$_6$H$_5$CH=CHCHO	C$_6$H$_5$(CH=CH)$_2$CON (pyrrolidine)	8	90
	C$_6$H$_5$(CH=CH)$_2$CON (piperidine)	13	83
7 CH$_3$CH=CHCHO	CH$_3$(CH=CH)$_2$CON (piperidine)	12	88
8 n-C$_5$H$_{11}$CH=CHCHO	n-C$_5$H$_{11}$(CH=CH)$_2$CON (pyrrolidine)	7	82

[a]Reaction of **33** with **34a** was carried out in CH$_3$CN and with **34b** was carried out in THF. All the reactions were run at 25 °C.
[b]All compounds were characterized by ^{1}H NMR, IR, and mass spectroscopy; no Z stereoisomer was found by ^{1}H NMR.
[c]Isolated yields after column chromatography.

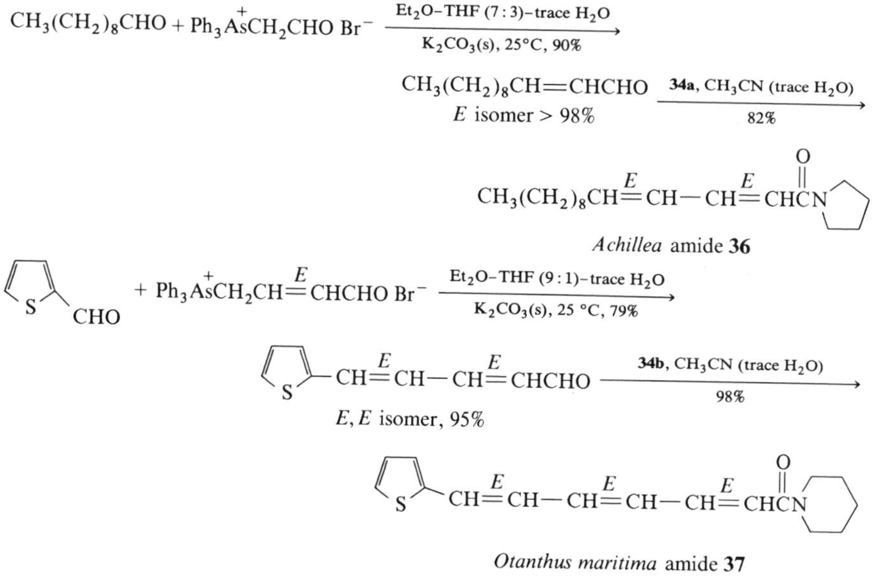

$$CH_3(CH_2)_8CHO + Ph_3\overset{+}{As}CH_2CHO\ Br^- \xrightarrow[\text{K}_2\text{CO}_3(s),\ 25°C,\ 90\%]{\text{Et}_2\text{O–THF (7:3)–trace H}_2\text{O}}$$

$$CH_3(CH_2)_8CH\!\!=\!\!CHCHO \xrightarrow[82\%]{\text{34a, CH}_3\text{CN (trace H}_2\text{O)}}$$
$$E\ \text{isomer} > 98\%$$

CH$_3$(CH$_2$)$_8$CH$\overset{E}{=}$CH—CH$\overset{E}{=}$CHCN (with O and pyrrolidine ring)

Achillea amide **36**

(thiophene)–CHO + Ph$_3\overset{+}{As}$CH$_2$CH$\overset{E}{=}$CHCHO Br$^-$ $\xrightarrow[\text{K}_2\text{CO}_3(s),\ 25\ °C,\ 79\%]{\text{Et}_2\text{O–THF (9:1)–trace H}_2\text{O}}$

(thiophene)–CH$\overset{E}{=}$CH—CH$\overset{E}{=}$CHCHO $\xrightarrow[98\%]{\text{34b, CH}_3\text{CN (trace H}_2\text{O)}}$

E, E isomer, 95%

(thiophene)–CH$\overset{E}{=}$CH—CH$\overset{E}{=}$CH—CH$\overset{E}{=}$CHCN (with O and piperidine ring)

Otanthus maritima amide **37**

Scheme 10.6

Several groups have reported successful syntheses of unsaturated isobutylamides [31, 34]. Recently, Crombie and co-workers [35] reported a hydrozirconation method. The interesting insecticidal activity of such lipid amides stimulated us to develop a more convenient and versatile synthetic approach to the above compounds. A general and highly stereoselective synthesis of both (2 E)- and (2 E,4 E)-unsaturated isobutylamides by means of the arsonium salt **42**, starting from readily available aldehydes **41**, has been achieved. The approach is outlined below and the results are summarized in Table 10.6.

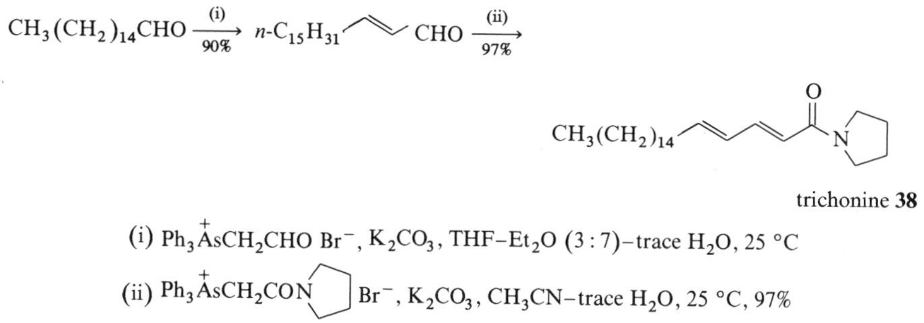

$$CH_3(CH_2)_{14}CHO \xrightarrow[90\%]{(i)} n\text{-}C_{15}H_{31}\diagup\!\!\!\diagdown CHO \xrightarrow[97\%]{(ii)}$$

CH$_3$(CH$_2$)$_{14}$—(chain)—C(=O)—N (pyrrolidine)

trichonine **38**

(i) Ph$_3\overset{+}{As}$CH$_2$CHO Br$^-$, K$_2$CO$_3$, THF–Et$_2$O (3 : 7)–trace H$_2$O, 25 °C

(ii) Ph$_3\overset{+}{As}$CH$_2$CON (pyrrolidine) Br$^-$, K$_2$CO$_3$, CH$_3$CN–trace H$_2$O, 25 °C, 97%

Scheme 10.7

Table 10.6 ▪ Conversion of Aldehyde 41 to (2*E*)- and (2*E*,4*E*)-Isobutylamide 43[a]

	Aldehyde	Solvent[b]	Reaction time (h)	Product[c]	Yield[d] (%)
1	C_6H_5CHO	A	19	$C_6H_5CH{=}CHCONHBu^i$	96
2	$p\text{-}ClC_6H_4CHO$	A	12	$p\text{-}ClC_6H_4CH{=}CHCONHBu^i$	95
3	$p\text{-}O_2NC_6H_4CHO$	A	7	$p\text{-}O_2NC_6H_4CH{=}CHCONHBu^i$	99
4	$CH_3(CH_2)_4CHO$	A	12	$CH_3(CH_2)_4CH{=}CHCONHBu^i$	95
5	$CH_3(CH_2)_8CHO$	A	11	$CH_3(CH_2)_8CH{=}CHCONHBu^i$	97
6	$p\text{-}CH_3OC_6H_4CHO$	A	21	$p\text{-}CH_3OC_6H_4CH{=}CHCONHBu^i$	60
		B	24		92
7	$C_6H_5CH{=}CHCHO$	A	21	$C_6H_5CH{=}CHCH{=}CHCONHBu^i$	61
		B	17		93
8	$CH_3CH{=}CHCHO$	A	15	$CH_3CH{=}CHCH{=}CHCONHBu^i$	57
		B	18		78

[a] All reactions were run at 25 °C.
[b] Solvent A, $CH_3CN—H_2O$ (200 : 1); solvent B, $CH_3CN—HCONH_2$ (100 : 1).
[c] All compounds were characterized by ^1H NMR, IR, mass spectroscopy, and elemental analysis. No *Z* stereoisomer was found in any case.
[d] Isolated yields after flash column chromatography.

$$\underset{\mathbf{41}}{R(CH{=}CH)_nCHO} + \underset{\mathbf{42}}{Ph_3\overset{+}{As}CH_2\overset{O}{\overset{\|}{C}}NHBu^i\ Br^-} \xrightarrow[CH_3CN,\ -Ph_3AsO]{K_2CO_3/\text{trace } H_2O,\ 25\ ^\circ C} \underset{\mathbf{43}}{R(CH\overset{E}{=}CH)_{n+1}\overset{O}{\overset{\|}{C}}NHBu^i}, \quad n = 0,1$$

In order to demonstrate the use of the present methodology, we have synthesized an insecticide, pellitorine (**40**). Formylolefination of **44** with reagent **2** was achieved according to our procedure, affording the α,β-unsaturated aldehyde **45** in 81% yield (97% pure by gas chromatography). Compound **45** was reacted with the reagent **42** in the solvent B to give pellitorine (**40**) in 79% yield. (See Scheme 10.8.) The ^1H NMR analysis of our synthetic **40** indicated the absence of *Z* double-bond isomer [36].

Our synthetic method reported here is a highly steroselective and versatile route to a wide range of lipid isobutylamides.

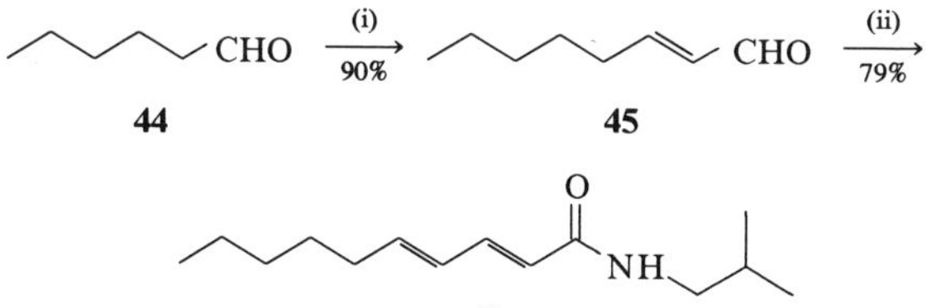

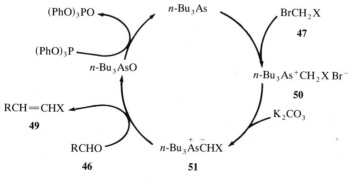

$$\text{44} \xrightarrow[\text{90\%}]{(i)} \text{45} \xrightarrow[\text{79\%}]{(ii)}$$

40

(i) $Ph_3As^+CH_2CHO\ Br^-$, K_2CO_3, Et_2O–THF (7:3)–trace H_2O, 25 °C

(ii) $Ph_3As^+CH_2CONHBu^i\ Br^-$, K_2CO_3, CH_3CN–trace $HCONH_2$, 25 °C

(iii) $Ph_3As^+CH_2CH=CHCONHBu^i\ Br^-$, K_2CO_3, CH_3CN–trace H_2O, 25 °C

Scheme 10.8

10.8. The First Example of a Catalytic Wittig-type Reaction: Tri-*n*-butylarsine–Catalyzed Olefination in the Presence of Triphenylphosphite [37]

The simplicity of our methodology and the ease of reconverting triphenylarsine oxide into triphenylarsine [38] stimulated us to consider the possibility of a catalytic Wittig-type reaction. To our knowledge, although the Wittig-type reaction and its alternatives have been widely used for many years, no catalytic process has ever appeared in the literature. Herein we wish to report the first example in this respect, as summarized in Scheme 10.9 and Table 10.7. A mixture of aldehyde (1 mmol), potassium carbonate (1.2 mmol), methyl bromoacetate (1.2 mmol), 0.5 ml of THF, 4 ml of CH_3CN, tributylarsine (0.2 mmol), and triphenylphosphite (1.2 mmol) were stirred in a reaction tube under nitrogen at room temperature. After the reaction was completed (monitored by thin-layer chromatography), ethyl acetate was added. The resulting mixture was passed through a short column of silica gel to remove the inorganic salt. The desired product was obtained by flash chromatography.

$(PhO)_3PO$ ← $n\text{-}Bu_3As$ $BrCH_2X$
                                           47

$(PhO)_3P$

         $n\text{-}Bu_3AsO$ $n\text{-}Bu_3As^+CH_2X\ Br^-$
                                                 50

$RCH=CHX$ ← K_2CO_3

49

        $RCHO$ $n\text{-}Bu_3\overset{+}{As}\overset{-}{CHX}$

           46   **51**

Scheme 10.9

Table 10.7 ▪ Reaction of Various Aldehydes with Methyl Bromoacetate (47a) and with ω-Bromoacetophenone (47b)[a]

	46 R	47	Reaction time (h)	Isolated yield (%)	$E:Z$[b]
1	p-ClC$_6$H$_4$—	a	18	87	98:2
2	C$_6$H$_5$—	a	30	86	99:1
3	o-ClC$_6$H$_4$—	a	12	81	98:2
4	p-Tolyl	a	24	80	98:2
5	n-C$_5$H$_{11}$—	a	31	64	100
6	2-Furyl	a	16	80	99:1
7	2-Thiophenyl	a	17	75	97:3
8	2-Pyridyl	a	17	68	99:1
9	PhCH=CH—	a	18	61	$E,E > 97$
10	p-ClC$_6$H$_4$—	b	24	75	> 98
11	2-Furyl	b	18	86	> 98
12	n-C$_4$H$_9$—	b	12	80	> 98

[a]All the products were characterized by ^{1}H NMR.
[b]The ratio of $E:Z$ isomers was determined by capillary gas chromatography.

$$RCHO + BrCH_2X + (PhO)_3P \xrightarrow[\text{THF–CH}_3\text{CN, rt}]{n\text{-Bu}_3\text{As(cat.), K}_2\text{CO}_3\text{(s)}} RCH=CHX + (PhO)_3PO$$

46   **47**   **48**   **49**

47a: X = CO$_2$CH$_3$
47b: X = C(O)Ph

The reaction path is proposed as shown in Scheme 10.9.

Reaction of tri-n-butylarsine with bromo compound **47** forms arsonium salt **50**, which, in the presence of potassium carbonate, generates **51** in situ. Ylide **51** reacts with the aldehyde rapidly to afford the desired olefin **49**, and the tri-n-butylarsine is regenerated by reduction of tri-n-butylarsine oxide with triphenylphosphite.

10.9. Summary

A facile methodology for the syntheses of unsaturated aldehydes, ketones, amides, and related natural products via arsonium salts has been reported. This method appears to be especially suitable for unstable substrate aldehydes in the synthesis of natural products. Furthermore, all the arsonium salts mentioned in this paper are stable and can be stored for a long time, in contrast with the corresponding ylides.

Acknowledgments. We are grateful for financial support from the National Natural Science Foundation of China and from Academia Sinica.

References

1. (a) Huang, Y. Z.; Shen, Y. C. *Adv. Organomet. Chem.*, **1982**, *20*, 115, and references cited therein. (b) Shi, L. L.; Xiao, W. J.; Ge, Y. T.; Huang, Y. Z. *Acta Chimica Sinica*, **1986**, *44*, 421.
2. Shao, M. C.; Jin, X. L.; Tang, Y. Q.; Huang, Q. C.; Huang, Y. Z. *Tetrahedron Lett.*, **1982**, *23*, 5343.
3. Trippett, S.; Walker, D. M. *J. Chem. Soc.*, **1961**, 1266.
4. Bestmann, H. J.; Vostrowsky, O. P.; Saluu, H.; Billmann, W.; Stransky, W. *Tetrahedron Lett.*, **1977**, 121.
5. Cresp, T. M.; Sargent, M. V.; Vogel, P., *J. Chem. Soc.*, *Perkin Trans. 1*, **1974**, 37.
6. Nagata, W.; Hayase, Y. *Tetrahedron Lett.*, **1968**, 4359.
7. Wollenberg, R. H.; Albizati, K. F.; Peries, R. *J. Am. Chem. Soc.*, **1977**, *99*, 7365.
8. Huang, Y. Z.; Shi, L. L.; Yang, J. H. *Tetrahedron Lett.*, **1985**, *26*, 6447.
9. Doolittle, K. E.; Roclofs, W. L.; Solomon, J. D.; Carde, R. T.; Beroza, M. *J. Chem. Ecol.*, **1976**, *2*, 399.
10. Ando, T.; Va, M. H.; Yoshida, S.; Takahashi, N.; Tatsuki, S.; Katagiri, K.; Yamane, A.; Ikeda, T.; Yamazaki, S. *Agric. Biol. Chem.*, **1982**, *46*, 709.
11. Meng, X. Z. *Kuxue Tongbao*, **1979**, *24*, 1004 (in Chinese).
12. Underhill, E. W.; Chisholm, M. D.; Steck, W. *Can. Entomol.*, **1980**, *112*, 629.
13. Chisholm, M. D.; Steck, W. F.; Bailey, B. K.; Underhill, E. W. *J. Chem. Ecol.*, **1981**, *7*, 159.
14. Ando, T.; Kurotsu, Y.; Kaiya, M.; Uchiyama, M. *Agric. Biol. Chem.*, **1985**, *49*, 141.
15. Bestmann, H. J.; Koschatzky, K. H.; Platz, H.; Suss, J.; Vostrowsky, O.; Knant, W.; Burghardt, G.; Schneider, I. *Liebigs Ann. Chem.*, **1982**, 1359.
16. Huang, Y. Z.; Shi, L. L.; Yang, J. H.; Cai, Z. W. *J. Org. Chem.*, **1987**, *52*, 3558.
17. Yang, J. H.; Shi, L. L.; Xiao, W. J.; Wen, X. Q.; Huang, Y. Z. *Heteroatom Chem.*, **1990**, *1*, 75.
18. Wang, Y. F.; Li, J. C.; Wu, Y. L.; Huang, Y. Z.; Shi, L. L.; Yang, J. H. *Tetrahedron Lett.*, **1986**, *27*, 4583.
19. Huang, Y. Z.; Shi, L. L.; Li, S. W. *Synthesis*, **1988**, 975.
20. Sakakibara, M.; Matsui, M. *Agric. Biol. Chem.*, **1979**, *43*, 117.
21. Font, J.; March, P. de *Tetrahedron*, **1981**, *37*, 2391.
22. Camps, J.; Font, J.; March, P. de *Tetrahedron*, **1981**, *37*, 2493.
23. Shi, L. L.; Xiao, W. J.; Yang, J. H.; Wen, X. Q.; Huang, Y. Z. *Tetrahedron Lett.*, **1987**, *28*, 2155.
24. (a) Greger, H.; Grenz, M.; Bohlmann, F. *Phytochem.*, **1981**, *20*, 2579 and **1982**, *21*, 1071. (b) Bohlmann, F.; Granzer, M.; Kruger, M.; Nordhoff, E. *Tetrahedron*, **1983**, *39*, 123.
25. Bohlmann, F.; Huhn, C. *Chem. Ber.*, **1977**, *110*, 1183.
26. Trost, B. M.; Lautens, M.; Peterson, B. *Tetrahedron Lett.*, **1983**, *24*, 4525.
27. (a) Mandai, T.; Moriyama, T.; Tsujimoto, K.; Kawaba, M.; Otera, J. *Tetrahedron Lett.*, **1986**, *27*, 603. (b) Moriyama, T.; Mandai, T.; Kawada, M.; Otera, J.; Trost, B. M. *J. Org. Chem.*, **1986**, *51*, 3896.
28. Huang, Y. Z.; Shi, L. L.; Yang, J. H.; Zhang, J. T. *Tetrahedron Lett.*, **1987**, *28*, 2159.
29. Greger, H.; Zdero, C.; Bohlmann, F. *Phytochem.*, **1984**, *23*, 1503.
30. Shi, L. L.; Yang, J. H.; Li, M.; Huang, Y. Z. *Liebigs Ann. Chem.*, **1988**, 377.
31. (a) Bloch, R.; Hassen-Gonzoles, D. *Tetrahedron*, **1986**, *42*, 4975. (b) Crombie, L.; Fischer, D. *Tetrahedron Lett.*, **1985**, *26*, 2477.
32. Jacobson, M. *Chem. and Ind.*, **1957**, 50.
33. Crombie, L., *J. Chem. Soc.*, **1955**, 999.
34. Crombie, L.; Denman, R. *Tetrahedron Lett.*, **1984**, *25*, 4267. Blade, R. J.; Robinson, J. E. *Tetrahedron Lett.*, **1986**, *27*, 3209.

35. Crombie, L.; Hobbs, A. J. W.; Horsham, M. A. *Tetrahedron Lett.*, **1987**, *28*, 4875. Crombie, L.; Horsham, M. A.; Blade, R. J. *Tetrahedron Lett.*, **1987**, *28*, 4879.
36. Shi, L. L.; Yang, J. H.; Wen, X. Q.; Huang, Y. Z. *Tetrahedron Lett.*, **1988**, *29*, 3949.
37. Shi, L. L.; Wang, W. B.; Wang, Y. C.; Huang, Y. Z. *J. Org. Chem.*, **1989**, *54*, 2027.
38. (a) Xing, Y.; Hou, X.; Huang, N. *Tetrahedron Lett.*, **1981**, *22*, 4927. (b) Lu, X. Y.; Wang, Q. W.; Tao, X. C.; Sun, J. H.; Lei, G. X. *Acta Chim. Sinica*, **1985**, *43*, 450. (c) Huang, Y. Z.; Lu, J., unpublished.

Efficient Asymmetric Syntheses of Chiral Sulfoxides and Some Synthetic Applications

Henri B. Kagan

Laboratoire de Synthèse Asymétrique (URA 255)
Université Paris-Sud, 91405-Orsay
France

11.1. Introduction

11.1.1. Chiral Sulfoxides

Chiral sulfoxides are useful auxiliaries in asymmetric synthesis. Many reactions can be stereocontrolled by the sulfinyl moiety that later is removable by reductive methods or by β-elimination. Several reviews are devoted to the use of chiral sulfoxides in synthesis [1–4]. For example, interesting organic chemistry can be performed by condensation of sulfoxide carbanions, by 1,4-addition of organometallics on vinylsulfoxides, or by Diels–Alder cycloaddition on α,β-unsaturated sulfoxides. Another reason for interest in chiral sulfoxides is that biological properties of some natural compounds are often correlated to the absolute configuration at sulfur. For all these reasons there is a growing interest to devise methods giving access to many types of chiral sulfoxides.

11.1.2. Main Routes to Chiral Sulfoxides

There are some possibilities to resolve racemic sulfoxides or to make selective chemical transformations on chiral sulfoxides. The main route to enantiomerically pure sulfoxides remains the Andersen method [5,6] based on the separation of epimeric sulfinates derived from an optically active alcohol (Scheme 11.1). In the specific case of menthyl p-toluenesulfinate, an in-situ epimerization combined with a crystallization produces one epimer in high yield [7]. The substitution step occurs with inversion of configuration at sulfur with many organometallics.

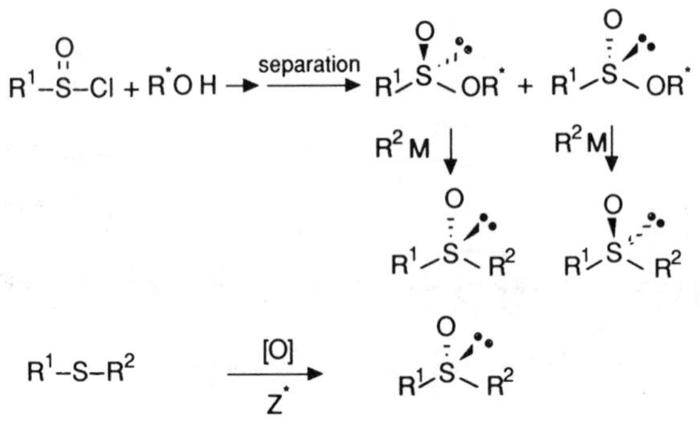

Scheme 11.1

The limitation of the Andersen method lies in the difficulty of separation of epimeric sulfinates. However, there is no problem in preparing *p*-tolyl sulfoxides because of the easy access to the sulfinate precursor [7]. Asymmetric oxidation by enzymatic reactions has been applied to various sulfides (for a review see [8]). In comparison, for a long time chemical systems gave disappointingly low enantiomeric excesses (ee). Chiral oxaziridines emerged as a useful family of chiral oxidants of sulfides, with continuous improvements giving ee up to 90% [9]. In 1984 we discovered a quite general method of oxidation of prochiral sulfides [10], based on the use of a modification of the Sharpless reagent for asymmetric epoxidation. A similar reagent has been independently described by Modena and co-workers [11]. We will discuss and summarize the main features of our oxidant system, which allows us in some cases to reach very high levels of enantiomeric excesses. Efficient asymmetric oxidation is not possible for some classes of sulfoxides. This is why we investigated a new route to chiral sulfoxides based on a modification of the Andersen method. This route proved to be especially valuable in preparing various types of alkyl sulfoxides. In a later section some uses of chiral sulfoxides in synthesis will be presented, especially when the sulfoxides are available by the above methods.

11.2. Asymmetric Oxidation of Sulfides

11.2.1. Introduction

After many attempts at asymmetric oxidation with chiral peroxomolybdenum complexes, we found by serendipity that chiral titanium complexes could be very useful [10]. The Sharpless reagent [Ti(O*i*-Pr)$_4$/(+)-diethyl tartrate = 1 : 1] and *t*-BuOOH gave racemic methyl *p*-tolyl sulfoxide by oxidation of methyl *p*-tolyl sulfide. This was anticipated because the sulfide has no hydroxyl group for anchoring on titanium. However, when the same reaction was performed in the presence of water (one mol eq. with respect to titanium), a new organometallic species is generated that is very efficient for asymmetric oxidation of sulfides. This species is inactive for epoxidation of allylic alcohols, in agreement with the observation of Hanson and Sharpless [12] that drying with molecular sieves is necessary for achieving catalytic reactions.

Our reagent is conveniently prepared at room temperature in dichloromethane under argon by the sequential addition of Ti(Oi-Pr)$_4$, (+)-DET, and water (1:2:1). A yellow solution is obtained, to which the sulfide is added. After cooling at -20 °C, 1.1 mol eq. of t-BuOOH is introduced. The reaction time is usually a few hours. After hydrolysis and filtration of titanium dioxide, the sulfoxide is recovered by flash chromatography in excellent yield. The amount of sulfone is negligible.

The amount of water to get the optimum ee is around 1 mol eq. (with respect to titanium). Under strictly anhydrous conditions, racemic sulfoxide has been recovered [10,13]. An excess of water is not beneficial. In the Modena procedure [11] an excess of diethyl tartrate (4 mol eq.) is used without water addition. The difference between the two reagents remains unclear. We developed the use of the water-modified reagent in the last four years [13–19]. The main features of our reagent will be summarized here.

11.2.2. Oxidation by the Water-Modified Reagent

The reagent Ti(Oi-Pr)$_4$/(+)-DET/H$_2$O/t-BuOOH (1:2:1:1.1) proved to be very valuable for the asymmetric oxidation of aryl methyl sulfides. The corresponding sulfoxides are obtained with 80–90% ee and the (R) configuration. Some representative results are indicated in Scheme 11.2. Chemical yields are good to excellent at the 5-mmol scale. In Scheme 11.3 are presented some asymmetric syntheses of aryl alkyl sulfoxides. The enantioselectivity decreases when the methyl group is replaced by an ethyl and then an n-butyl group. Asymmetric oxidation of Ar—S—R, where R = CH$_2$Cl, CH$_2$CN, vinyl, or COCH$_3$, is also less efficient than for R = Me (Scheme 11.3). Especially promising is the oxidation of phenyl cyclopropyl sulfide (95% ee). Oxidation of dialkyl sulfides has also been investigated [10,14]. R—S—Me

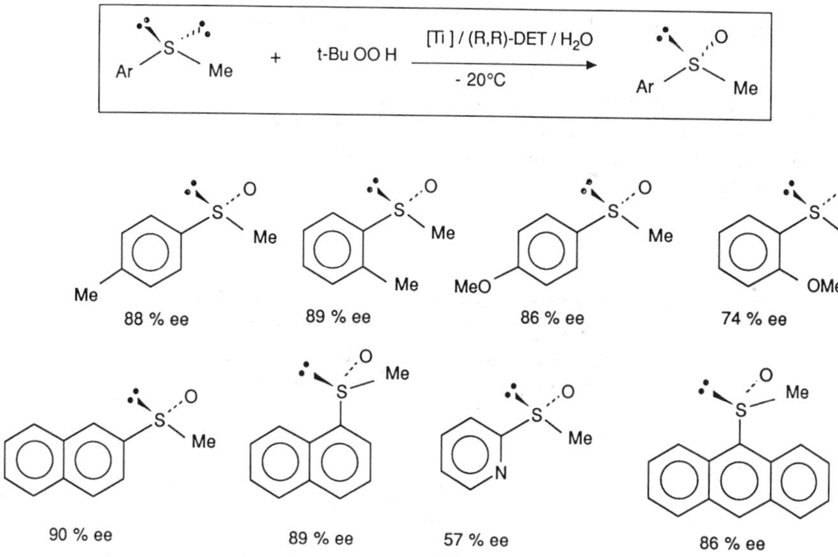

Scheme 11.2 ■ (See references [10,14].)

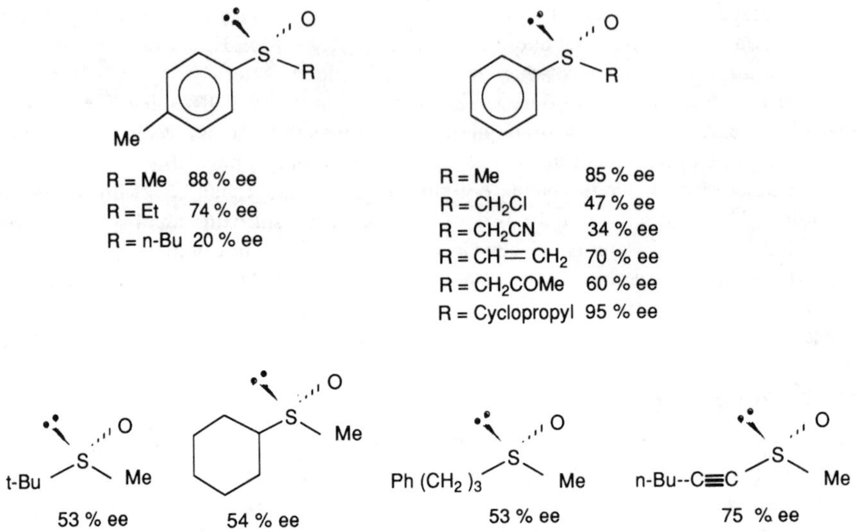

R = Me   88 % ee
R = Et   74 % ee
R = n-Bu 20 % ee

R = Me   85 % ee
R = CH$_2$Cl   47 % ee
R = CH$_2$CN   34 % ee
R = CH=CH$_2$ 70 % ee
R = CH$_2$COMe   60 % ee
R = Cyclopropyl 95 % ee

t-Bu 53 % ee

54 % ee

Ph (CH$_2$)$_3$ 53 % ee

n-Bu--C≡C 75 % ee

Scheme 11.3 ▪ (See references [10, 13, 14, 19].)

are transformed into the corresponding sulfoxides with ee in the range of 50–55% (R = t-Bu, n-octyl, cyclohexyl). Asymmetric oxidation of some sulfur derivatives R—S—X into R—S(O)—X (where X = S-alkyl, NR^{1}R^2, or OMe) has been achieved [15], but enantiomeric excesses remain between 25 and 50%.

11.2.3. Improvements by the Use of Cumene Hydroperoxide

We found that replacement of t-BuOOH by Ph$_2$C(Me)OOH (cumene hydroperoxide) in our standard reagent gives a significant increase in the enantiomeric excess [13]. Many sulfoxides were recovered with ee's above 90% (up to 96%). Some results are listed in Table 11.1. Trityl hydroperoxide has also been checked; it gives a much lower ee than t-BuOOH.

Table 11.1 ▪ Enantiomeric Excess (%) in Asymmetric Oxidation by Various Hydroperoxides in the Presence of [Ti(Oi-Pr)$_4$ ⪯(R,R)-DET ⪯H$_2$O = 1 : 2 : 1)]^a

	Sulfide	Cumene hydroperoxide	t-BuOOH	Ph$_3$COOH
1	Me—S—(p-tolyl)	96	89	16
2	Me—S—(o-anisyl)	93	74	
3	Me—S-phenyl	93	88	
4	Me—S—(n-octyl)	80	53	32
5	Me—S-benzyl	61	35	

^{a}Reactions performed at −20 °C in CH$_2$Cl$_2$ [13].

The use of cumene hydroperoxide allows convenient preparation of optically pure sulfoxides, when there are crystalline compounds. Thus, sulfoxides with 100% ee, corresponding to entries 1–4 of Table 11.1, were isolated after two or three crystallizations in 76, 80, 60, and 40% overall yields, respectively [13]. A procedure with cumene hydroperoxide, working at multigram scale, has been exemplified with synthesis of (S)-methyl p-tolyl sulfoxide [19].

11.2.4. Catalytic Oxidation

A careful investigation has been made [13] for achieving catalytic processes (with respect to titanium complex). When 0.5 mol eq. of titanium complex is used, with t-BuOOH or cumene hydroperoxide, the reaction (oxidation of methyl p-tolyl sulfide) works well, giving the same ee as in the stoichiometric reaction. If the catalyst ratio is decreased to 0.2–0.25, the ee is 50% (t-BuOOH) or 84% (cumene hydroperoxide). An optimization was achieved with cumene hydroperoxide as oxidant and 0.2 mol eq. of titanium complex. An 88% ee and almost quantitative yield are obtained when the reaction is performed in the presence of molecular sieves (pellets). It seems that the molecular sieves regulate the amount of water in the medium. This result appears to be the best up to now for catalytic sulfide oxidation (except for enzymatic reactions).

11.2.5. Oxidations of 1,3-Dithianes

Early attempts to oxidize 1,3-dithianes by t-BuOOH and our titanium complex gave low ee (20%) [15]. We reexamined this problem by using cumene hydroperoxide as oxidant. A careful investigation of the experimental conditions showed that the ee can reach 80% when the oxidation is performed at −40 °C (instead of at −20 °C) [19]. In this way, various 1,3-dithianes 1-oxides have been prepared in excellent yields (Scheme 11.4). For the 2,2-disubstituted monosulfoxides only trans isomer (with respect to Ph or CO_2Et) was produced. With 2-monosubstituted 1,3-dithianes, a mixture of cis–trans monosulfoxides (of the same ee) is formed. It was established that this lack of diastereoselectivity is the result of the in-situ formation of titanium enolates (because of the acidity of the monosulfoxides) and protonation during the workup. Modena and

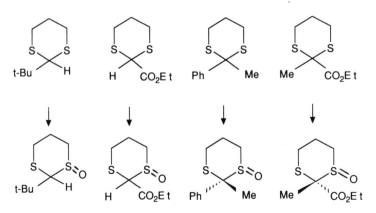

Scheme 11.4 ■ (See reference [18].)

co-workers [20] have studied the monoxidation of 1,3-dithiolanes with their reagent (ee < 83%).

11.2.6. Stereochemistry and Mechanism

11.2.6a. Asymmetric Induction. Absolute configuration of the sulfoxides could be correlated with the configuration of the tartrate by a simple picture (Scheme 11.5) based on "large" (L) and "small" (S) classification of the groups bound to sulfur [10,13]. This scheme has a good predictive value if one takes aryl > alkyl or alkyl (+ Me) > Me. A triple bond behaves as an aromatic ring, but gives lower ee. This shows that asymmetric induction is here the result of a delicate balance of electronic and steric factors.

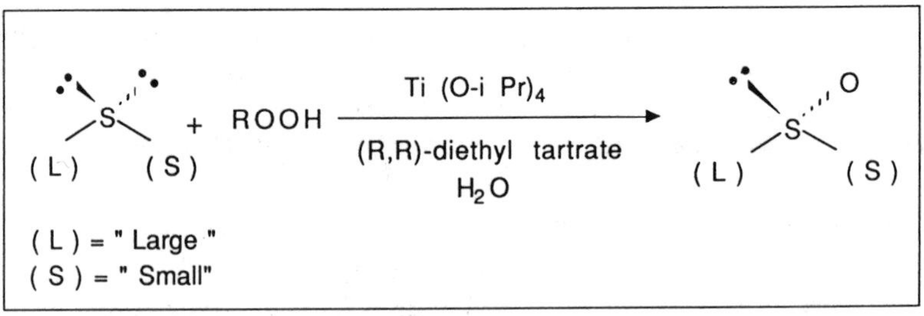

Scheme 11.5

11.2.6b. Mechanism. Mechanistic details are still missing for explaining the various steps of the oxidation process. The structure of the Sharpless reagent itself remains unclear, although some chiral titanium complexes have been analyzed by X-ray crystallography, and some correlations between optical yield and ligand structures have been investigated [21]. We believe that the presence of water, which is deleterious to the Sharpless reagent, builds Ti—O—Ti bridges. We proposed structure **A** as the active species for the oxidation, without establishing whether sulfoxide is produced by an attack on sulfur or by coordination of sulfur to titanium prior to the oxidation (case **B** or **C**) [10,17]. One argument for the choice of the open structure **A** and not the structure **D** is the extended X-ray absorption fine structure (EXAFS) and X-ray absorption near-edge structure (XANES) spectroscopy performed in solution or in the solid state [22].

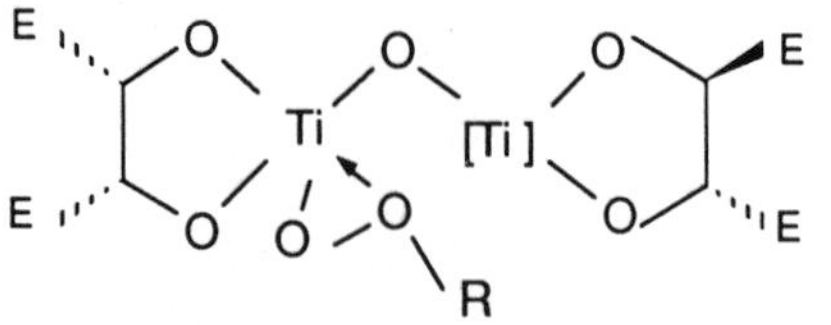

A 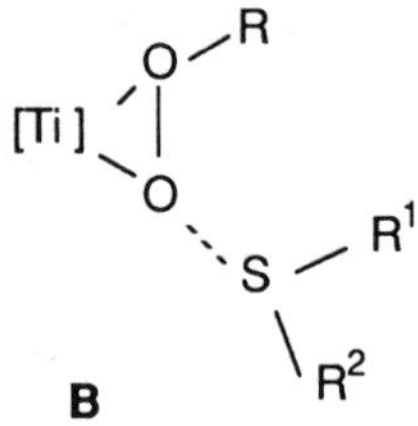   B

This study has shown that the main species in solution either has not two Ti atoms in close vicinity or is highly mobile. XANES shows the permanence of distorted TiO_6 octahedron. More work is needed to elucidate the mechanistic details of this system in order to improve both enantioselectivity and catalytic activity. Some deactivation could occur by complexation to titanium of the produced sulfoxides. A screening of the ee as a function of conversion extent does not indicate any significant variation during asymmetric oxidation of methyl p-tolyl sulfide [13].

11.2.7. Limitations

The main limitation, apart from the moderate catalytic activity of the titanium complex, is the weak ee obtained as soon as there is no aromatic ring directly connected to sulfur. In view of the widespread interest for many types of chiral sulfoxides, we devised an alternative method that avoids oxidation and is described in the next section.

11.3. A Chiral Cyclic Sulfite for Preparation of Sulfoxides

11.3.1. Introduction

The Andersen method is not very convenient for the preparation of several classes of sulfoxides. For example, t-butyl alkyl sulfoxides give interesting stereoselective 1,4-additions but are difficult to get enantiomerically pure [23] with the Andersen method or by asymmetric oxidation. We recently devised a very simple method, which starts from (S)-ethyl lactate, one of the cheapest chiral materials [24]. Reaction with phenylmagnesium bromide produces in good yield diol (S)-1 (Scheme 11.6). This diol reacts with $SOCl_2$ and NEt_3 to give a mixture of diastereomers 2 (95 : 5) from which the stereochemically pure cyclic sulfite 2 was isolated by crystallization (overall yield 60%). Trans stereochemistry has been assigned to 2, based on the subsequent transformations. This compound is an excellent starting material for synthesis of various kinds of optically pure sulfoxides, through substitution reactions.

11.3.2. Transformation of 2 into Chiral Sulfoxides [23]

The basis of the process is described in Scheme 11.6. One can expect that 2 will react under suitable conditions with organometallics R^1M, leading to sulfinates 3 or 4. If this reaction is fully regioselective, then an asymmetric sulfur atom is generated, because

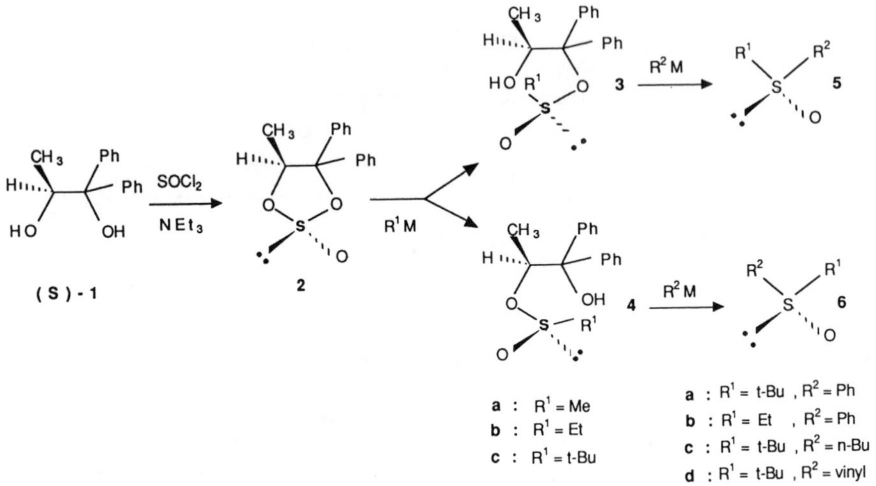

Scheme 11.6 ■ (See reference [24].)

substitution reactions at sulfur usually occur with inversion of configuration [2]. Addition of the second organometallics R^2M will produce optically active sulfoxide (again with inversion at sulfur). Absolute configuration and ee of the final product will be fully connected to the regioselectivity of the ring opening of cyclic sulfite 2. Another essential condition to reach our goal is the necessity to get clean monosubstitution in the first step [to avoid formation of symmetrical R^1—S(O)—R^1 sulfoxide].

We were delighted to see that the Grignard reagent from t-butyl bromide cleanly reacts at room temperature in THF with sulfite 2, giving predominant formation of 4c by cleavage at the more-hindered side (4c:3c = 95:5). The minor isomer 3c could be removed by crystallization. The yield of purified 4c is close to 60%. Grignard reagents from methyl iodide or ethyl bromide react at −78 °C in THF to give pure crystalline 3a and 3b in 55% yield (Table 11.2). These alkylsulfinates are very good starting material for preparation of optically pure sulfoxides. We developed especially the

Table 11.2 ■ Synthesis of Chiral Alkylsulfinates from Sulfite (S)-2

	R^1M^a	3:4 ratio[b]	Isolated yield of sulfinate (major isomer)
1	MeMgI	80:20	3a, 56%
2	MeLi	75:25	3a, 55%
3	EtMgBr	92:8	3b, 57%
4	t-BuMgBr	5:95	4c, 60%
5	t-BuMgCl	10:90	4c, 70%

[a]Reaction performed in THF, at −78 °C (except entry 4, 25 °C).
[b]Measured by NMR on the crude product.

Table 11.3 ■ Synthesis of Chiral *t*-Butyl Sulfoxides from Sulfinate 4c

Sulfinate	R²M	Isolated yield	Configuration and ee
4c	PhLi	6a, 99%	(S) 100%
4c	*n*-BuLi	6c, 99%	(R) 100%
4c	Vinyl-MgCl	6d, 99%	(R) 100%

synthesis of *t*-butyl sulfoxides (Table 11.3), which are obtained in quantitative yields and 100% ee by action of various organometallic reagents. The stereochemistry at sulfur of chiral sulfite **2** is deduced from the absolute configuration of sulfoxides (assuming overall retention of configuration at sulfur).

11.3.3. Conclusions

The two-step asymmetric synthesis of alkyl sulfoxides described here [24] is a useful complement to the existing methods [25–27]. We are actively investigating the scope and the development of the reaction to other classes of sulfoxides.

11.4. Synthetic Applications

It is impossible to describe here the huge number of asymmetric reactions performed with chiral sulfoxides (for some review see [1–4]). We will focus the discussion on some syntheses using the routes to chiral sulfoxides presented in Sections 11.2 and 11.3.

Beckwith and Boate [28] wanted to study stereochemistry of radical substitution at sulfur atom in sulfoxides in cyclization reactions. For this purpose *o*-bromophenylethyl methyl sulfoxide **7** and dihydrobenzothiophene 1-oxide **8** (Scheme 11.7) were prepared with high ee (≥ 96 and ≥ 94%, respectively) by using our water-modified titanium reagent and *t*-BuOOH. The high ee observed in the formation of **7** is surprising (compared with oxidation of $Ph(CH_2)_3SMe$ in Scheme 11.3). It could be the result of a beneficial effect of the bromine in ortho position (weak coordination on Ti). Absolute configuration of **7** was assigned by the model described in Scheme 11.5, and confirmed by X-ray crystallography. The authors treated (R)-**7** with *n*-Bu₃SnH and a trace of AIBN, and converted it into (R)-**8**, with complete inversion of configuration.

The procedure with cumene hydroperoxide had been recently used in some synthetic applications. For example (Scheme 11.8), the Syntex Company wanted to prepare the two enantiomers of **10** for clinical evaluation as cardiovascular drug [29]. The two enantiomers of *p*-anisyl methyl sulfoxide **9** were obtained with greater than 98% ee by asymmetric oxidation. Each enantiomer was converted in four steps into one enantiomer of **10a**. The synthetic scheme involves an asymmetric synthesis controlled by the sulfinyl moiety (asymmetric Michael reaction). Once the dihydropyridine ring has been formed the sulfoxide is oxidized into the final sulfone **10b**. The initial chiral sulfoxide had played a dual role: transient chiral inducer and building block.

A second example of application (Scheme 11.9) is the total synthesis of sulfoxide **12** (itomanindole A), which has been isolated from *Alga okinawa* [30]. This compound is not optically pure (51% ee) and its absolute configuration could not be assigned by

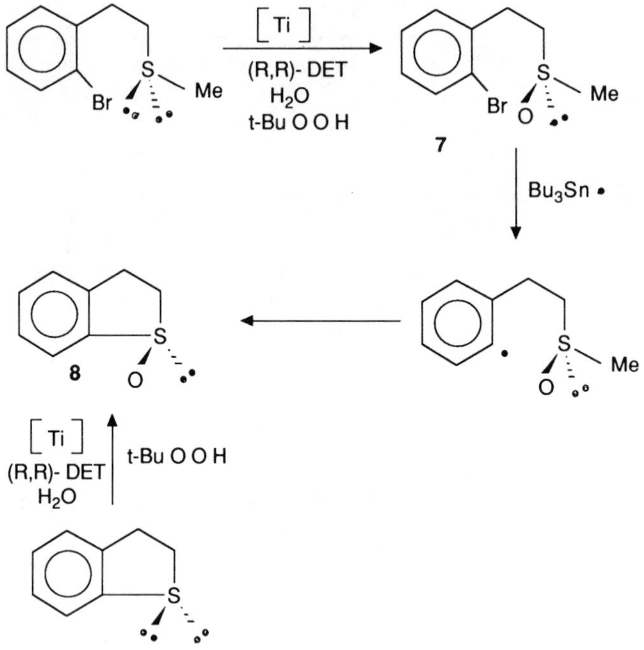

Scheme 11.7 ■ (See reference [28].)

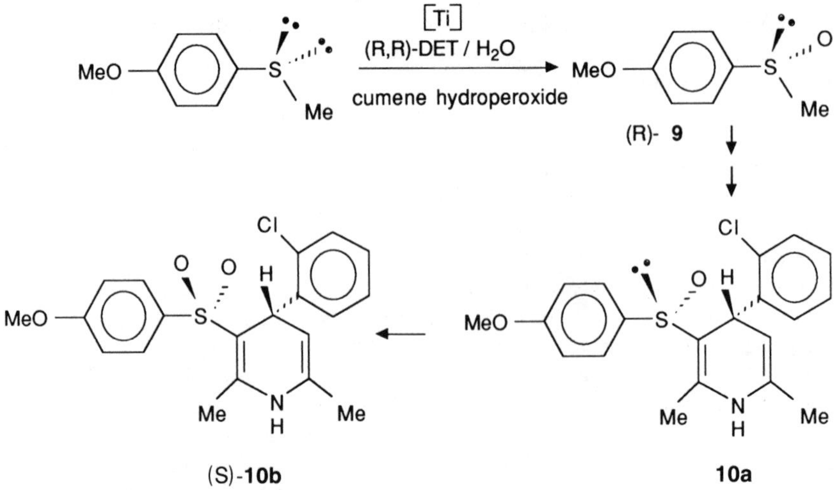

Scheme 11.8 ■ (See reference [29].)

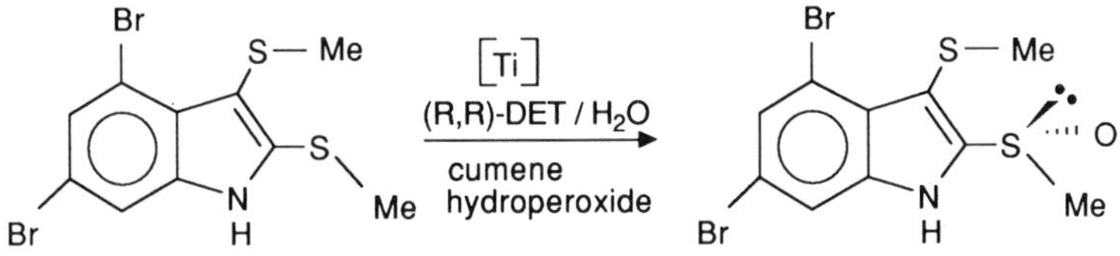

Scheme 11.9 ■ (See reference [30].)

X-ray crystallography because recrystallization yielded a racemic mixture. A fast synthesis of (+)-**12** (90% ee) was achieved by regioselective oxidation of indolic disulfide **11** by our reagent Ti(Oi-Pr)$_4$/(R,R)-DET/H$_2$O (1:2:1). The (R) configuration was assigned on the basis of the model in Scheme 11.5. Interestingly, the oxidation is entirely regioselective, presumably thanks to complexation of titanium by the nitrogen atom.

It is too early to quote applications of chiral alkyl sulfoxides prepared by our chiral cyclic sulfite (Scheme 11.6). However, one can anticipate some interesting developments on the line of the work of Casey, Manage, and Nezhat [31]. These authors used the anion of racemic t-butyl sulfoxide **13** (Scheme 11.10) in some highly stereoselective 1,4-additions to conjugated esters. The t-butyl group seems to be important due to its bulkiness, which prevents attack on the carbonyl carbon and introduces a strong steric control in the transition state. The relative stereochemistry of the products was established by a combination of spectroscopic and X-ray crystallographic methods. Sulfoxide **13** should be easily available from our t-butyl sulfinate **3c** (Scheme 11.6) and suitable organometallics.

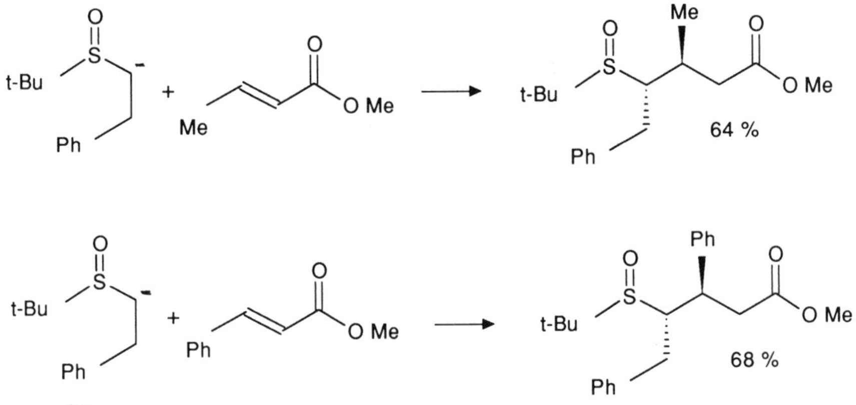

Scheme 11.10 ■ (See reference [31].)

11.5. Conclusion

Substantial progress has been achieved in recent years in asymmetric oxidation of sulfides into sulfoxides by organometallic reagents [10–20] or organic reagents [9]. Our titanium reagent Ti(Oi-Pr)$_4$/DET/H$_2$O allows in many cases the realization of ee's higher than 90% [13,14,19,28,29]. Unfortunately, these high enantioselectivities are observed almost exclusively when an aryl group is directly connected to sulfur. Hopefully, the asymmetric synthesis of many alkyl sulfoxides is possible by using an easily available chiral cyclic sulfite derived from lactic acid [24]. Various t-butyl sulfoxides were prepared and a method is now under development in our laboratory for the synthesis of other types of alkyl sulfoxides. A limitation of all these processes is that the chiral auxiliary is employed in stoichiometric amount. Our titanium reagent can be used in catalytic amount, especially in presence of cumene hydroperoxide, but the catalyst ratio is modest (20%). A goal for the future is to devise classes of specific chiral catalysts combining high catalytic activity and high enantioselectivity.

Acknowledgment. I thank my co-workers who performed the work reported herein: E. Dunach, P. Pitchen, B. Ronan, F. Rebière, O. Samuel, and S. Zhao. I also acknowledge the financial support of C.N.R.S.

References

1. Solladié, G. *Synthesis*, **1981**, 185.
2. Mikolajczyk, M.; Drabowicz, J. *Top. Stereochem.*, **1982**, *13*, 333.
3. Barbachyn, M. R.; Johnson, C. R. in *Asymmetric Synthesis*, Morrison, J. D., ed., Vol. 4, Academic Press, New York, 1985, p. 227.
4. Posner, G. *Acc. Chem. Res.*, **1987**, *20*, 72.
5. Andersen, K. K. *Tetrahedron Lett.*, **1962**, 93.
6. Mislow, K.; Green M. M.; Laur, P.; Melillo, J. T.; Simmons, T.; Ternay, A. L., Jr. *J. Am. Chem. Soc.*, **1965**, *85*, 1958.
7. Mioskowski, C.; Solladié, G. *Tetrahedron*, **1980**, *31*, 227.
8. Holland, H. L. *Chem. Rev.*, **1988**, *88*, 473.
9. Davis, F. A.; McCauley, J. P., Jr.; Chattopadhyay, S.; Harakal, M. E.; Towson, J. C.; Watson, W. H.; Tavanaiepour, I. *J. Am. Chem. Soc.*, **1987**, *109*, 3370; Davis, F. A.; ThimmaReddy, R.; Weismiller, M. C. *J. Am. Chem. Soc.*, **1989**, *111*, 5964.
10. Pitchen, P.; Deshmukh, M.; Dunach, E.; Kagan, H. B. *J. Am. Chem. Soc.*, **1984**, *106*, 8188.
11. Di Furia, F.; Modena, G.; Seraglia, R. *Synthesis*, **1984**, 325.
12. Hanson, R. M.; Sharpless, K. B. *J. Org. Chem.*, **1986**, *51*, 1922.
13. Zhao, S.; Samuel, O.; Kagan, H. B. *Tetrahedron*, **1987**, *43*, 5135.
14. Dunach, E.; Kagan, H. B. *New J. Chem.*, **1985**, *9*, 1.
15. Nemecek, E.; Dunach, E.; Kagan, H. B. *New J. Chem.*, **1986**, *10*, 761.
16. Kagan, H. B.; Dunach, E.; Nemecek, C.; Pitchen, P.; Samuel, O.; Zhao, S. *Pure Appl. Chem.*, **1985**, *57*, 1911.
17. Kagan, H. B. in *Stereochemistry of Organic and Bioorganic Transformations*, Bartmann, W., Sharpless, K. B., eds., Vol. 17, VCH, Weinheim, 1987, p. 31.
18. Ronan, B.; Samuel, O.; Kagan, H. B. *J. Organomet. Chem.*, **1989**, *370*, 43.
19. Samuel, O.; Zhao, S.; Kagan, H. B. *Organic Synth.*, **1989**, *68*, 49.
20. Bortolini, O.; Di Furia, F.; Licini, G.; Modena, G.; Rossi, M. *Tetrahedron Lett.*, **1986**, *27*, 6257. Also see Conte, V.; Di Furia, F.; Licini, G.; Modena, G. *Tetrahedron Lett.*, **1989**, *30*, 4859.

21. Burns, C. J.; Martin, C. A.; Sharpless, K. B. *J. Org. Chem.*, **1989**, *54*, 2826 and references quoted therein.
22. Verdaguer, M.; Cartier, C.; Dunach, E.; Kagan, H. B., unpublished results.
23. Casey, M.; Manage, A. C.; Nezhat, L. *Tetrahedron Lett.*, **1988**, *29*, 3370.
24. Rebière, F.; Kagan, H. B. *Tetrahedron Lett.*, **1989**, *30*, 3659.
25. Wudl, F.; Lu, T. B. K. *J. Am. Chem. Soc.*, **1973**, *95*, 6349.
26. Hiroi, K.; Sato, S.; Kitayama, R. *Chem. Lett.*, **1980**, 1595.
27. Andersen, K. K.; Bujnicki, B.; Drabowicz, J.; Mikolajczyk, M.; O'Brien, J. B. *J. Org. Chem.*, **1984**, *42*, 4070.
28. Beckwith, A. L. J.; Boate, D. R. *J. C. S. Chem. Commun.*, **1986**, 189.
29. Davis, R.; Kern, J. R.; Kurz, L. J.; Pfister, J. R. *J. Am. Chem. Soc.*, **1988**, *110*, 7873.
30. Tanaka, J.; Higa, T. *Tetrahedron Lett.*, **1988**, *29*, 6091.
31. Casey, M.; Manage, A. C.; Nezhat, L. *Tetrahedron Lett.*, **1988**, *29*, 5821.

Original Syntheses of Molecules Containing Adjacent Quaternary Carbons

A. Krief, J. L. Laboureur, M. Hobe, and P. Barbeaux

Department of Chemistry
Facultés Universitaires Notre-Dame de la Paix
61 rue de Bruxelles, 5000, Namur
Belgium

The synthesis of compounds bearing a quarternary carbon is often a difficult task [1]. Most of the classical methods for carbon–carbon bond formation are in general not applicable to this specific case due to steric hindrance around the reaction site that precludes the formation of a new carbon–carbon bond and also due to competing side reactions, such as elimination, which in fact can predominate. Although some significant progress has been made in this area over the last decade, there still remain significant goals to be achieved and new methods to be invented. Among these, the problem of constructing adjacent quaternary centers increases to a large extent the difficulties already mentioned. Not only are the two carbon units to be linked now both crowded but also one must try to solve the stereochemical problems that undoubtedly will be encountered.

We have been active in this field over the past decade, having developed two different approaches that allow the synthesis of products possessing such structural features and that have proved particularly efficient for the synthesis of carbocycles highly substituted with alkyl or aryl groups (Figure 12.1).

Both of these approaches involve carbonyl compounds as the starting material and α-selenoalkyllithiums or α-selenobenzyllithiums as the key intermediates [2]. The first procedure employs the reaction of these organometallics with carbonyl compounds and leads to β-hydroxyalkyl selenides. Further reaction of these pinacol-type derivatives with soft electrophiles produces homologated carbonyl compounds. The whole process

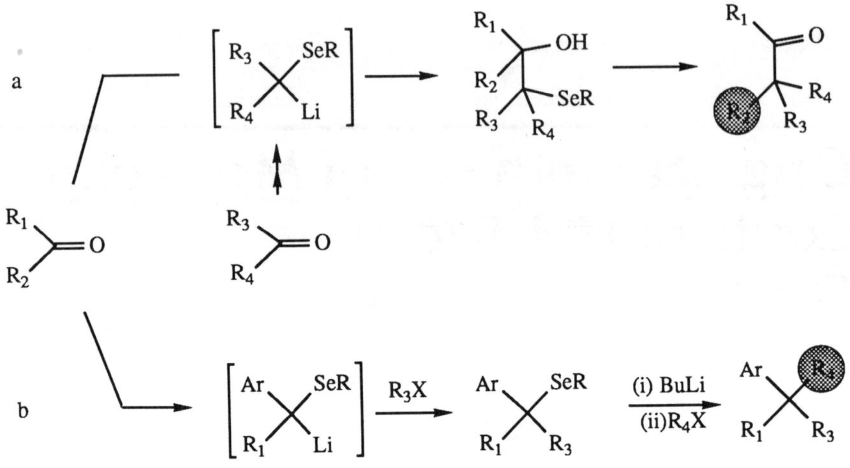

Figure 12.1

involves the insertion of a ketone in between the carbonyl group of aldehydes and ketones and their alpha carbon, thereby allowing the ring enlargement of cyclic ketones (Figure 12.1a).

The second procedure takes advantage of the alkylation of α-selenobenzyllithiums. The resulting compounds on further reaction with butyllithium produce benzyllithiums that can themselves undergo alkylation (Figure 12.1b). The whole process allows the geminal dialkylation of the carbonyl group of aromatic aldehydes and ketones.

We have applied these methods to the synthesis of (d,l)-cuparene and of α- and β-cuparenones, which both possess two vicinal quaternary centers (Figure 12.2). Various syntheses of such compounds exist. We have described some that employ the selective formation of the type a, b, or c bond at the final stage of the synthesis. We have been particularly interested in the last approach because it requires the linkage of the two fully alkyl- or aryl-substituted carbons.

Organolithium reagents bearing a seleno moiety on their carbanionic center are, as already mentioned, the key intermediates in these transformations. They are readily available from aldehydes and ketones by a two-step sequence that involves the synthesis of selenoacetals [3] from selenols under acid catalysis and their cleavage with butyllithiums [4,5]. The acetal synthesis is easier [3]

1. with aldehydes rather than ketone,
2. with methylseleno rather than phenylseleno derivatives,
3. with aliphatic rather than aromatic compounds,
4. when $TiCl_4$ is used instead of $ZnCl_2$ (although the later catalyst is the more commonly employed),
5. when the reaction is carried out in nitromethane rather than in methylene chloride or carbon tetrachloride (Figure 12.3).

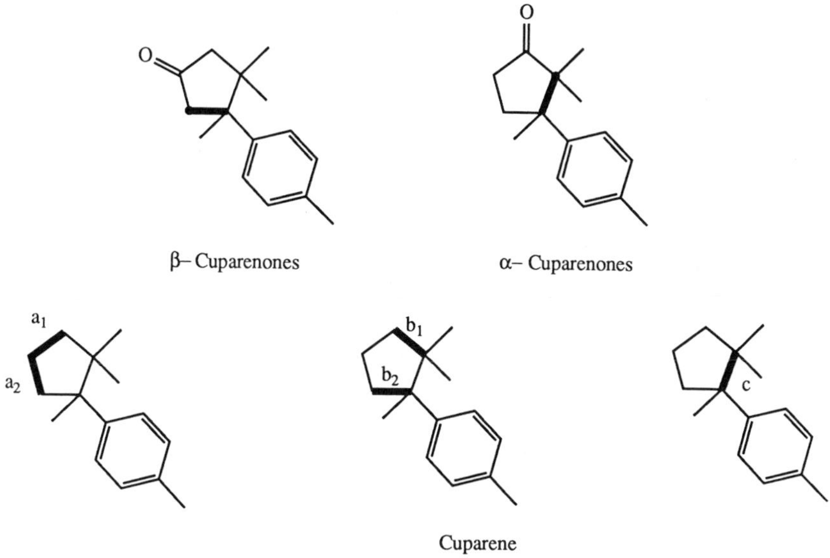

β–Cuparenones   α–Cuparenones

Cuparene

Figure 12.2

The synthesis of α-selenoorganolithiums from selenoacetals is easier [4,5]

1. with those derived from aldehydes rather than from ketones,
2. with phenylseleno rather than methylseleno compounds,
3. with aromatic rather than aliphatic selenoacetals,
4. with *tert*- or *sec*-butyllithium rather than *n*-butyllithium,
5. when the reaction is carried out in THF rather than in ether or alkanes.

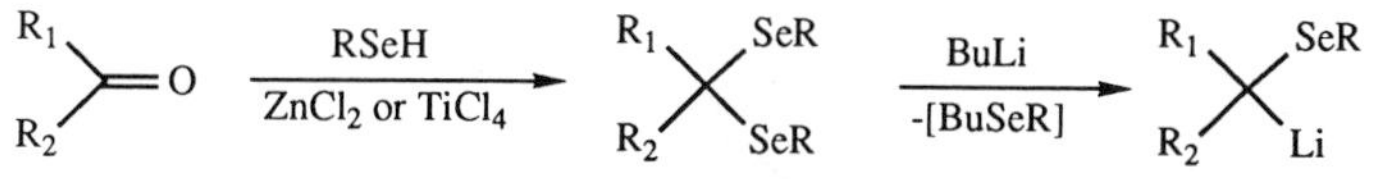

General for R = Me, Ph; R_1, R_2: H, Alkyl, Aryl

No metalation or further reaction with BuSeR, stable: $-78°C$

Easier with

Aldehydes > ketones (hindered)	Aldehydes > ketones
SeMe > SePh	SePh > SeMe
Aliphatic > Aromatic	Aromatic > Aliphatic
$TiCl_4$ > $ZnCl_2$	tBuLi > sBuLi > nBuLi
$MeNO_2$ > CH_2Cl_2 > CCl_4	THF > Ether > Alkanes

Figure 12.3

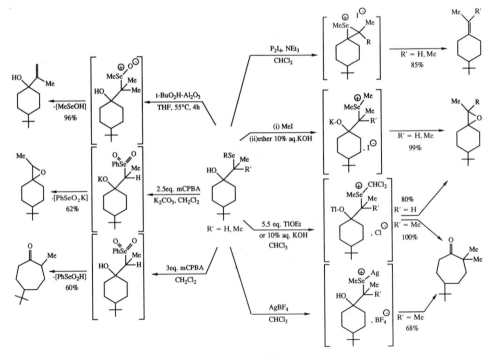

Figure 12.4

Figure 12.5

	n = 1	n = 2	n = 3	n = 4	n = 12
5.5 equiv. TlOEt - CHCl$_3$, 20°C, 5-8h	-	56 (Me)*	71 (Me)	90 (Me)	79 (Me)
10% KOH - CHCl$_3$, PhCH$_2$N$^+$Et$_3$Cl$^-$, 20°C	63 (Me)*	70 (Me)	61 (Me)	74 (Me)	73 (Me)
AgBF$_4$ - Al$_2$O$_3$, CH$_2$Cl$_2$, 50°C	-	70 (Ph)	68 (Me)	-	-

* refers to the nature of the R group

Figure 12.6

In the latter solvent only selenoacetals derived from aromatic carbonyl compounds can be cleaved (Figure 12.3).

The high nucleophilicity of these organometallics towards carbonyl compounds is one of the interesting features of the first process because it allows the synthesis in high yield of a large variety of β-hydroxyalkyl selenides [2, 4, 6–8], including those derived from the highly enolizable deoxybenzoin [6] or highly hindered 2,2,6,6-tetramethyl-cyclohexanone [7, 8a], 2,2,6-trimethylcyclohexanone [7, 8a], and di-*tert*-butyl ketone [8a] (Figure 12.4).

β-Hydroxyalkyl selenides in turn proved to be versatile intermediates that lead selectively to a large variety of selenium-free compounds [2]. They can be viewed as super-pinacols bearing hard (OH) and soft (SeR) moieties that can be selectively activated on reaction with suitable reagents belonging to one of these two classes. Olefins [7] are selectively produced on reaction with thionyl chloride or diphosphorus

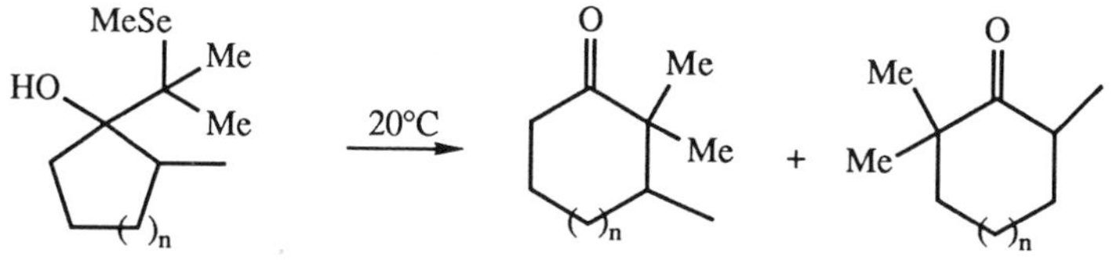

		n = 1	n = 2	n = 3
		Yield A / B	Yield A / B	Yield A / B
a	TlOEt - CHCl$_3$	76% (90 / 10)	77% (60 / 40)	65% (78 / 22)
b	10% KOH - CHCl$_3$	69% (92 / 08)	58% (62 / 38)	56% (84 /16)
c	AgBF$_4$, Al$_2$O$_3$ - CH$_2$Cl$_2$	74% (90 / 10)	54% (22 / 78)	49% (72 / 28)

Figure 12.7

tetraiodide, whereas epoxides [7, 8] are obtained on reaction of β-hydroxyalkyl selenides with an alkylating agent (such as MeI, Me_2SO_4, or $MeSO_3F$) followed by reaction
of the resulting β-hydroxyalkyl selenonium salt with a base (such as 10% aq KOH).
Other typical reactions of β-hydroxyalkyl selenides are listed in Figure 12.5 [2, 8].

Of particular interest for our purpose is the reaction with dichlorocarbene or silver
tetrafluoroborate. These reagents interact selectively with the soft selenium atom and
provide, if the carbon atom bearing the seleno moiety if *fully alkyl-substituted*,
homologated carbonyl compounds [9, 10, 12–20]. When applied to cycloalkanones, both
reagents result in ring enlargement regardless of the size of the cycle (from 4 up to 15).
As expected, the transformation of cyclobutanones to cyclopentanones proved [10–19]
to be faster than the others due to the strain released during the transformation
(Figure 12.6).

Thallous ethoxide in chloroform [12–19] [5.6 eq, $CHCl_3$, 20 °C, 10–20 h (Tl
conditions)] was the most efficient and versatile among the various sources of carbene.
Although the reaction rate is slower than under phase transfer conditions [13, 14] [10%
aq KOH, $CHCl_3$, cat. $PhCH_2NEt_3^+$ Cl^-, 20 °C, 1–2h (PTC conditions)], the resulting

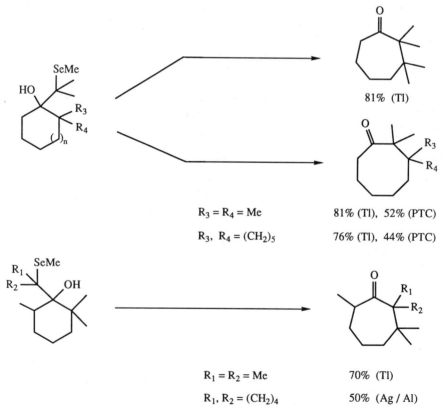

$R_3 = R_4 = Me$	81% (Tl), 52% (PTC)
$R_3, R_4 = (CH_2)_5$	76% (Tl), 44% (PTC)

$R_1 = R_2 = Me$	70% (Tl)
$R_1, R_2 = (CH_2)_4$	50% (Ag / Al)

Figure 12.8

ketone is usually produced very cleanly and does not have to be separated from various compounds derived from dichlorocarbene oligomerization when phase transfer conditions are used instead [13,14]. The reaction with silver tetrafluoroborate [9,13] (Ag conditions) often leads to the concomitant formation of olefins resulting from the interaction of hard fluoroboric acid formed in the medium with the hard oxygen of the hydroxyl group of the β-hydroxyalkyl selenide. This side reaction can be avoided if the reaction is performed in the presence of basic alumina (Ag/Al conditions). Other bases proved [9] unsuitable because they inhibit the ring-enlargement reaction.

When the reaction is applied to β-hydroxyalkyl selenides derived from unsymmetrically α-substituted cycloalkanones, it is found that the more highly alkyl-substituted carbon generally possesses the highest tendency for migration [9,13] (Figure 12.7). This is particularly evident with β-hydroxyalkyl selenides derived from cyclopentanones but is less pronounced with those derived from cyclohexanones [13] (Figure 12.7). In the latter case the reverse tendency has even been found when the reaction is performed with silver tetrafluoroborate (Figure 12.7). Interesting, the migration of the more alkyl-substituted carbon is observed almost exclusively if the starting ketone is itself α,α-dialkyl–substituted (Figure 12.8). The set of reactions reported above therefore allows the synthesis of cycloalkanones bearing two vicinal quaternary carbons. Detailed information on the intimate mechanism of such ring-enlargement reactions has been reported recently [15].

This reaction has been successfully applied by Fitjer, Scheuermann, and Wehle [20] to the transformation of permethylcyclobutanone to permethylcyclopentanone and permethylcyclohexanone, respectively (Figure 12.9), and has been used for the synthesis

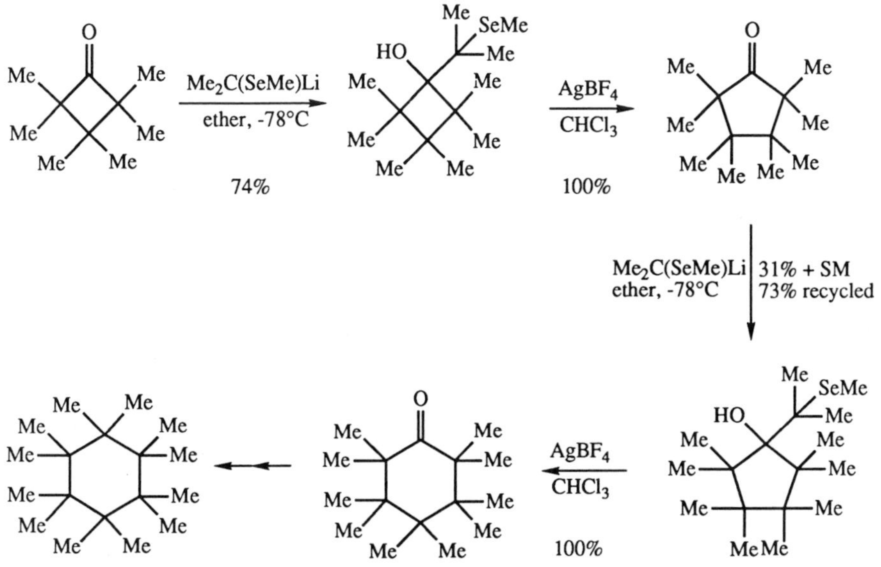

Figure 12.9

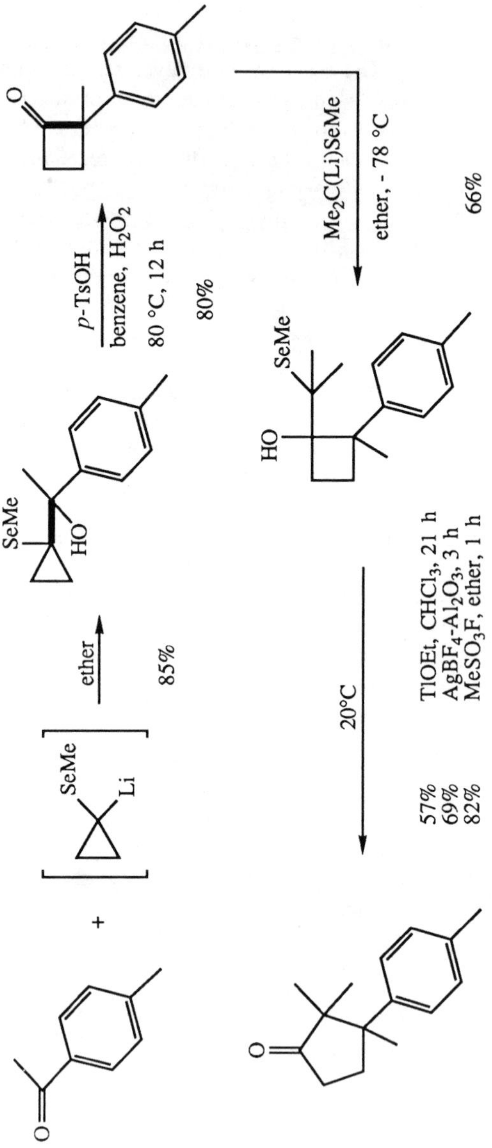

Figure 12.10

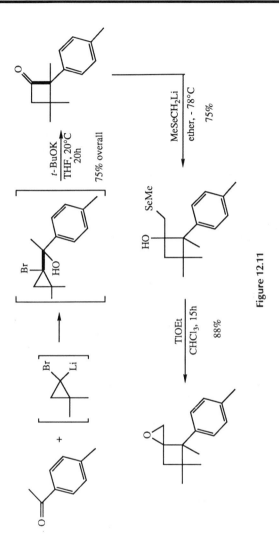

Figure 12.11

of α-cuparenone [10] from 2-methyl-2-p-tolyl cyclobutanone, which was in turn generated from p-methyl acetophenone (Figure 12.10). All these ketones are sterically
congested compounds bearing two quaternary centers in close proximity.

We have also synthesized [10] some other regioisomeric cuparenones from the same
acetophenone by other routes, which employ the reaction of methylselenomethyllithium
with 2-p-tolyl-2,3,3-trimethylcyclobutanone. Treatment of the resulting β-hydroxyalkyl
selenide with thallous ethoxide in chloroform does not lead to a cyclopentanone but
instead to an oxaspirohexane (Figure 12.11). The epoxide formation is almost always
encountered if the reaction is carried out on β-hydroxyalkyl selenides in which the
carbon atom bearing the selenium atom is not fully alkyl-substituted.

We took advantage of the ring strain present in the oxaspirohexane to perform [10]
the ring-expansion reaction that leads once more to cyclopentanones (Figure 12.12).

Reaction of the oxaspirohexane shown in Figure 12.12 with lithium chloride in
methylene chloride or lithium iodide in benzene leads [10] to an 80:20 mixture of
isomeric cuparenones in which the product resulting from the migration of the more
highly alkyl-substituted carbon prevails (Figure 12.12). Interestingly, a much higher
regiocontrol in favor of the latter compound is achieved [10] when the reaction is
instead carried out in dioxane. An almost complete regioselectivity is observed if the
reaction is performed in the presence of 18-crown-6 (Figure 12.12a). Furthermore, the
synthesis of the previously mentioned minor regioisomer arising from the migration of
the less-substituted carbon has been selectively performed [10] in a two-step sequence
that involves the intermediary formation of a chlorohydrin (BeCl$_2$, 20 °C, 20 h) and its
rearrangement promoted by silver tetrafluoroborate (Figure 12.12b). The synthesis of
cuparene from these ketones then requires a further reduction step.

The second approach to cuparene is much shorter. It takes advantage of the easy
alkylation of selenobenzyllithiums [21,22] with suitably functionalized alkyl halides
and of the high propensity of the resulting benzyl selenides to produce benzyllithiums
[22,23] on further reaction with butyllithiums. Although the alkylation of α-selenoal-

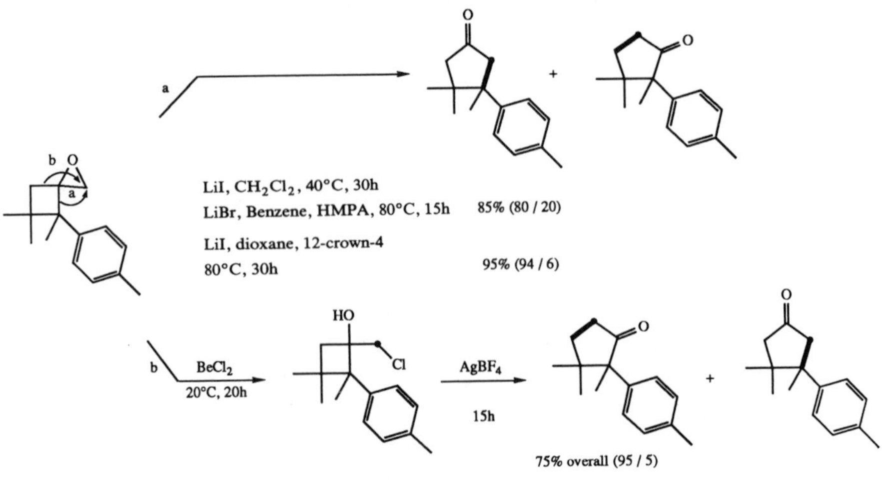

LiI, CH$_2$Cl$_2$, 40°C, 30h
LiBr, Benzene, HMPA, 80°C, 15h   85% (80 / 20)

LiI, dioxane, 12-crown-4
80°C, 30h   95% (94 / 6)

BeCl$_2$
20°C, 20h

AgBF$_4$
15h

75% overall (95 / 5)

Figure 12.12

Figure 12.13

n-BuLi, THF or *s*-BuLi, ether or *s*-BuLi, ether - HMPA	80%	20	80
s-BuLi, pentane	73%	81	19

Figure 12.14

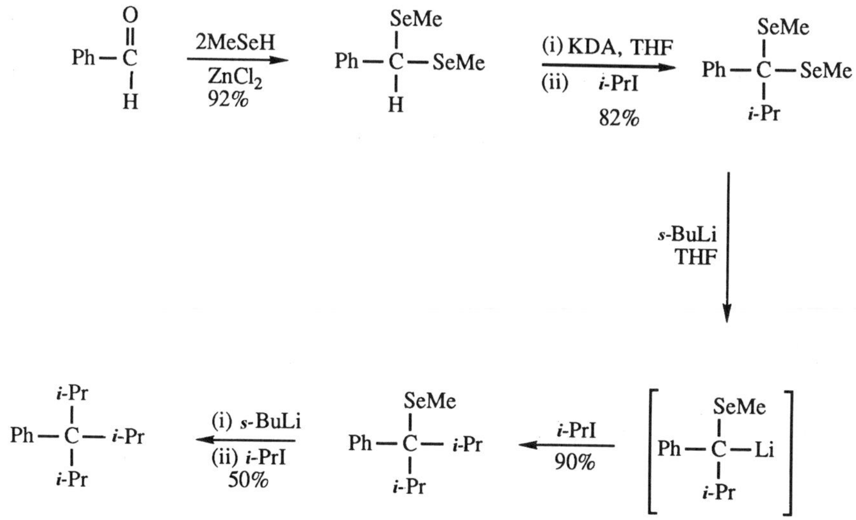

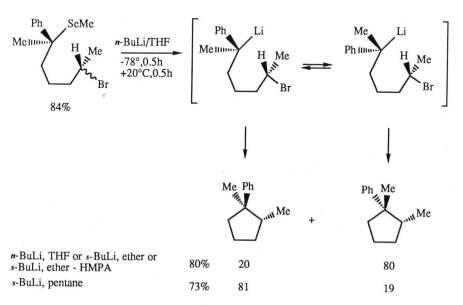

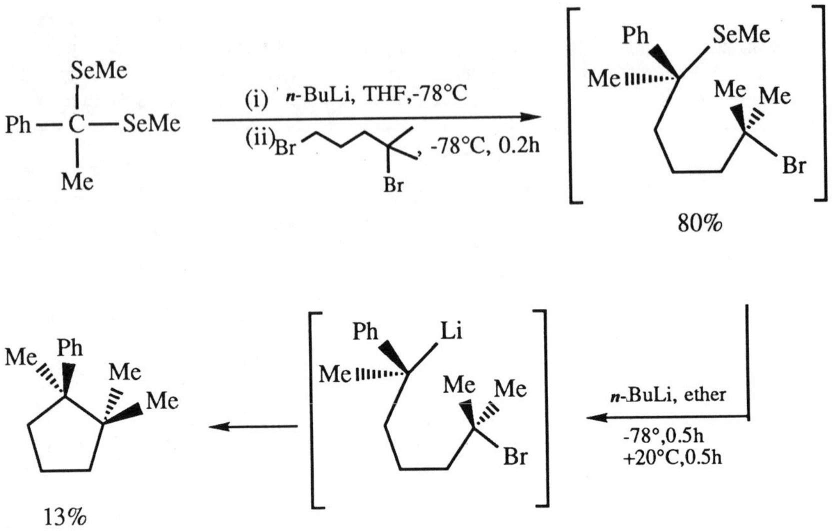

Figure 12.15

kyllithiums [2,19] is strictly limited to primary alkyl halides and some allyl halides, α-selenobenzyllithiums are commonly alkylated with a variety of alkylating agents [2,21–23], including secondary alkyl halides [2,21,23]. Benzyllithiums derived from these selenides also proved valuable as nucleophilic species toward alkyl halides (Figure 12.13). They have been efficiently alkylated even with secondary alkyl halides, but unfortunately not with the tertiary ones.

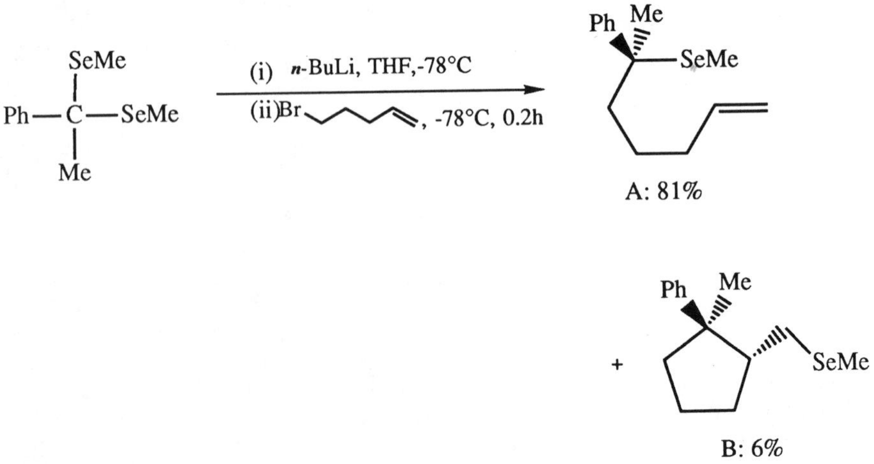

Figure 12.16

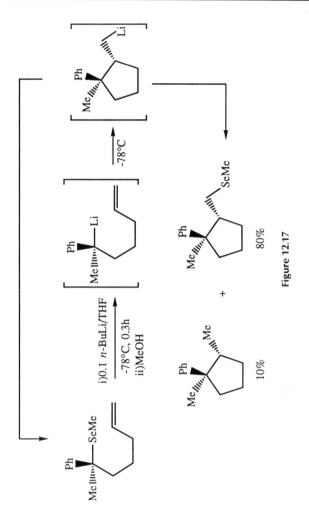

Figure 12.17

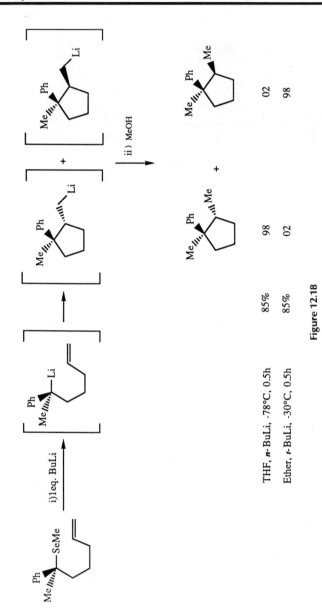

THF, *n*-BuLi, -78°C, 0.5h   85%   98   02

Ether, *t*-BuLi, -30°C, 0.5h   85%   02   98

Figure 12.18

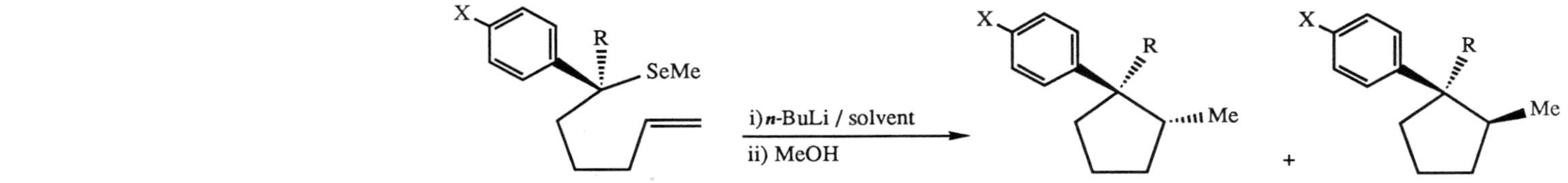

	X	R				
a	H	i-Pr	ether-Hexane, -30°C, 2h	84%	00	100
b	H	H	ether-Hexane, 0°C, 1h	88%	00	100
c	MeO	H	ether-Hexane, 0°C, 1h	94%	03	97
d	H	i-Pr	THF-Hexane, -78°C, 2h	86%	100	00
e	H	H	THF-Hexane, -30°C, 1h	70%	55	45
f	MeO	H	THF-Hexane, -30°C, 1h	77%	58	42

Figure 12.19

We wished to apply this set of reactions to the construction of aryl cycloalkanes and ultimately to the synthesis of cuparene. This transformation requires at one stage the alkylation of a benzyllithium with a tertiary alkyl halide. In a model experiment we have sequentially reacted in THF the methylseleno acetal derived from acetophenone with n-butyllithium, 1,4-dibromopentane, and n-butyllithium again (Figure 12.14).

We obtained in 80% overall yield 1,2-dimethyl-1-phenylcyclopentane as an 2:8 mixture of cis and trans stereoisomers (relative to methyl groups). This ratio is different from the stereoisomeric ratio of the intermediately formed 6-bromo-2-methylseleno-2-phenylheptane, which, in fact, has been isolated [22] in 84% yield as a 1:1 mixture of stereoisomers when the last reagent is omitted. These results lead us to suspect that the benzyllithium does not retain its original stereochemistry and that some control of the relative stereochemistry at C-1 and C-2 on the resulting cyclopentane derivative could be achieved by the proper choice of experimental conditions. A similar stereoisomeric ratio, however, is observed when the last step is carried out in ether (in this case the use of sec- or $tert$-butyllithium instead of n-butyllithium is required to cleave the C—Se bond) or in ether–HMPA. The reversed ratio of stereoisomers, however, is found [24] if the reaction is instead performed in pentane at 20 °C.

With these results in hand, we attempted [24] the synthesis of the cuparene analogue missing the p-methyl group on the phenyl substituent. The alkylation of 1-phenyl-1-ethyllithium with 2,5-dibromo-2-methylpentane proceeded in very good yield but, as expected, the cyclization of the tertiary bromide was not completely successful. The highest yield of 1,2,2-trimethyl-1-phenyl cyclopentane obtained when this reaction was performed in ether unfortunately did not exceed 13% (Figure 12.15).

We then turned our attention toward the cyclization of 2-methylseleno-2-phenyl-6-heptene. It was known [25] when we started this work that alkenyl selenides do cyclize to cyclopentane and cyclohexane derivatives when reacted with tin hydrides and that this reaction involves a single electron transfer. The synthesis of the starting material was readily achieved [22] from the previously mentioned benzyllithiums and 5-bromo-1-pentene but the reaction was not always reproducible (Figure 12.16).

Often, 1-methyl-1-phenyl-2-methylselenomethylcyclopentane was concomitantly produced [22] and in some cases this even became the major product. We later proved [22] that even a trace of n-butyllithium catalyzes the transformation of 2-methyl-

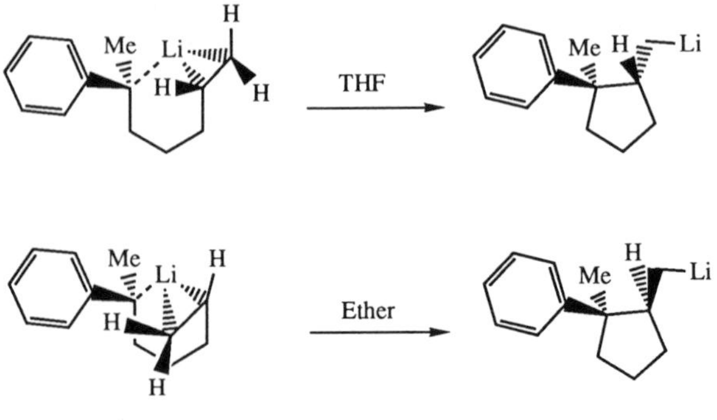

Figure 12.20

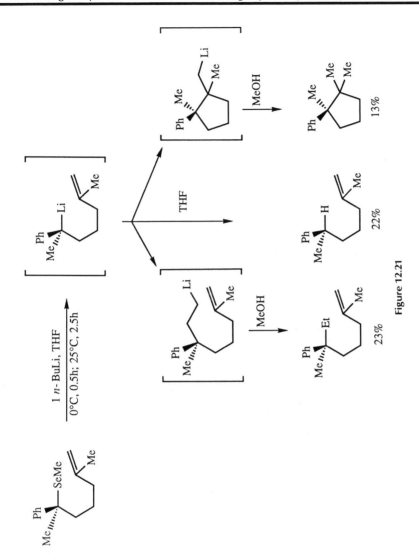

Figure 12.21

seleno-2-phenyl-6-heptene to 1-methyl-1-phenyl-2-methylselenomethylcyclopentane (Figure 12.17).

The reaction involves the intermediate formation of 2-lithio-2-phenyl-6-heptene, which immediately cyclizes to 2-lithiomethyl-1-methyl-1-phenylcyclopentane. This further reacts with the starting benzylselenide to produce 1-methyl-1-phenyl-2-methyl-selenomethylcyclopentane and 2-lithio-2-phenyl-6-heptene which perpetuates the cycle until none of the starting material remains.

We took advantage of this interesting observation to perform the selective synthesis of 2-lithiomethyl-1-methyl-1-phenylcyclopentane. For this purpose, 2-methylseleno-2-phenyl-6-heptene was chemoselectively prepared at -78 °C from 1,1-bis-(methylseleno)-1-phenylethane, stoichiometric amounts of n-butyllithium in THF, and 5-bromo-1-pentene. It was then reacted at -78 °C in THF with an equimolar amount of n-butyllithium or, preferably, with sec- or tert-butyllithium (addition of the selenide to the butyllithium solution in alkanes maintained at -78 °C) and the resulting mixture was then quenched with methanol at this temperature to deliver regioselectively- and stereoselectively cis-1,2-dimethyl-1-phenylcyclopentane (relative to the two methyl groups). Even more interesting is the observation that its trans 1,2-dimethyl isomer was generated [22] highly selectively if the reaction was instead performed in ether at -30 °C (Figure 12.18).

This cyclization reaction proved to be quite general. It allows, from benzylselenides in which the benzylic carbon is fully alkyl-substituted, the regioselective synthesis of various 1-aryl-2-methylcyclopentane [22, 26] (Figure 12.19) and 1-aryl-2-methyl-cyclohexane [22] derivatives, with complete stereochemical control of the type reported above (Figures 12.19d and 12.19e).

In the case of analogous compounds whose carbanionic center is not fully alkyl-substituted, the cyclopentane derivative possessing cis stereochemistry between the methyl and aryl groups can still be produced [26] completely stereoselectively if the reaction is carried out in ether at -30 °C (Figures 12.19b and 12.19c), but a 1:1 mixture of cis and trans stereoisomers is formed [26] if the reaction is performed in THF (Figure 12.19e and 12.19f). None of the various conditions we tried allowed the selective synthesis of trans-1-aryl-2-methylcyclopentanes. We suspect that the selective formation of cyclopentane derivatives possessing an aryl and a methyl group cis to one another is due to internal chelation of the lithium cation that is sandwiched between the aryl group and the carbon–carbon double bond (Figure 12.20). The reasons for the reverse selectivity found in THF remain unclear at present.

The cyclization reaction does not proceed [26] under the previously mentioned conditions if carried out on unsaturated benzylselenides bearing two methyl sub-

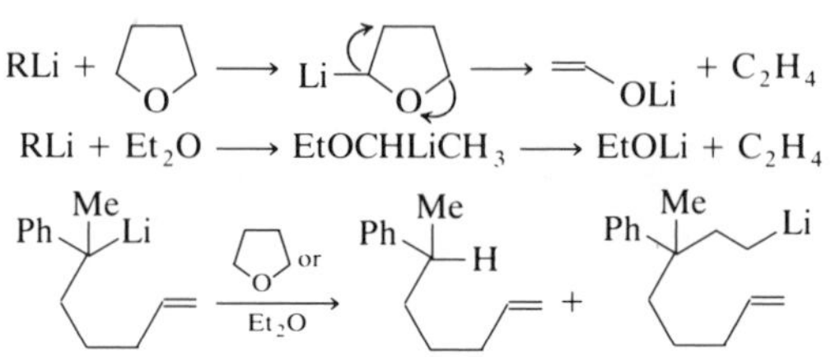

Figure 12.22

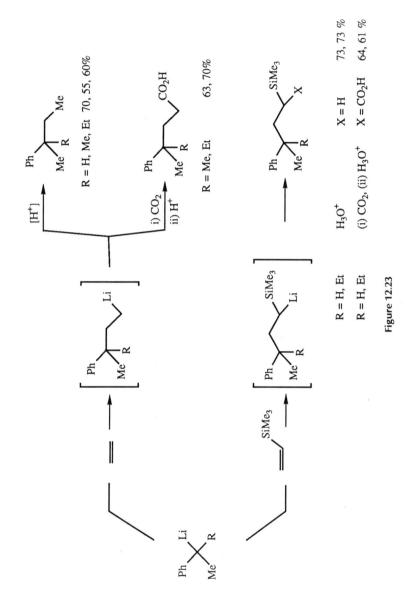

Figure 12.23

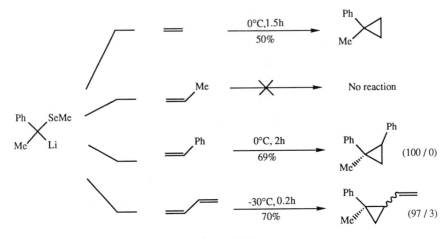

Figure 12.24

stituents at the terminus of the carbon–carbon double bond and takes place very slowly at room temperature with 6-methyl-2-methylseleno-2-phenyl-6-heptene. It provides [26], although in very low yield (13%), the cyclized product 1,2,2-trimethyl-1-phenylcyclopentane, which bears two quaternary centers on vicinal carbons, as well as 6-methyl-2-phenyl-6-heptene and 3,7-dimethyl-3-phenyl-7-octene resulting, respectively, from the protonation and the ethylation of 2-lithio-6-methyl-2-phenyl-6-heptene formed intermediately (Figure 12.21).

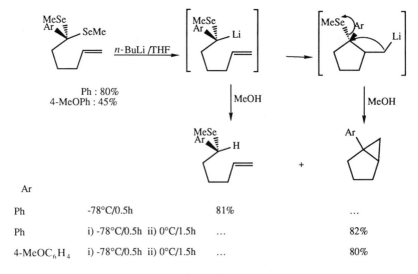

Ar			
Ph	-78°C/0.5h	81%	...
Ph	i) -78°C/0.5h ii) 0°C/1.5h	...	82%
4-MeOC$_6$H$_4$	i) -78°C/0.5h ii) 0°C/1.5h	...	80%

Figure 12.25

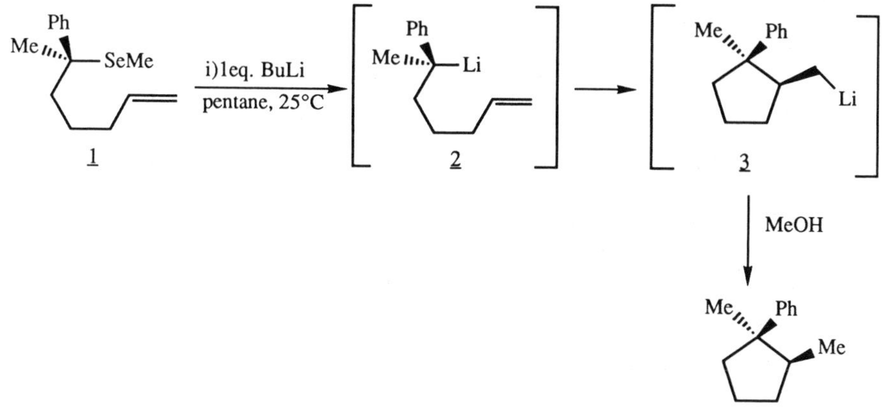

n	Conditions		1	4
	s	t		
a	*n*-BuLi	24h	51%	13%
b	*s*-BuLi	0.25h	84%	16%
c	*s*-BuLi	0.75h	43%	57%
d	*t*-BuLi	0.75h	10%	90%

Figure 12.26

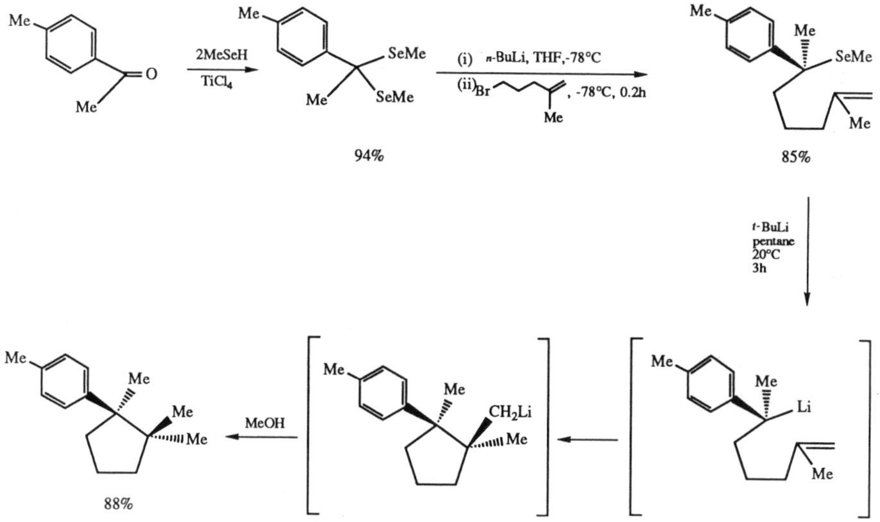

Figure 12.27

The concomitant formation of these two products has been rationalized in the following way. The benzyllithium resulting from the C—Se bond cleavage of the benzyl selenide metalates the solvent at this temperature. Ethylene resulting from the well-known decomposition of 1-lithioethyl ethyl ether or 2-lithiotetrahydrofuran further reacts with the benzyllithium to provide after hydrolysis the observed products (Figure 12.22). These results emphasize the high propensity for benzyllithiums to react with ethylene.

In a separate experiment we have indeed found [26] that the previous reaction is particularly facile with benzyllithiums bearing two alkyl substituents on the carbanionic center. It proceeds rapidly at room temperature when ethylene is bubbled into the medium, providing a novel organolithium compound that is then quenched with water or with carbon dioxide. Benzyllithiums also react with trimethylsilylethylene, bearing on the carbon–carbon double bond a substituent capable of stabilizing the incipient carbanionic center (Figure 12.23).

The reaction also proceeds with α-selenobenzyllithiums, but the organometallic resulting from the addition of the α-selenobenzyllithium to the olefinic compound further reacts to produce [26] instead cyclopropane derivatives with remarkable stereocontrol (Figure 12.24).

The intramolecular version of this reaction is even more spectacular because it provides, in a two-pot–three-step process from benzaldehyde, the bicyclo[3.1.0]hexane derivative shown in Figure 12.25.

We have, in the meantime, tried to avoid the metalation of the solvent by the benzyllithium in order to get a higher yield of 1,2,2-trimethyl-1-phenylcyclopentane. On using 2-methylseleno-2-phenyl-6-heptene as a model, we have found [26] that the cleavage of the C—Se bond can be carried out in pentane if *sec*- or *tert*-butyllithium is employed and that the resulting benzyllithium smoothly adds at room temperature to the built-in carbon–carbon double bond (Figure 12.26).

Under these conditions, 6-methyl-2-methylseleno-2-phenyl-6-heptene smoothly cyclizes [26] to 1,2,2-trimethyl-1-phenylcyclopentane, which is produced in very good yield. Its *p*-methyl-phenyl analogue provides cuparene very efficiently after hydrolysis (Figure 12.27). Furthermore, carbonation of the resulting organometallic leads to only one stereoisomer of 1,2-dimethyl-2-carboxymethyl-1-phenylcyclopentane, clearly indicating that the addition reaction proceeds [26] with high stereoselectivity.

In conclusion, we have shown that the synthesis of cycloalkanes bearing quaternary centers in the vicinity of one another is feasible by two different routes, employing either the homologation of substituted cycloalkanones or cyclization of benzyllithiums containing a carbon–carbon double bond. Both the carbonyl group of the former compounds and the aryl group on the latter derivatives can be further transformed to various functional groups. Although the former approach is quite general and allows the synthesis of cycloalkanones of various sizes (4- to 16-membered and even larger rings), the second method is currently limited to the formation of three-, five- and six-membered rings.

References

1. Martin, S. F. *Tetrahedron*, **1980**, *36*, 419.
2. Krief, A. in *The Chemistry of Selenium and Tellurium Compounds*, Patai, S. Rappoport, Z., eds., vol. 2, John Wiley, Chichester, 1987, p. 675.

3. Clarembeau, M.; Cravador, A.; Dumont, W.; Hevesi, L.; Krief, A.; Lucchetti J.; Van Ende, D. *Tetrahedron*, **1985**, *41*, 4793.

4. Krief, A.; Dumont, W.; Clarembeau, M.; Bernard, G.; Badaoui, E. *Tetrahedron*, **1989**, *45*, 2005.

5. Krief, A.; Dumont W.; Clarembeau M.; Badaoui, E. *Tetrahedron*, **1989**, *45*, 2023.

6. Van Ende, D.; Dumont W.; Krief, A. *Angew. Chem., Int. Ed.*, **1975**, *14*, 700.

7. Labar, D.; Krief, A.; *J. Chem. Soc., Chem. Commun.*, **1982**, 564.

8. (a) Krief, A.; Dumont, W.; Van Ende, D.; Halazy, S.; Labar, D.; Laboureur J. L.; Lee, Q. T. *Heterocycles*, **1989** *28*, 1203. (b) Laboureur, J. L.; Dumont W.; Krief, A. *Tetrahedron Lett.*, **1984**, *25*, 4569.

9. Labar, D.; Laboureur, J. L.; Krief, A. *Tetrahedron Lett.*, **1982**, *23*, 983.

10. Halazy, S.; Zutterman, F.; Krief, A. *Tetrahedron Lett.*, **1982**, *23*, 4385.

11. Halazy, S.; Krief, A. *J. Chem. Soc., Chem. Commun.*, **1982**, 1200.

12. Laboureur, J.; Krief, A. *Tetrahedron Lett.*, **1984**, *25*, 2713.

13. Krief, A.; Laboureur, J. L. *Tetrahedron Lett.*, **1987**, *28*, 1545.

14. Krief, A.; Laboureur, J. L. *Tetrahedron Lett.*, **1987**, *28*, 1549.

15. Krief, A.; Laboureur, J. L.; Evrard, G.; Norberg, B.; Guittet, E. *Tetrahedron Lett.*, **1989**, *30*, 575.

16. Schmit, C.; Sahraoui-Taleb, S.; Differding, E.; Dehasse-De Lombaert, C. G.; Ghosez, L. *Tetrahedron Lett.*, **1984**, 5043.

17. Paquette, L. A.; Peterson, J. R.; Ross, R. J. *J. Org. Chem.*, **1985**, *50*, 5200.

18. Krief, A.; Laboureur, J. L. *J. Chem. Soc., Chem. Comm.*, **1986**, 702.

19. Krief, A. in *Topics in Current Chemistry*, De Mejiere, A., ed., vol. 135, Springer-Verlag, Heidelberg, 1987.

20. Fitjer, L.; Scheuermann H. H.; Wehle, D. *Tetrahedron Lett.*, **1984**, *25*, 2329.

21. Clarembeau M.; Krief, A. *Tetrahedron Lett.*, **1986**, *27*, 1719.

22. Krief, A.; Clarembeau, M.; Barbeaux, P. *J. Chem. Soc., Chem. Commun.*, **1986**, 457.

23. Krief A.; Barbeaux, P. *J. Chem. Soc., Chem. Commun.*, **1987**, *16*, 1214.

24. Hobe, M., unpublished results from our laboratory.

25. Set, L.; Cheschire D. R.; Clive, D. L. J. *J. Chem. Soc., Chem. Commun.*, **1985**, 1205.

26. Barbeaux, P., unpublished results from our laboratory.

A Review of Multidentate Lewis Acid Chemistry

Henry G. Kuivila

Department of Chemistry
State University of New York at Albany
Albany, New York 12222

13.1. Introduction

Multidentate Lewis bases, among which ethylenediamine is the prototype, have long been familiar tools in the inventory of the coordination chemist. In the course of the past two decades acyclic and, particularly, cyclic polyethers have emerged as important and rapidly expanding classes of donor species. This development began with the discovery of the cyclic polyethers by Pedersen in 1967 [1] and has achieved a high degree of sophistication in a little over two decades, especially as a result of the work of Cram [2] and Lehn [3]. The counterpart multidentate Lewis acids have received very little attention until recently, and the development of their chemistry can be said to be at a comparatively primitive stage.

Some of the reasons for the current undeveloped state of the field with respect to the main-group elements are quite evident. In order to construct a multidentate Lewis acid one needs a chain of atoms, such as carbons, to join the acidic centers. This eliminates the alkali metals because they are univalent. In principle, the alkaline earths could be introduced into chains, but no significant synthetic efforts have been reported. The same also holds true for members of group XIII. An important drawback with these groups is the sensitivity of the carbon–metal bonds to oxidation and/or hydrolysis. The aromatic carbon–boron bond can be an exception in being stable to air. The group XIV elements can provide air- and water-stable derivatives that are easily manipulated, although they vary considerably in the reactivity of the carbon–metal bonds. Among other divalent metals, mercury forms easily manipulated and stable carbon derivatives.

When a Lewis base enters into bond formation, whether electrostatic or covalent, little change in the geometry about the donor center is required because the unshared electron pair of the nonbonding orbital projects out from the donor atom. In contrast, when a Lewis acid enters into covalent bond formation, more deep-seated changes are usually observed. If one assumes that donor–acceptor bonds with some covalent character are formed, the hybridization changes sp–sp^2, sp^2–sp^3, sp^3–dsp^3, or sp^3–d^2sp^3 will tend to occur and these will be accompanied by substantial changes in geometry and concomitant changes in conformations. If a donor–dipole interaction occurs, this will also be facilitated by any change in geometry that enables the donor to approach close to the positive end of the acceptor dipole. Conformation changes can be especially dramatic when the metal atom is in a ring. For example, a group XIV metal with sp^3 hybridization in an unstrained ring will have all intraanular C — M — C bonds at the tetrahedral angle. When the coordination number increases to five, with dsp^3 hybridization and the preferred trigonal bipyramidal geometry, the intraanular bonds could be 90°, 120°, or conceivably even 180°. The message is that a model of the uncoordinated Lewis acid will be of limited value in an attempt to analyze the potential ease of forming an unstrained multidentate complex with a donor. In contrast, models have been very useful in analyzing the complexation behavior of multidentate Lewis bases [4].

Another important characteristic of multidentate Lewis acid chemistry stems from the fact that the unshared electrons of the donor lie in discrete orbitals. This results in a tendency toward formation of a defined number of covalent bonds with fixed spatial orientations rather than simple electrostatic interactions with minimal directional requirements. Thus, compounds normally bond to only one Lewis acid through nitrogen and to one or two through oxygen. A halide ion might interact with three Lewis acid centers to form pyramidal bridging units. It will be of interest to learn how spherical donors like halide ions, because of their spherical geometry, actually interact with molecules with four or more Lewis acid centers.

13.2. Acyclic Multidentate Lewis Acid

13.2.1. Examples Involving Boron

The first explicit search for a multidentate effect appears to be that of Shriver and Biallas in 1967 [4]. They prepared 1,2-bis-(difluoroboryl)ethane ($F_2BCH_2CH_2BF_2$, **1**) and examined its behavior with oxygen donors such as ethers, a ketone carbonyl, and an amine oxide, but all gave complexes with one donor molecule per boron. However, they used an ingenious method for obtaining methoxide ion without use of a strong base. Trityl methyl ether was allowed to react with **1** resulting in formation of the trityl cation and methoxide complex **1OCH$_3^-$** as shown in eq. (13.1).

$$Ph_3COCH_3 + \begin{array}{c} \diagdown BF_2 \\ \\ \diagup BF_2 \end{array} \longrightarrow \left[\begin{array}{c} \diagdown BF_2 \\ OCH_3 \\ \diagup BF_2 \end{array} \right]^- + Ph_3C^+ \qquad (13.1)$$

$$\textbf{1} \qquad\qquad \textbf{1OCH}_3{}^-$$

A competition experiment showed that **1** formed a much more stable complex with methoxide than did BF_3. Thus, although a strongly basic donor was used in order to reveal the bidentate potential of **1**, this led to a more stable complex than does BF_3, which is a stronger acid than a single BF_2 unit of **1**.

Later, Shore and co-workers [5] reported the preparation of a compound with geminal borons bridged by hydride as in **2** and tetramethylene diboron hydride complexes bearing the bridging hydride in a seven-membered ring as in **3**. This latter compound is interesting because the entropy decrease needed for formation of the ring is more than compensated for by the driving force for formation of the hydride bridge.

A rigid diborane system designed to provide high selectivity has been prepared by Katz by placing borons in the peri positions of naphthalene, as in **4**, which formed complexes with hydride (**4H**⁻, shown in Figure 13.1), fluoride (**4F**⁻), and hydroxide (**4OH**⁻) [6]. It was shown that the hydride in **4H**⁻ is so tightly bound that it is unreactive towards acetic acid and benzaldehyde. Hence, **4** was given the trivial name "hydride sponge" by analogy with the name "proton sponge" for 1,8-bis-(dimethyl-amino)naphthalene, which is an exceptionally strong Bronsted base.

The affinity for the hydride was also revealed by the ability of **4** to form **4H**⁻ by abstracting hydride from the 1-naphthyldimethylborohydride ion.

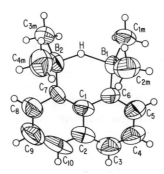

Figure 13.1 ■ ORTEP drawing of **4H**⁻ in crystalline **4** · KH(dioxane)₃, with atoms represented by thermal ellipsoids at 50% of electron density.

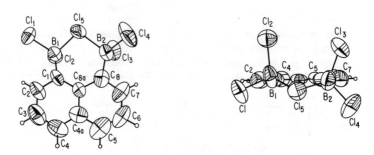

Figure 13.2 ■ Perspective drawings of **5Cl⁻** in crystalline $Ph_2P{=}N{=}PPh_2{}^+C_{10}H_6B_2Cl_6$: (a) top view; (b) view from end looking toward bridged chlorine.

The bis-dichloroboron analog of **4** has also been prepared and has been shown to be an effective agent for coordinating chloride ion, forming **5Cl⁻**, in which the chloride bridge is symmetrical (Figure 13.2). This constitutes the first example of a compound in which two borons are symmetrically bridged by chloride [7].

A novel system in which the bidentate Lewis acid is a species with two different acid centers is provided by the (8-silyl-1-naphthyl)boranes, prepared by the following scheme [8]:

The structure of **6F⁻** shown in Figure 13.3, as determined by X-ray diffractometry, shows that the fluoride forms an unsymmetrical bridge between boron and silicon, as expected. It is less strongly bonded to the silicon than to the boron, as shown by the Si—F bond length of 2.714 Å as compared to the sum of the covalent radii of Si and F of 3.4 Å. The B—F bond length is in the normal range at 1.475 Å.

13.2.2. Examples Involving Tin

Most of the studies on organotins have involved 1,n-distannylalkanes that could also serve as building blocks for the cyclic compounds. 1,1-Bis-distannylmethanes have received the greatest study and they have provided useful information concerning

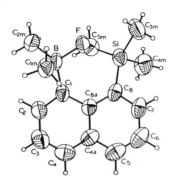

Figure 13.3 ■ ORTEP drawing of **6F⁻** with atoms represented by thermal ellipsoids of 50% electron density.

bidentate complexation with neutral ligands in particular. We have prepared several derivatives with the general structure **7** and studied their complexation with

$$R_2C \Big\langle \begin{matrix} SnMe_xCl_{3-x} \\ \\ SnMe_xCl_{3-x} \end{matrix}$$

7

dimethylsulfoxide (DMSO) [9]. Three different types of structures that crystallized readily from solutions, **8**, **9**, and **10**, are shown in Figures 13.4, 13.5, and 13.6, respectively. In **8**, DMSO bridges the tins symmetrically, and two DMSO molecules bridge in the same way in **9**. On the other hand, when three chlorines are present on each tin, an open structure **10** is isolated with two DMSO molecules on each tin, which assume a distorted octahedral configuration with the methylene group serving as a common apex. Of particular note in this structure is the 130° Sn—C—Sn angle, which is required to accommodate the bulk of the ligands on the two tins. Proton NMR studies in acetone indicate clearly that a fairly stable intermediate coordinated with a single DMSO is a precursor to **9**. Solubility problems frustrated a similar search for precursors to **10**.

Jurkschat and Gielen studied analogs of **8** in which the methyl groups were replaced by phenyls [10]. They observed that the uncomplexed bis-(diphenylchlorostannyl)-methane structure comprises chlorine-bridged units **11**, shown in Figure 13.7, that are aligned in linear polymeric strands joined by tin–chlorine bridges. No structures for methyl analogs of **11** have been reported. The complex of **11** with HMPA ($[(CH_3)_2N]_3P = O$) as the donor has a structure (shown as **12** in Figure 13.8) with a chlorine bridge rather than an oxygen bridge. It seems unlikely that this should be due to any difference between the oxygens of the sulfoxide and phosphine oxide. It is more likely that the large bulk of the phenyl groups compared to methyl on the bis-(halostan-nyl)methane and the three dimethylamino groups on the phosphorus of the donor create a sterically strained environment such that the chlorine-bridged structure is the more stable.

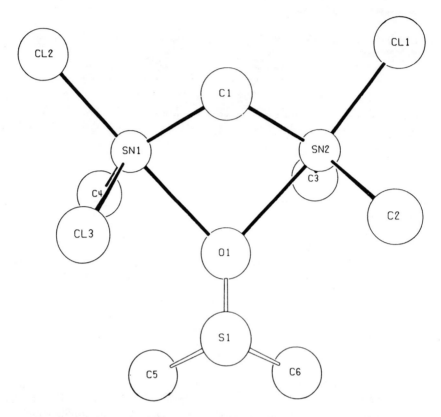

Figure 13.4 ■ Schematic of X-ray structure of **8** [(CH$_3$)$_2$ClSnCH$_2$SnCH$_3$Cl$_2$][DMSO].

Structures such as **8** and **9** on the one hand, and **12** on the other, represent two ways in which bidentate Lewis acids have been shown to interact with donors. The former group represents classical chelate binding to the donor, whereas in the latter the donor bonds to one acceptor center and a ligand on this center forms a bridge to the second acid center. In this case the donor–acceptor bond strength should be enhanced over that in a simple monodentate Lewis acid because the bridging of a ligand on the first Lewis acid site to the second depletes electron density at the first site, thus increasing its acidity. It is convenient to use the term *cooperative binding* to refer to the cause of enhanced Lewis acidity in multidentate species when the actual structure of the complex has not been established.

If 1,2-bis(dimethylchlorostannyl)methane is mixed in various proportions with DMSO the only complex which crystallizes out is the 1 : 2 complex with one DMSO coordinated to each tin in the extended structure **13** (Figure 13.9) [11]. With 1,2-bis-(chlorodiphenylstannyl)ethane, however, Jurkschat isolated the first bridged anion complex; this involves chloride bridging between two tins (**14**) (Figure 13.10). Ph$_3$P $=$ N $-$ PPh$_3$$^+$ was the counterion [12]. No bridged complexes have been reported with longer chain ditins.

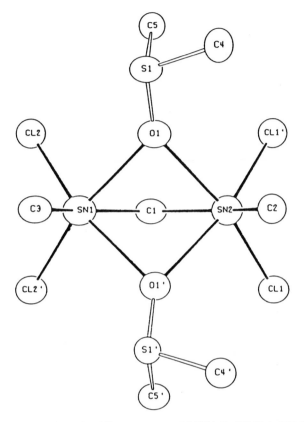

Figure 13.5 ■ Schematic of X-ray structure of **9** $[CH_2(SnCH_3Cl_2)_2][DMSO]_2$.

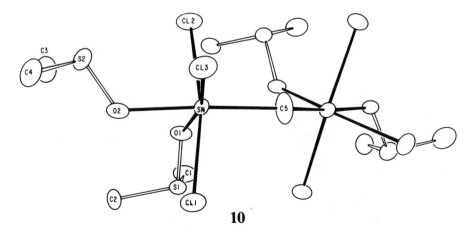

10

Figure 13.6 ■ Schematic of X-ray structure of **10** $[CH_2(SnCl_3)_2][DMSO]_4$.

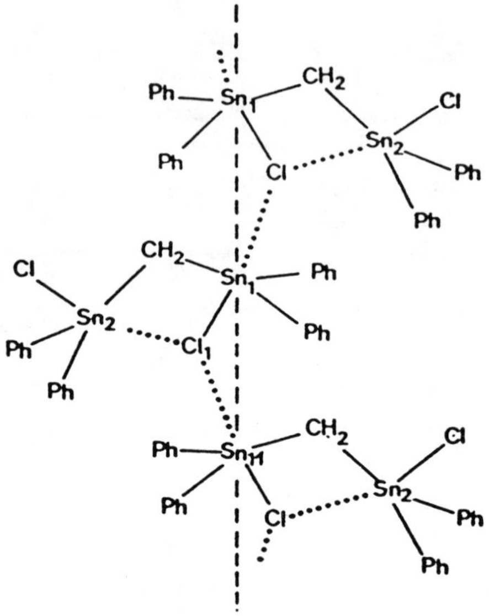

11

Figure 13.7 ■ Schematic of polymeric chain of **11** [$CH_2(SnPh_2Cl)_2$].

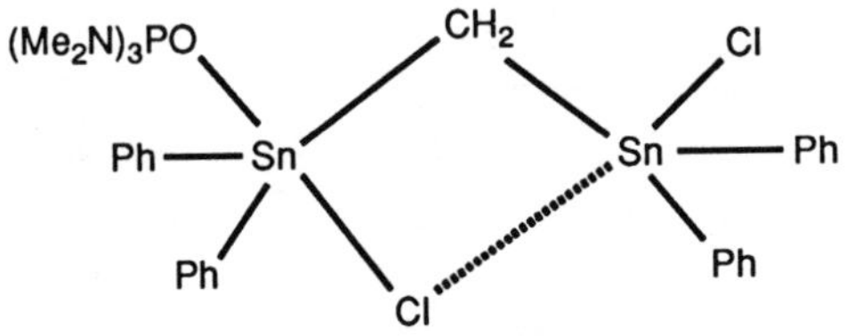

12

Figure 13.8 ■ Schematic drawing of **12** [$(Ph_2ClSn)_2CH_2$][HMPA].

13.2.3. Examples Involving Mercury

Wuest has found that organomercurials show some very interesting complexation behavior [13]. For example, phenylmercuric chloride (**15**) reacts with tetraphenylphosphonium chloride *via* disproportionation [eq. (13.2)]. On the other hand, dichloro-1,2-phenylenedimercury (**16**) forms a 2 : 1 complex (**17**) [eq. (13.3)]. Thus, the isolation of a stable complex with chloride

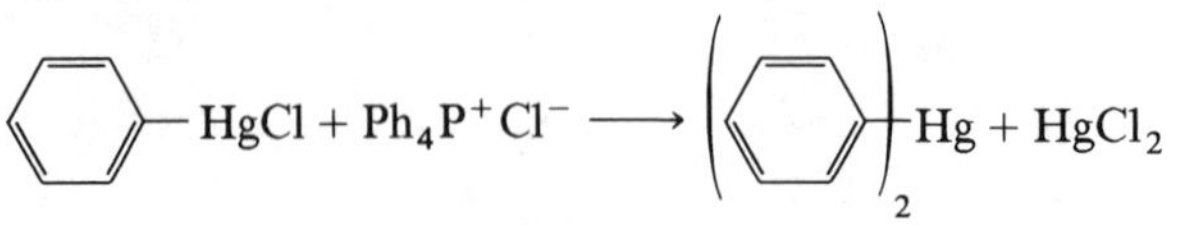

$$(13.2)$$

15

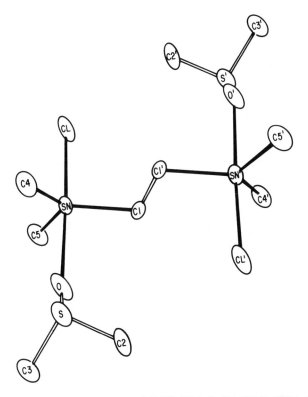

Figure 13.9 ■ Schematic of X-ray structure of **13** $[Cl(CH_3)_2SnCH_2CH_2Sn(CH_3)_2Cl][DMSO]_2$.

$$2 \quad \begin{array}{c} \text{HgCl} \\ \text{HgCl} \end{array} + Ph_4P^+Cl^- \longrightarrow \left[\left(\begin{array}{c} \text{HgCl} \\ \text{HgCl} \end{array} \right)_2 Cl \right]^- \qquad (13.3)$$

16   **17**

requires a bidentate Lewis acid. Presumably, complex **17** is too stable to react to form the disproportionation products.

The structure of **17** is complex. ^{199}Hg NMR studies indicate that a stable $1:1$ complex is formed with chloride in solution. Its structure is very probably **18** as shown in Figure 13.11. The solid $2:1$ complex shows interaction of this unit with a phenylenedimercury unit in two similar nonequivalent structures, one of which is **19** (Figure 13.12). When chloride and bromide ions are allowed to compete for **16**, bromide is incorporated into a $2:1$ complex, thus revealing high selectivity of **16** among halide ions.

The bis-trifluoroacetate analog of **8** facilitates reduction of thiones and leads to a different course of reaction from that observed when a monodentate mercurial is used,

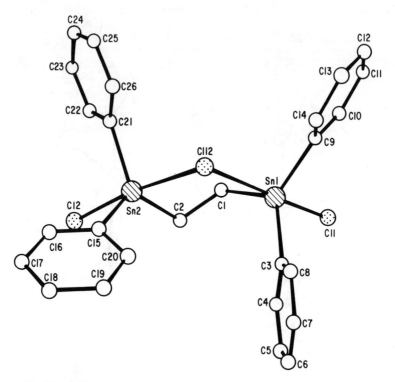

Figure 13.10 ■ Schematic of X-ray structure of the chloride complex **14** $[(\text{ClPh}_2\text{SnCH}_2\text{CH}_2\text{SnPh}_2\text{Cl})\text{Cl}]^- [\text{Ph}_2\text{P}=\text{N}=\text{PPh}_2]^+$.

eqs. (13.4) and (13.5) [14]. A dimeric intermediate which is particularly susceptible to reduction was proposed for the reaction facilitated by **16**.

$$\text{⟨⟩—HgOTf} + \text{Ar}_2\text{C}=\text{S} \xrightarrow{\text{(Red)}} \underset{(\text{major})}{(\text{Ar}_2\text{CH})_2\text{S}} \qquad (13.4)$$

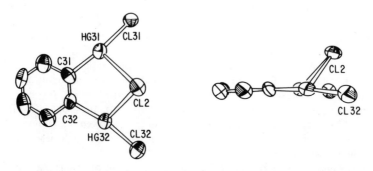

Figure 13.11 ■ Proposed structure for the 1:1 complex **18** $[(\text{C}_6\text{H}_4\text{Hg}_2\text{Cl}_2)\text{Cl}]^-$.

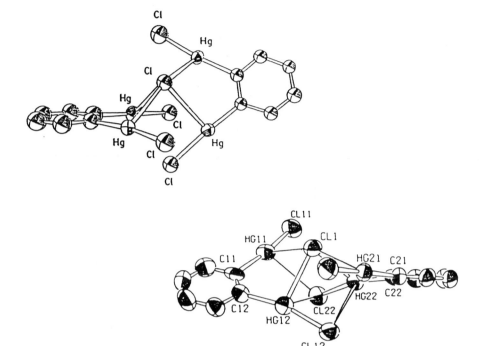

Figure 13.12 ■ (a) ORTEP drawing of one perspective of the complex **19** $[(C_6H_4Hg_2Cl_2)_2Cl]^-$, with ellipsoids representing 50% probability. (b) Simplified ORTEP drawing with weaker Hg—Cl; interactions not shown.

$$\underset{\text{HgOTf}}{\overset{\text{HgOTf}}{\bigcirc}} + Ar_2C{=}S \xrightarrow{\text{(Red)}} \underset{(\text{major})}{Ar_2CH_2} \qquad (13.5)$$

$$Red = \underset{}{\bigcirc}NCH_2Ph$$

Carbonyl coordination by **16** was examined using dimethylformamide as the donor [15]. A stable 1 : 1 complex was isolated; its bidentate structure is shown as **20a** and **20b** in Figure 13.13. The oxygen forms a symmetrical bridge between the mercury atoms. One novel characteristic of mercury coordination chemistry is shown in **20b**. The aryl carbon–mercury–chloride bond angle is greater than 170°, giving the mercury a T-shaped configuration. Detailed examination of the unit cell in the structure of **20**, as well as that of **11**, shows chlorine–mercury interactions between the individual complexes.

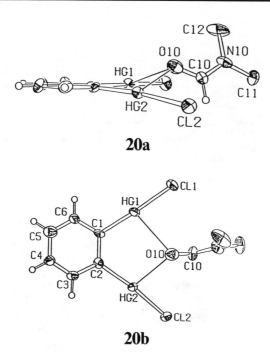

20a

20b

Figure 13.13 ■ Two views of the complex **20** [C$_6$H$_4$(HgCl)$_2$][DMSO], with ellipsoids at 50% probability. Note nearly linear aryl C—Hg—Cl arrangement.

13.3. Cyclic Multidentate Lewis Acids

The presence of two or more Lewis acid centers on a cyclic structure would be expected to facilitate cooperative binding to donors because of the favorable entropy effect. Examples to be considered in this section show that this is not always realized.

13.3.1. Examples Involving Tin

We have initiated studies on cyclic multitins (**21**) with initial emphasis on simple examples bearing two tins in the ring. The general synthetic method is shown in eq. (13.6) [14, 16]. With $n = 4$ and $m = 1$, 2, 3, or 4, the products were obtained in yields of 25–40% without using high dilution methods.

$$BrMg(CH_2)_nMgBr + Cl(CH_3)_2Sn(CH_2)_mSn(CH_3)_2Cl \longrightarrow$$

$$
(CH_3)_2Sn \underset{\diagdown (CH_2)_m \diagup}{\overset{\diagup (CH_2)_n \diagdown}{}} Sn(CH_3)_2
$$

21   (13.6)

Each ditin could be separated from oligomers containing four or more tins by distillation at low pressure. The compound with $n = m = 3$ was also prepared, albeit in yields of only 8–10%. Replacement of methyl groups by chlorines could be effected by reaction with tin tetrachloride or mercuric chloride, eq. (13.7) [16].

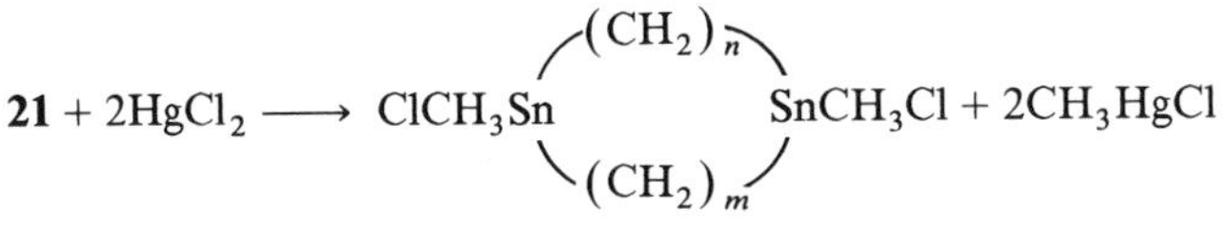

$$21 + 2HgCl_2 \longrightarrow ClCH_3Sn \overset{(CH_2)_n}{\underset{(CH_2)_m}{\diagup\diagdown}} SnCH_3Cl + 2CH_3HgCl \qquad (13.7)$$

21Cl

The reaction is straightforward with larger rings, but complications arise with the smaller ones. For example, cleavage of a seven-membered ring in **22** occurred according to eq. (13.8). Little, if any, cleavage of the tetramethylene bridge was observed.

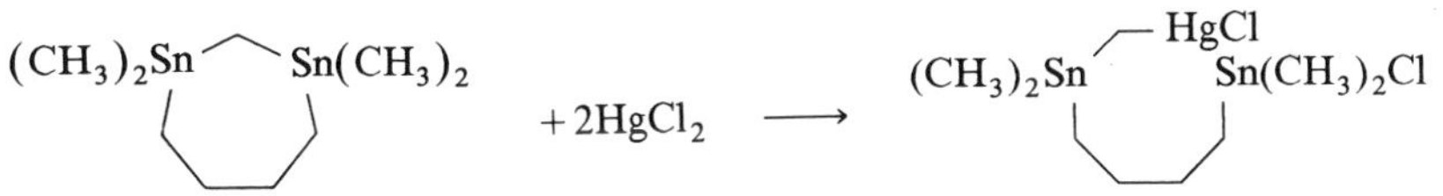

$$(CH_3)_2Sn \frown Sn(CH_3)_2 + 2HgCl_2 \longrightarrow (CH_3)_2Sn \overset{-HgCl}{\frown} Sn(CH_3)_2Cl$$

22   **23**

$$(13.8)$$

Ring cleavage may also compete with replacement of the second methyl on each tin; thus **24** yields both cyclic **25** and acyclic **26** [eq. (13.9)]. The structure of **25** was easily assigned on the basis of $^{119}Sn-^{13}C$ coupling constants.

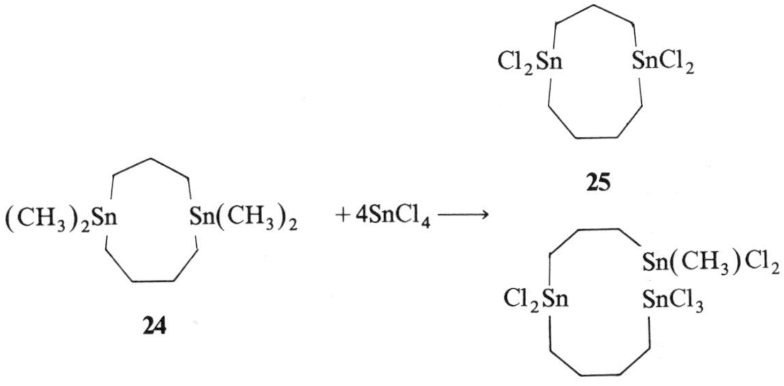

$$Cl_2Sn \frown SnCl_2$$

25 (13.9)

$$(CH_3)_2Sn \frown Sn(CH_3)_2 + 4SnCl_4 \longrightarrow$$

24

$$Cl_2Sn \overset{Sn(CH_3)Cl_2}{\frown} SnCl_3$$

26

We have studied the complexation of DMSO with acyclic ditins and analogous cyclics using ^{119}Sn NMR as a probe. Results are presented in Table 13.1. Trimethyltin chloride (**27Cl**) can be taken as one reference and 1,6-bis-(chlorodimethylstannyl)hexane (**28Cl**) as a ditin that would not be expected to display cooperative binding. The chemical shifts for these two compounds are very similar and move upfield to the same degree upon addition of DMSO. 2,2-Bis(chlorodimethylstannyl)propane (**29Cl**) shows a

Table 13.1 ▪ Effects of DMSO on ^{119}Sn Chemical Shifts

Compound	Moles of DMSO per mole of organotin in CDCl$_3$			
	0	1.0	2.0	5.0
Me$_3$SnCl **27Cl**	171.9	141	117.2	74.7
ClMe$_2$Sn(CH$_2$)$_6$SnMe$_2$Cl **28Cl**	167.9	144.9	124.5	74.6
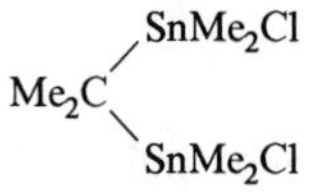 Me$_2$C(SnMe$_2$Cl)(SnMe$_2$Cl) **29Cl**	166.6	93.1	71.6	59.5
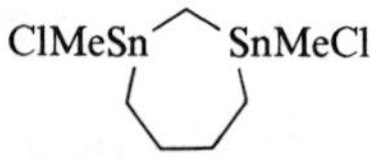 ClMeSn⌒SnMeCl (ring) **30Cl**	166	135	115	83.4
ClMeSn(CH$_2$)$_2$SnMeCl **31Cl**	159.6	101.3	74.6	52.9
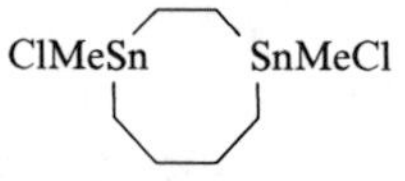 ClMeSn⌒SnMeCl (ring) **32Cl**	163.9	119.4	99.5	75.8
ClMeSnCH$_2$CH$_2$CH$_2$SnMeCl **33Cl**	162.8	133.1	110.7	72.4
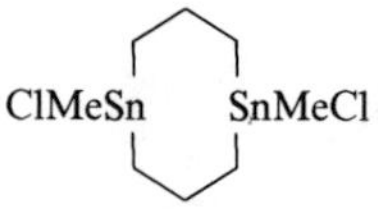 ClMeSn⌒SnMeCl (ring) **34Cl**	159.2	131.8	113.4	82.5
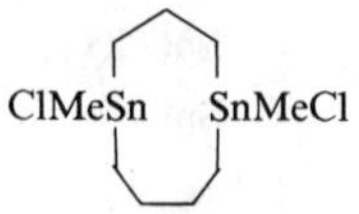 ClMeSn⌒SnMeCl (ring) **35Cl**	149.8	76.1	49.1	31.3

normal chemical shift, but addition of 1 mol of DMSO causes an upfield shift of 73.5 ppm as compared to 29.1 and 23.0 ppm for **27Cl** and **28Cl**, respectively. This indicates formation of a 1 : 1 complex that is quite stable, undoubtedly due to the bridging that is seen in the solid state of the analog **9**. If one joins the two tins of **27Cl** by a 4-carbon chain to form **30Cl**, the cooperative effect of the two tins is largely lost, as indicated by the upfield shift of only 31 ppm upon addition of 1 mol of DMSO.

In 1,2-bis-(chlorodimethylstannyl)ethane **31Cl** the chemical shift is at a higher field than those considered thus far, suggesting some chlorine–tin interaction. Upon addition of 1 mol of DMSO, an upfield shift of 58.3 ppm results. This is large enough to indicate cooperative binding analogous to that in the chloride complex **13**. Upon joining the two tins by a 4-carbon chain in **32Cl**, a slight (4.3 ppm) downfield shift occurs. Addition of 1 mol of DMSO causes an upfield shift of 44.5 ppm, suggesting cooperative binding; however, this is less than that for the acyclic **31Cl**. Thus, the entropic benefit gained by including the tins in the ring is more than counterbalanced by enthalpic cost created by joining the two tins with the 4-carbon chain.

When the tin atoms are separated by the 3-carbon chain in **33Cl**, the chemical shift trend is similar to that of **27Cl**, providing very little indication of cooperative binding. Closing these tins into a ring with a 4-carbon chain leads to a slight change in the magnitude of the trend as seen by the parameters for **34Cl**. However, when the tins are joined by two 3-carbon links as in **35Cl**, dramatic changes occur. First, the chemical shift moves upfield to 149.8 ppm, suggesting substantial chlorine–tin interaction in this compound. Support for this is provided by its structure, which was determined by X-ray diffraction to be as shown in **36a** and **36b**. An intramolecular bridge joins the tins of each unit as in **36a** and these are joined by intermolecular chlorine bridges between tins to form a linear polymeric chain, **36b**.

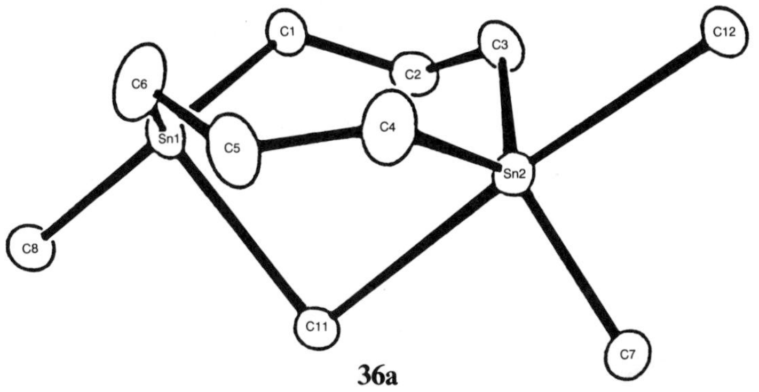

36a

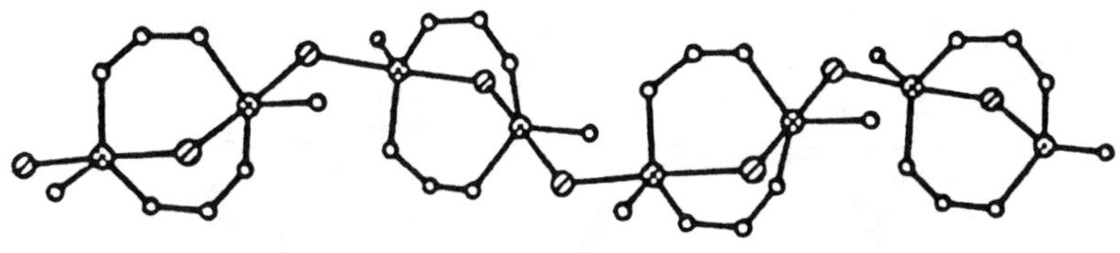

36b

Second, a remarkable difference from its higher homolog also appears when 1 mol of DMSO is added to a solution: The chemical shift moves upfield by 73.9 ppm, and this is carried through even when 5 mol of DMSO is added, whence the shift moves up to 31.3 ppm. This implies a higher degree of coordination than shown by any of the other members of this group, making **35Cl** by far the most powerful Lewis acid. The striking difference in behavior between **34Cl** and **35Cl** is attributable to greater accessibility in **35Cl** to the configurations and conformations most conducive to stability of structures with coordination numbers larger than 4.

Three-carbon chains have been used to construct the cyclic tritin **37**, which was converted to the tris-dichlorotin derivative (**37Cl$_2$**) using the reactions of eqs. (13.10) and (13.11) [17,18]. We were able to isolate the complex of **37Cl$_2$** formed with $Ph_3P{=}N{=}PPh_3{}^+Cl^-$ and to show the anion $C_9H_{18}Cl_7Sn_3{}^-$ to have the structure **38** (Figure 13.14).

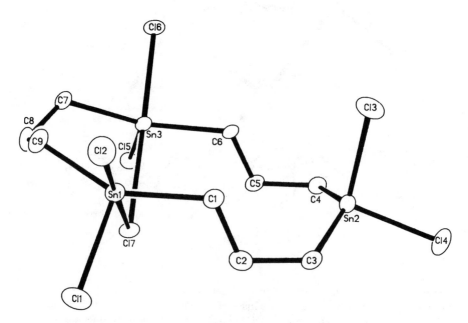

Figure 13.14 ■ ORTEP view of the anion **38** of 1,1,5,5-tetrachloro-1,5-distannacyclododecane in the complex $[C_9H_{18}Cl_7Sn_3]^-[Ph_3P{=}N{=}PPh_3{}^+]$.

It has pseudo mirror symmetry because one Sn—Cl bond length involving the bridging chlorine is 2.906(2) Å and the other is 2.700(2) Å. Thus, the three tins are different from each other, and symmetrical bridging of all three by the chlorine is not realized. The solution NMR spectrum at room temperature shows a single tin signal, indicating rapid exchange of coordinated chlorine.

Newcomb's group has been exploring the utility of macrocyclic multitins as selective hosts for anions. They have prepared ditins **39** with $n = 4, 5, 6, 8, 10, 12$ and tetratins **40** with the same values of n [19].

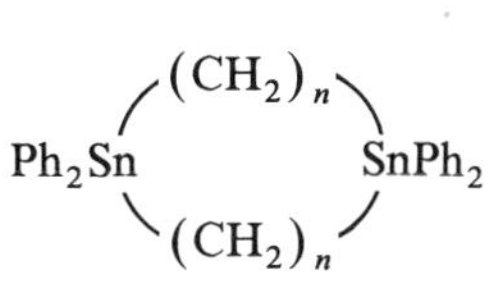

$$Ph_2Sn \overset{\displaystyle /(CH_2)_n\diagdown}{\underset{\diagdown(CH_2)_n/}{}} SnPh_2$$

39

$$\begin{array}{c} \overset{\frown}{}(CH_2)_n - SnPh_2 \overset{\frown}{} \\ Ph_2Sn \qquad\qquad\qquad (CH_2)_n \\ | \qquad\qquad\qquad\qquad | \\ (CH_2)_n \qquad\qquad\qquad SnPh_2 \\ \underset{\smile}{} Ph_2Sn - (CH_2)_n \underset{\smile}{} \end{array}$$

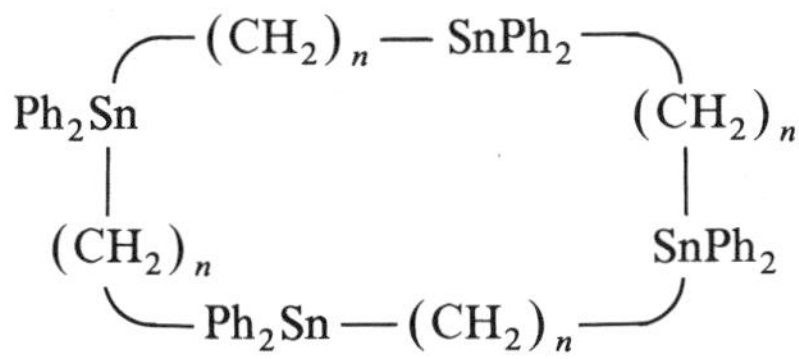

40

These could be readily converted into the monochlorotins or dichlorotins by treatment with HCl [20]. Complexation between chloride and the dichloroditins derived from **39** with $n = 8$, 10, and 12 was studied by ^{119}Sn NMR [19a]. Cooperative binding was observed with the magnitude for $n = 8$ larger than that for $n = 10$ or 12. The equilibrium constants for binding one chloride ion were in the range 700–800 M^{-1} in acetonitrile, up to four times as large as that for dibutyltin dichloride. Constants for addition of a second chloride were of the same magnitude.

Bicyclic compounds **41** were shown to bind chloride when $n = 8$, 10, or 12 with constants in the range 30–70 M^{-1} in chloroform at $-20\,°C$.

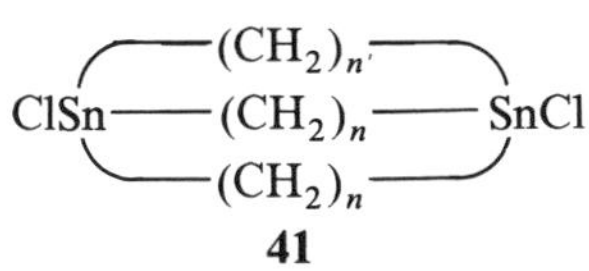

$$ClSn \underset{\smile(CH_2)_n\smile}{\overset{\frown(CH_2)_{n'}\frown}{-(CH_2)_n-}} SnCl$$

41

The cavity is large enough in these to accommodate the chloride. The compound with $n = 6$ is remarkable because it displays no significant Lewis acidity toward chloride. On the other hand this molecule binds fluoride ion with an equilibrium constant of the order of 100 M^{-1}, a value similar to that for a linear bis-monochlorotin [21]. This constitutes evidence that the bicyclic molecule complexes the fluoride by encapsulation.

13.3.2. Examples Involving Mercury

If the dichloro-1,2-phenylenedimercury **16** is converted into the oxide by treatment with base and the oxide treated with a carboxylic acid, the carboxylates are formed. With perfluorosuccinic acid, the product is an insoluble polymeric species. In contrast, the perfluoroglutarate yields a crystalline complex when prepared in THF [22]. This contains a 2:1 complex (**42**) in which one THF molecule is bonded to the mercury atoms of each 1,2-phenylenemercury unit; one is perpendicular to and above the plane of the macrocycle and the other is perpendicular to and below, as seen in the structures shown as **42a** and **42b** in Figure 13.15.

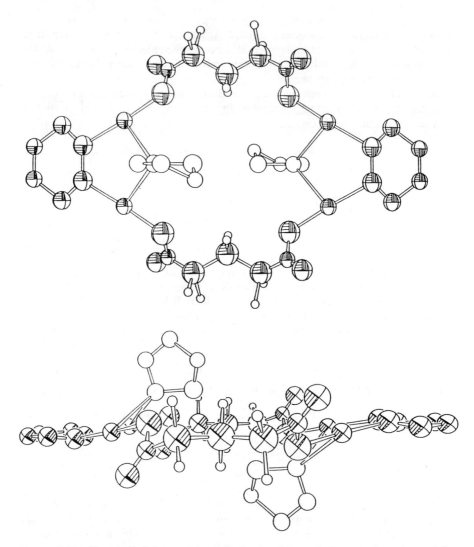

Figure 13.15 ▪ Two views of the structure of **42**, showing the overall planar configuration and the location of the THF molecules above and below the plane.

The cavity in the macrocycle is about 12 Å long by 7 Å wide and should be able to accommodate selectively a molecule with two appropriately disposed basic sites, as the authors suggest.

The difference between the perfluorosuccinate and the perfluoroglutarate is of interest. Structure **42** shows that the macrocyclic ring is essentially planar and the carbon chain is close to the extended staggered conformation. With perfluorosuccinate, the preferred antiperiplanar conformation of the chain would have to be abandoned in order for two succinates to join two 1,2-phenylenedimercury units; therefore, a polymeric structure proves to be the more stable.

13.4. Dynamic Aspects

13.4.1. Lewis Acids as Ionophores

Little attention has been paid to the potential for using multidentate Lewis acids as ionophores or, conversely, using ionophoric behavior as a test for complexation. One example has been reported by Jung and Xia [23]. They prepared the trisiladodecane **43** and conducted liquid membrane experiments using methylene chloride as the organic phase and tested for transport of halide from saturated solutions of tetramethylammonium halides into water. They observed transport of chloride and bromide, but not of iodide. This result is highly remarkable because tetraalkylsilicon should display very little Lewis acidity. The multidentate structure is necessary because tetramethylsilane does not promote transport.

43   44

We have carried out similar preliminary experiments using bis-(chlorodiphenyl)stannane (**44**) as the ionophore [24]. Halides and several anions were included in the study, but only nitrate and perchlorate showed transport. This was attributed to very slow transport of strongly bound anions such as the halides. These results along with those for the trisilane suggest that transport studies may serve as sensitive probes for weak complexation.

13.4.2. Lewis Acids as Catalysts

Cooperative binding would be expected to enhance the effectiveness of a Lewis acid as a catalyst. In order to test this we have begun studies on the catalytic efficiency of the bidentate organotin Lewis acids by using the acid-catalyzed Diels–Alder reaction between methyl vinyl ketone and isoprene as a probe [16]. The effectiveness changes with the nature of the electronegative ligand in the order Cl < mesylate < Cl_2 < triflate, with the chlorides being very poor, and the triflates being very good, catalysts. A comparison of the effects of acyclic and cyclic triflates on the reaction is shown in Table 13.2. The numbers in the second column represent the times required for 20% of adduct formation and those in the last column provide an approximation of the relative rates at which this extent of reaction is achieved in the presence of the catalysts.

Trimethyltin triflate **27TfO** is a modest catalyst and the bis-stannylmethane **29TfO** is quite effective. When the tins are separated by two carbons in **31TfO** the effectiveness goes down, but increases when an additional methylene group separates the tins in **43TfO**. Further lengthening of the chain as in **44TfO** results in a decrease in the

Table 13.2 ■ Catalysis of the Isoprene / Methyl Vinyl Ketone Reaction
by Triflates at 0 °C in CH_2Cl_2 with
MVK = 1.5 M, isoprene = 1.7 M, catalyst = 0.03 M

Catalyst	t_{20}	Relative "rate"
None	194	1
Me₃SnTfO **27Cl**	7.7	25
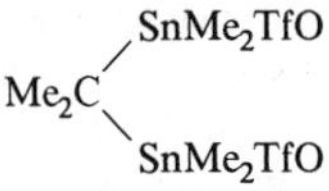 Me₂C(SnMe₂TfO)(SnMe₂TfO) **29TfO**	0.12	1620
TfOMe₂Sn(CH₂)₂SnMe₂TfO **31TfO**	3.2	60
TfOMe₂Sn(CH₂)₃SnMe₂TfO **43TfO**	0.56	346
TfOMe₂Sn(CH₂)₅SnMe₂TfO **44TfO**	4.3	45
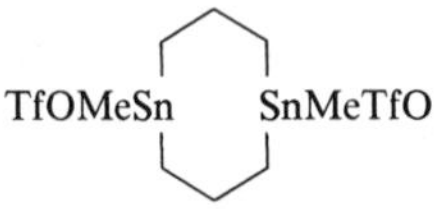 TfOMeSn ... SnMeTfO **34TfO**	0.04	4850
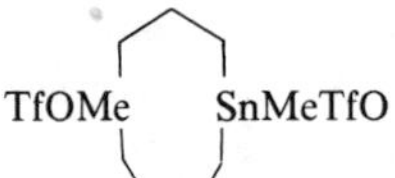 TfOMe ... SnMeTfO **35TfO**	2.9	67
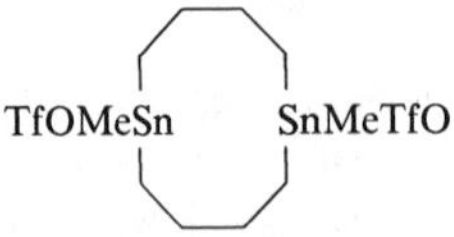 TfOMeSn ... SnMeTfO **45TfO**	7.0	28

catalytic efficiency. The most effective catalyst is the distannacyclooctane **34TfO**. Enlarging the ring by lengthening one chain segment by one methylene to form **35TfO** or each chain segment by a methylene to form **45TfO** brings about a substantial decrease in catalytic efficiency. Thus, the most effective catalyst is the triflate analog of the chloride that proved to be the most effective of the chlorides in coordinating to DMSO (Table 13.1).

Acknowledgments. It is a pleasure to acknowledge contributions of my collaborators K. Swami, D. Farah, R. Krishnamurti, K. Jurkschat, and C. Vu and to acknowledge the X-ray structural studies by J. A. Zubieta and his collaborators, J. Hyde, J. Hutchinson, N. Shaik, and C. Liu. We are grateful for support by the National Science Foundation and the donors of the Petroleum Research Fund administered by the American Chemical Society.

References

1. Pedersen, C. J. *J. Am. Chem. Soc.*, **1967**, *89*, 2495, 7017.
2. Cram, D. J. *Science*, **1988**, *240*, 760.
3. Lehn, J. M. *Angew. Chem., Int. Ed. Engl.*, **1988**, *27*, 89.
4. Schriver, D. F.; Biallas, M. J. *J. Am. Chem. Soc.*, **1967**, *89*, 1078.
5. Saturnino, D. J.; Yamauchi, M.; Clayton, W. R.; Nelson, R. W.; Shore, S. G. *J. Am. Chem. Soc.*, **1975**, *97*, 6063.
6. (a) Katz, H. E. *J. Am. Chem. Soc.*, **1985**, *107*, 1420. (b) Katz, H. E. *J. Org. Chem.*, **1985**, *50*, 5027.
7. Katz, H. E. *Organometallics*, **1987**, *6*, 1134.
8. (a) Katz, H. E. *Organometallics*, **1986**, *5*, 2308. (b) Katz, H. E. *J. Am. Chem. Soc.*, **1986**, *108*, 7640.
9. (a) Hyde, J. R.; Karol, T. J.; Hutchinson, J. P.; Kuivila, H. G.; Zubieta, J. A. *Organometallics*, **1982**, *1*, 404. (b) Karol, T. J.; Hutchinson, J. P.; Hyde, J. R.; Kuivila, H. G.; Zubieta, J. A. *Organometallics*, **1983**, *2*, 106.
10. (a) Jurkschat, K.; Gielen, M. *Bull. Soc. Chim. Belg.*, **1982**, *91*, 803. (b) Gielen, M.; Jurkschat, K. *Bull. Soc. Chim. Belg.*, **1982**, *93*, 153. (c) Gielen, M. Jurkschat, K.; Meunier-Piret, J.; van Meerssche, M. *Bull. Soc. Chim. Belg.*, **1984**, *93*, 379. (d) Meunier-Piret, J.; van Meerssche, M.; Jurkschat, K.; Gielen, M. *J. Organomet. Chem.*, **1985**, *288*, 139.
11. Swami, K. Ph.D. dissertation, State University of New York at Albany, 1984.
12. Jurkschat, K., private communication.
13. (a) Wuest, J. D.; Zacharie, B. *Organometallics*, **1985**, *4*, 410. (b) Beauchamp, A. L.; Olivier, M. J.; Wuest, J. D.; Zacharie, B. *J. Am. Chem. Soc.*, **1986**, *108*, 73.
14. Wuest, J. D.; Zacharie, B. *J. Am. Chem. Soc.*, **1985**, *107*, 6121.
15. Beauchamp, A. L.; Olivier, M. J.; Wuest, J. D.; Zacharie, B. *Organometallics*, **1987**, *6*, 153.
16. Farah, D. Ph.D. dissertation, State University of New York at Albany, 1987.
17. Jurkschat, K.; Kuivila, H. G.; Liu, S.; Zubieta, J. A. *Organometallics*, **1989**, *8*, 2755.
18. Compound **3** has been previously detected by GC—MS but not isolated: Seetz, J. W. F. L.; Schat, G.; Akkerman, O. S.; Bickelhaupt, F. *J. Am. Chem. Soc.*, **1983**, *105*, 3336.
19. (a) Newcomb, M.; Azuma, Y.; Courtney, A. R. *Organometallics*, **1983**, *2*, 175. (b) Azuma, Y.; Newcomb, M. *Organometallics*, **1984**, *3*, 9.
20. (a) Newcomb, M.; Blanda, M. T.; Azuma, Y.; Delord, T. J. *J. Chem. Soc., Chem. Commun.*, **1984**, 1159. (b) Newcomb, M.; Madonik, A. M.; Blanda, M. T.; Judice, J. D. *Organometallics*, **1987**, *6*, 145. (c) Newcomb, M.; Horner, J. H.; Blanda, M. T. *J. Am. Chem. Soc.*, **1987**, *109*, 7878.
21. Newcomb, M.; Blanda, M. T. *Tetrahedron Lett.*, **1988**, *29*, 4261.
22. Wuest, J. D.; Zacharie, B. *J. Am. Chem. Soc.*, **1987**, *109*, 4714.
23. Jung, M. E.; Xia, M. *Tetrahedron Lett.*, **1988**, *29*, 297.
24. Vu, C., unpublished observations.

The Interaction of Group IVB (14) Elements with Positive Charge on a Beta Carbon

Joseph B. Lambert, Gen-tai Wang, and Erik C. Chelius

Department of Chemistry
Northwestern University
2145 Sheridan Road
Evanston, Illinois 60208

14.1. Introduction

The ability of group IVB (14) elements [M in eq. (14.1)] to activate the electrofugal departure of groups at a beta position has been known since 1937 [1].

$$Me_3M-CH_2-CH_2-X \longrightarrow Me_3M-CH_2-CH_2^+ + X^-$$
$$\longrightarrow CH_2{=}CH_2 + Me_3M-X \quad (14.1)$$

For the case of silicon, Sommer, Whitmore, and co-workers [2] provided seminal studies in the 1940s, although the mechanism was not well defined. Some 20 years later, Davis and co-workers [3] made similar observations for tin. The beta effect was firmly established in a wider context through studies of protiodemetalation of $ArMR_3$ [4], electronic spectroscopy, and dehydrometalation [5].

In protiodemetalation, the relative rates for $Ar-MMe_3$ (M = Si, Ge, Sn, Pb) were found to be $1:40:4 \times 10^5:2 \times 10^8$ going down the series [4]. For dehydrometalation $(M-C-C-H \rightarrow M-C-C^+)$ of Et_4M, the relative rates were $1:70:10^5:10^7$ [5].

The common thread in all these reactions is the development of positive charge on a carbon beta to the metal in the transition state. The metal exercises a stabilizing influence to an increased extent the larger the metal. Several factors can contribute to this phenomenon.

1. The atoms below carbon are much more electropositive and therefore can stabilize positive charge by induction.
2. Hyperconjugative stabilization increases going down the column, because the polarizability of the $M-C$ bond and the ability of M to stabilize positive charge increase (1).

$$1 \qquad\qquad\qquad\qquad 2$$

3. For similar reasons, anchimeric assistance leading to a three-membered ring of the type **2** increases down the column.

Hyperconjugative stabilization has been termed *vertical* by Traylor because it does not involve movement of the nuclei, and anchimeric assistance has been termed *nonvertical* because the nuclei, of necessity, must move to form the three-membered ring [5]. The relative importance of these mechanisms of stabilization was not clear from the early work. Moreover, the maximum value of the beta effect had never been measured because previous studies had all used conformationally flexible systems. Hyperconjugation and neighboring-group participation are closely tied to conformation. Consequently, we have prepared a series of conformationally biased systems that have enabled us to measure large rate accelerations and to distinguish some of the mechanistic possibilities.

14.2. Silicon

All early work on the beta effect of silicon was carried out on open-chain systems, for example, $Me_3SiCH_2CH_2X$ [6]. These systems not only demonstrated a significant rate enhancement (up to 10^6 or 10^7 for SiCCX over HCCX) but also enabled the favored stereochemistry between Si and X in the $Si-C-C-X$ fragment to be determined. Placement of substituents on the intervening carbon atoms created stereochemistry that could be followed from starting material to product alkene [7]. In this way, the stereochemistry between Si and X in the transition state was shown to be antiperiplanar, thus disproving, for example, a synperiplanar elimination mechanism. The antiperiplanar arrangement is optimal for hyperconjugative overlap (1) but also is optimal for neighboring-group participation leading to **2**.

Study of conformationally mobile open-chain systems omitted a number of important considerations. (1) The maximum kinetic acceleration could not be measured. In freely rotating systems, reaction may occur most readily from a different conformation from that favored at equilibrium. In simple ethanes with polar substituents, gauche conformations are more the rule than anti conformations. The situation for silyl systems is not specifically known, but, if a conformational equilibration must precede reaction, Curtin–Hammett considerations become important. Only by constraining a system to the antiperiplanar conformation can the full kinetic acceleration be measured with certainty. (2) The beta effect may be exhibited in geometries other than the antiperiplanar. If nonvertical participation is the exclusive mechanism, alternative geometries such as gauche or syn should exclude participation. Hyperconjugation on the other hand falls off slowly with movement away from the antiperiplanar geometry (as the square of the cosine of the dihedral angle), so that some participation is still possible even in a gauche conformation, none in a perpendicular arrangement of Si — C — C — X, and strong participation in the eclipsed syn geometry.

In order to measure the maximum beta acceleration of silicon and to obtain mechanistic insight from alternative geometries, we have prepared a series of conformationally constrained systems containing silicon beta to a leaving group. We carried out these studies in solvents of low nucleophilicity, such as trifluoroethanol, to promote the k_c reaction (E1) that leads to the carbocation intermediate ($Si—C—C—X \rightarrow Si—C—C^+$). There has been some evidence for a nucleophilic component (possibly attack at either Si or X) in solvents of higher nucleophilicity [8]. Thus, mechanistic tests such as variation of solvent nucleophilicity are obligatory.

These studies focused on systems on the types **3** and **4** (M = Si) [9].

3   **4**

The cyclohexane structure restricts rotational freedom of the Si — C — C — X system. When R = H, there still are two conformations available to **3** and **4**, in which the leaving group is either axial or equatorial. In either conformation of the cis form **4** (R = H), the 60° dihedral angle between Si and X is poorly disposed for hyperconjugation or anchimeric assistance. In the trans form **3** (R = H), the reactive diaxial form shown (antiperiplanar Si — C — C — X geometry) interconverts with a poorly reactive 60° conformation (diequatorial). In this case the reactive conformation is clearly the thermodynamically less favored, by 2.0–2.5 kcal mol^{-1} according to A-value analysis. As with the open chain systems, reaction will take place from the most reactive conformation.

Solvolysis of **3** and **4** with R = H was carried out with trifluoroacetate as the leaving group X [9]. For the sake of comparison with a system lacking silicon, parallel measurements were made on cyclohexyl trifluoroacetate. Aqueous mixtures of trifluoroethanol and ethanol provided sufficient variation of solvent ionizing power and nucleophilicity to determine whether the reaction was k_c or k_s [10]. The unsilylated cyclohexyl system solvolyzed by a k_s mechanism, in which solvent serves as a nucleophile to displace the leaving group. Introduction of silicon at either the 60° or the 180° dihedral angle (**4** and **3**) changed the mechanism to k_c. These experiments confirm that the transition state leads to a carbocation. For standard conditions of 97% trifluoroethanol at 25 °C, the cis substrate **4** solvolyzed about 10^4 times faster than cyclohexyl and the trans substrate **3** solvolyzed about 10^9 times faster than cyclohexyl [9].

These accelerations showed for the first time that the beta effect can operate even at a dihedral angle as poor as 60°. Such a result is consonant with a hyperconjugative mechanism of stabilization, which falls off slowly as the dihedral angle deviates from

180°. The kinetic acceleration for the trans form was about a hundred times faster than any measured in an open-chain system. Ring constraint therefore favors silicon participation, even though the reactive diaxial form is not the most stable form.

These results encouraged us to synthesize the fully constrained systems, 3 and 4, in which R = tert-butyl [9]. The presence of this large group biases the conformational preference by at least 5 kcal mol^{-1} in favor of the axial-X/equatorial-Si conformer for 4 and the diaxial conformer for 3. Thus, the leaving group is constrained to be axial in both cases and the favored trans conformer is also the most reactive arrangement. For the cis case incorporation of the tert-butyl group had little or no effect on the rate of solvolysis under the standard conditions: The acceleration was still about 10^4 compared with cyclohexyl trifluoroacetate [9]. The biased trans form (R = tert-butyl), however, solvolyzed several hundred times faster than the unbiased form (R = H), so that the total acceleration was about 10^{12} compared with cyclohexyl. This value is one of the largest in solvolysis chemistry, comparable to anchimeric acceleration by the double bond in norborn-2-en-7-yl tosylate.

The additional factor of several hundred is too large to explain in terms of conformational proportions alone. If for 3 (R = H) the diequatorial form is about 50 times more populous than the diaxial form, then only a factor of 50 is gained by taking populations into account. It is possible that these systems comprise an exception to the Curtin–Hammett principle, which requires that the conformational interconversion be faster than the rate of reaction of the separate conformers. If the reaction of the diaxial form is faster than the conformational equilibration, then when R = H the rate of the conformational process contributes to the observed rate. Simply put, the favored diequatorial form may not be able to feed the reactive diaxial form fast enough, so that the overall rate for R = H is less than maximal.

If hyperconjugation provides the major mechanism of stabilization, then the process is mathematically analogous to H/D secondary kinetic isotope effects and to the vicinal (three-bond) coupling constant in NMR spectroscopy. The mathematical function that best describes an interaction that is maximal at 0 and 180° and minimal at 90° is the square of the cosine of the dihedral angle, expressed as eq. (14.2) by analogy with the isotope effect [11].

$$\log(k_{Si,\theta}/k_H) = \cos^2\theta \left[\log(k_{Si,\,max}/k_H)\right]^V + \log(k_{Si}/k_H)^I \qquad (14.2)$$

In eq. (14.2) k_{Si} is the rate for the silicon system (3 or 4), k_H is the rate for the nonsilylated cyclohexyl system, θ is the Si — C — C — X dihedral angle, V indicates the vertical or hyperconjugative contribution, and I indicates the inductive contribution that is independent of the dihedral angle. Substitution of the observed values of k and θ yields a hyperconjugative acceleration of about 10^{10} and an inductive acceleration of about 10^2. These two factors combine to form the 10^{12} factor observed for the biased trans system. The same inductive factor should apply to the cis system, so that the 10^4 acceleration observed in that case is composed of about 10^2 induction and 10^2 residual hyperconjugation.

This mathematical model has two drawbacks. First, it assumes the vertical mechanism; the results are consistent with this model but cannot prove it. Second, the simple form of eq. (14.2) assumes that the cosine-squared function is symmetrical about 90°. In fact, synperiplanar overlap is not thought to be as effective as antiperiplanar overlap (0 versus 180°). Both questions would be resolved in large part by experiments on a system with a 0° dihedral angle.

Five-membered rings (**5** and **6**) provide considerably different dihedral angles from those of six-membered rings [12].

Although exact geometries cannot be specified, the Si — C — C — X dihedral angle in the trans form **5** should be less than that in the biased trans six-membered ring **3**, and the dihedral angle for the cis form **6** also should be less than that of the six-membered ring **4**. A decrease from 180° should bring about a slower rate, but a decrease from 60° should bring about a faster rate. In the former case the cosine-squared function is at a maximum at 180° and must decrease with a smaller angle, but in the latter case the function is relatively low for 60° and will increase on approaching the 0° maximum that would occur for a planar cyclopentyl system.

Solvolysis of the trifluoroacetates generally corroborated these expectations and hence substantiated the hyperconjugative model [12]. Under the standard conditions, the trans cyclopentyl trifluoroacetate **5** solvolyzed about 10^7 times faster than cyclopentyl (this point of reference is intended to allow for differences in angle strain), compared with the 10^9 increase for the trans six-membered ring (R = H) with respect to cyclohexyl. On the other hand, the cis cyclopentyl trifluoroacetate **6** was 10^5 times faster than cyclopentyl, compared with 10^4 for the six-membered pair. The trans cyclopentyl system **5** was only 76 times faster than the cis cyclopentyl system **6**, whereas the six-membered trans : cis ratio was about 10^5. These results closely follow the expectations based on the analysis of dihedral angles within a hyperconjugative model. They do not, however, exclude the possibility of nonvertical contributions at 180°.

14.3. Germanium

Despite the long history of the beta effect of silicon, no previous work was reported on an analogous role for germanium in the solvolytic setting. The studies on protiodemetalation [4] and dehydrometalation [5] clearly indicated a major interaction. In order to investigate germanium in a stereochemically defined system and to measure the kinetic acceleration in a k_c reaction, we prepared the germanium analogues of **3** and **4** (M = Ge) [13]. For synthetic simplicity, we used the unbiased version (R = H). In the unbiased silicon systems, substantial but not maximal accelerations were observed. Moreover, in the cis silicon system, it made no difference whether R was H or *tert*-butyl, as there apparently was no conformational change when the system was biased.

The unbiased cis germanium system [4, M = Ge, R = H, X = O(CO)CF$_3$] solvolyzed about 10^5 times faster than cyclohexyl trifluoroacetate under the standard conditions (97% trifluoroethanol at 25 °C) [13]. In agreement with Eaborn and Pande [4] and Traylor and co-workers [5] the beta effect of germanium is slightly larger than that of silicon. The cis system, it should be remembered, represents the unfavorable gauche or skew geometry (Ge—C—C—X dihedral angle of about 60°). This large acceleration for an unfavorable geometry argues against an E$_2$ mechanism or direct (nonvertical) participation to form a germanonium ion, both of which would require an antiperiplanar arrangement. Residual hyperconjugation in an E_1 (k_c) mechanism can explain the results. Solvent studies showed that the rate was dependent on solvent ionizing power rather than nucleophilicity.

The unbiased trans germanium system [3, M = Ge, R = H, X = O(CO)CF$_3$] reacted about 10^{11} times faster than cyclohexyl trifluoroacetate, or more than an order of magnitude faster than the unbiased silicon system. The dominance of the diequatorial form, which is poorly disposed for any type of participation, and Curtin–Hammett considerations suggest that the rate should be a thousand times faster or so for a system biased into the antiperiplanar form shown in 3. By analogy with silicon, the biased trans germanium form therefore should solvolyze 10^{13}–10^{14} times as fast as cyclohexyl. Again by analogy with silicon, this maximum acceleration should be made up of about 10^2 induction and 10^{11}–10^{12} hyperconjugation.

Germanium seems to exert a beta effect very similar to that of silicon. Because of its higher polarizability, the effect of germanium is slightly larger than that of silicon.

14.4. Tin

In the context of protiodemetalation [4] and dehydrometalation [5], tin was much more reactive than silicon or germanium. We found this also to be the case in the solvolytic setting [13]. In fact, the trifluoroacetates were far too reactive to be isolated. We had recourse to acetate as a leaving group in order to measure the rate for the unbiased cis tin compound [4, M = Sn, R = H, X = O(CO)CH$_3$]. The rate of the acetate could be converted to that of the trifluoroacetate by an appropriate factor, which we selected to be 10^6 [14]. The unbiased cis tin system [4, X = O(CO)CF$_3$] then must have solvolyzed about 10^{11} times faster than cyclohexyl trifluoroacetate. There is a good deal of approximation in fixing this figure, because of the use of different leaving groups, temperatures, and solvents, but the acceleration could not be much less than this value. Under any circumstance, the beta effect of tin is several orders of magnitude faster than that of silicon or germanium. Even at the skew geometry, the kinetic acceleration is one of the largest on record in organic chemistry.

Although we were able to prepare the unbiased trans tin alcohol (3, M = Sn, R = H, X = OH), its esters reacted in solution faster than we could measure a rate. From the lower limit we set for the rate ($\gg 10^5$ s^{-1}), we could calculate a rate acceleration for the trans trifluoroacetate [3, M = Sn, R = H, X = O(CO)CF$_3$] with respect to cyclohexyl of at least 10^{14}, and probably many orders of magnitude more, because biasing and Curtin–Hammett factors have not yet been considered. Therefore, tin is probably the strongest neighboring group participator currently known in solvolysis. The extremely high polarizability of tin, its low electronegativity, and its weak bonds to carbon all make tin able to hyperconjugate very effectively.

14.5. Future Directions

The remarkable accelerations observed for tin immediately suggest that lead might provide even larger factors. To date there are no solvolytic data for a beta effect of lead. It is very important to obtain a quantitative measure of the maximum acceleration for tin in the antiperiplanar arrangement. The rapid rate of the reaction would require use of fast kinetic methods. Even for silicon there still are several unexplored geometries of interest: synperiplanar (0°), with parallel orbitals but a theoretically weaker interaction than antiperiplanar; anticlinal (120°), with poor overlap but on the anti side of the cosine-squared plot; and orthogonal (90°), in which no hyperconjugation can occur but the weak inductive effect may still provide some beta acceleration. Exploration of these phenomena awaits the synthesis of suitable systems.

Acknowledgment. The authors are indebted to the National Science Foundation, the Petroleum Research Fund, and the Dow Corning Corporation for support of this work.

References

1. Ushakov, S. N.; Itenberg, A. M. *Zh. Obshch. Khim.*, **1937**, *7*, 2495.
2. Sommer, L. H.; Dorfman, E.; Goldberg, G. M.; Whitmore, F. C. *J. Am. Chem. Soc.*, **1946**, *68*, 488–489. Sommer, L. H.; Whitmore, F. C. *J. Am. Chem. Soc.*, **1946**, *68*, 485–487. Sommer, L. H.; Bailey, D. L.; Whitmore, F. C. *J. Am. Chem. Soc.*, **1948**, *70*, 2869–2872.
3. Davis, D. D.; Gray, C. E. *J. Organomet. Chem.*, **1969**, *18*, P1–P4. Davis, D. D.; Gray, C. E. *J. Org. Chem.*, **1970**, *35*, 1303–1307. Davis, D. D.; Jacocks, H. M., III. *J. Organomet. Chem.*, **1981**, *206*, 33–47.
4. Eaborn, C.; Pande, K. C. *J. Chem. Soc.*, **1966**, 1566–1571.
5. Hanstein, W.; Berwin, H. J.; Traylor, T. G. *J. Am. Chem. Soc.* **1970**, *92*, 829–836. Hanstein, W.; Berwin, H. J.; Traylor, T. G. *J. Am. Chem. Soc.*, **1970**, *92*, 7476–7477. Traylor, T. G.; Berwin, H. J.; Jerkunica, J. M.; Hall, M. L. *Pure Appl. Chem.*, **1972**, *30*, 599–606. Traylor, T. B.; Jerkunica, J. M. *J. Am. Chem. Soc.*, **1971**, *93*, 6278–6279. Hannon, S. J.; Traylor, T. G. *J. Org. Chem.*, **1981**, *46*, 3645–3650. Koermer, G. S.; Traylor, T. G. *J. Org. Chem.*, **1981**, *46*, 3651–3657.
6. For a review, see Lambert, J. B. *Tetrahedron*, **1990**, *46*, in press.
7. Jarvie, A. W. P.; Holt, A.; Thompson, J. *J. Chem. Soc. B*, **1969**, 852–855. Jarvie, A. W. P. *Organomet. Chem. Rev. A*, **1970**, *6*, 153–207. Hudrlik, P. E.; Peterson, D. *J. Am. Chem. Soc.*, **1975**, *97*, 1464–1468.
8. Vencl, J.; Hetflejs, J.; Cermak, J.; Chvalovský, V. *Coll. Czech. Chem. Commun.*, **1973**, *38*, 1256–1262.
9. Lambert, J. B.; Wang, G.-t.; Finzel, R. B.; Teramura, D. H. *J. Am. Chem. Soc.*, **1987**, *109*, 7838–7845.
10. Raber, D.; Neal, W. C., Jr.; Dukes, M. D.; Harris, J. M.; Mount, D. L. *J. Am. Chem. Soc.*, **1978**, *100*, 8137–8146. Harris, J. M.; Mount, D. L.; Smith, M. R.; Neal, W. C., Jr.; Dukes, M. D.; Raber, D. *J. Am. Chem. Soc.*, **1978**, *100*, 8147–8156.
11. Melander, L.; Saunders, W. H. *Reaction Rates of Isotopic Molecules*, Wiley, New York, 1980, pp. 174–180. Sunko, D. E.; Szele, I.; Hehre, W. J. *J. Am. Chem. Soc.*, **1977**, *99*, 5000–5005.
12. Lambert, J. B.; Wang, G.-t. *J. Phys. Org. Chem.*, **1988**, *1*, 169–178.
13. Lambert, J. B.; Wang, G.-t.; Teramura, D. H. *J. Org. Chem.*, **1988**, *53*, 5422–5428.
14. Noyce, D. S.; Virgillo, J. A. *J. Org. Chem.*, **1972**, *37*, 2643–2647.

Synthesis and Structures of Some Novel Organic Compounds Containing Boron and Phosphorus

Anthony G. Avent, Michael F. Lappert, and Brian Skelton

School of Chemistry and Molecular Sciences
University of Sussex
Brighton BN1 9QJ
United Kingdom

Colin L. Raston

Division of Science and Technology
Griffith University, Nathan
Brisbane, Queensland, Australia 4111

Lutz M. Engelhardt, Stephen Harvey, and Allan H. White

Department of Physical and Inorganic Chemistry
University of Western Australia
Perth, Western Australia 6009

15.1. Introduction

Silyl-substituted methyls of type $^-CH_{n-3}(SiMe_3)_n$ were introduced into transition-metal chemistry in 1969 [2]. Homoleptic transition-metal alkyls such as $[MR_3]$ [R = $CH(SiMe_3)_2$] became accessible, for example, M = Cr [3] or L [4]. As for main group chemistry, some novel thermally stable open-shell compounds were made, including $R_2M=MR_2$ [5], $\dot{M}R_3$ [6a] (M = Ge or Sn), and $\dot{M'}R_2$ (M' = P or As) [6b]. For reviews, see [7].

In the present work, we sought to explore the potential of such alkyl ligands, in particular $^-CH(SiMe_3)_2$ ($\equiv R^-$), in the context of boron chemistry. As the boron atom is particularly small, it was envisaged that the substantial steric requirements of the ligand R^- might give rise to some unusual compounds. It had previously been shown that the reaction between boron trichloride and an excess of LiR yielded R_2BCl, which upon hydrolysis afforded R_2BOH [8].

This paper is mainly concerned with aspects of the chemistry of a bulky dichlorobo-rane $RBCl_2$ [R = $CH(SiMe_3)_2$] (1), obtained by reaction of equimolar proportions of BCl_3 + LiR in hexane. The dichloride 1 was converted into a number of derivatives: $RB(Cl)Bu^t$ (2), $RBCl(NHAr)$ (Ar = $C_6H_2Bu^t_3$-2,4,6) (3), $RB(Cl)(NR'_2)$ (R' = $SiMe_3$) (4), $[RB(\mu\text{-}NR')]_2$ (5), $RB{=}NAr$ (6), $[RB(\mu-O)]_3$ (7), and $RBCl_2$ (py) (8). The X-ray structure of the boronic anhydride 7 shows a planar six-membered ring of alternating boron and oxygen atoms with $\langle B{-}O \rangle$ 1.39(1) Å $\langle B{-}O{-}B \rangle$ 124(1)°, and $\langle O{-}B{-}O \rangle$ 116(1)°. Further reactions of the dichloride 1 include those with (1) $Mg(An^x)(pmdeta)$ (pmdeta = $[Me_2NCH_2CH_2]_2NMe$), yielding 9,10-$RBAn^H$ (9) or 9,10-$RBAn^{R'}$ (10) (An^H = anthracene, $An^{R'}$ = 9-trimethylsilylanthracene), and (2) Li_2COT (COT = cyclooctatetraene), affording $RB(COT')$ (11) (COT' is an isomer of COT). Compounds 1–12 [12 = $ArBCl_2$, from $LiAr(OEt_2)$ + BCl_3] have been fully characterized. Variable temperature ^1H NMR spectra reveal that 11 undergoes a degenerate Cope rearrangement in toluene solution with $\Delta G^\ddagger_{238\ K}$ = 41.8 ± 1.3 kJ mol^{-1}. While at ambient temperature this process is fast, 11 in PhMe at $-90°C$ consists of a major and a minor isomer, the structures of which are assigned; they differ from one another by a B$-$C bond rotation and the free energy difference between them (ΔG) is 1.7 kJ mol^{-1}. Further aspects of chemistry that were discussed in a lecture by Lappert [1] relate to the following:

1. (results of W.-P. Leung) the reactions of 1,2-$C_6H_4(PPh)_2[Li(tmeda)]_2$ with compounds of the formula ML_nCl_2, where ML_n = RB, $ZrCp''_2$ [Cp'' = η-$C_5H_3(SiMe_3)_2$-1,3], or R_2Sn [20];
2. (results of G. Beck) sterically hindered boron compounds having the ligand $^-OBR_2$ [21];
3. (results of W.-P. Leung) aspects of the chemistry of transition metal complexes having a terminal ligand ^{2-}PAr [22].

15.2. Synthesis of Boron Compounds

Among the reaction types that are reported are the following general categories:

1. nucleophilic substitutions at boron;
2. elimination reactions with or without concomitant or subsequent oligomeriza-tion;
3. base induced eliminations with concomitant or subsequent trapping;
4. nucleophilic addition.

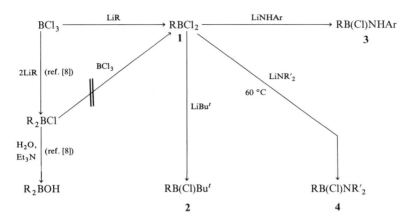

Scheme 15.1 ■ Synthesis of bis(trimethylsilyl)methyl(chloro)boranes and some of their nucleophilic displacement reactions. *Abbreviations*: $R = CH(SiMe_3)_2$; $Ar = C_6H_2Bu^t_3\text{-}2,4,6$; $R' = SiMe_3$. *Reaction conditions*: addition in *n*-hexane at $-78\ °C$ and slow warming to $25\ °C$, unless otherwise stated.

15.2.1. Nucleophilic Substitution at Boron

Starting with boron trichloride, routes to the alkyl(chloro)boranes $RB(Cl)X$ [$X = Cl$ (**1**), Bu^t (**2**), $NHAr$ ($Ar = C_6H_2Bu^t_3\text{-}2,4,6$) (**3**), or NR'_2 ($R' = SiMe_3$) (**4**)] are shown in Scheme 15.1.

There were no major surprises; the use of an excess of the bulky amide $LiNR'_2$ failed to displace the second Cl^- ligand of $RBCl_2$. The compound $ArBCl_2$ (**12**) was made from $LiAr(OEt_2)_2 + BCl_3$ in *n*-hexane.

15.2.2. Base-Induced Elimination Reactions

The object of making the bulky amino(chloro)boranes **3** and **4** was to explore their use as suitable substrates for either dehydrochlorination or de(trimethylchlorosilylation), respectively. In the event, both of these expectations were realized, but in the latter case rather severe conditions were required (Scheme 15.2). Perhaps for this reason a monomeric product of formula $RB{=}NR'$ was not isolated, but instead the diazaboretane $[RB(\mu\text{-}NR')]_2$ (**5**). We believe that the monomer $RB{=}NAr$ (**6**) was the product in the dehydrochlorination reaction. Compounds of type $(XBNY)_n$ are generally the trimers ($n = 3$) [9]; the first dimer ($n = 2$) was made in 1963 [10], whereas the first ambient temperature stable crystalline monomer ($n = 1$) dates from 1983 [9,11].

Our original purpose in converting the dichloroborane **1** into the boronic anhydride was to determine whether n was other than the normal $n = 3$. In the event, the trimer ($n = 3$) (**7**) was formed (Scheme 15.2). This was the first such compound to be crystalline at ambient temperature, and hence its X-ray structure was determined. There was no evidence for the formation of an $RB(OH)_2$ or $RB{=}O$ intermediate.

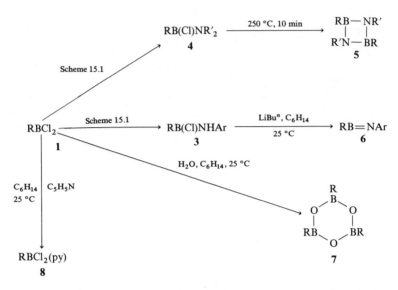

Scheme 15.2 ■ Some elimination reactions of chloroboranes **1**, **3**, and **4**, and an addition reaction of RBCl$_2$. *Abbreviations*: R = CH(SiMe$_3$)$_2$; Ar = C$_6$H$_2$Bu$'_3$-2,4,6; R' = SiMe$_3$.

15.2.3. Elimination–Trapping Reactions

Attempts were made to generate the boron equivalent R$\overset{..}{B}$ of a carbene by a suitable dichlorine elimination from RBCl$_2$ (**1**). These reactions, summarized in Scheme 15.3, led to the new compounds 9,10-RBAn^H (**9**) (An^H = anthracene), 9,10-RBAn$^{R'}$ (**10**) (An$^{R'}$ = 9-trimethylsilylanthracene), and RB(COT') (**11**) (COT' = an isomer of cyclooctatetraene). At this stage we are not in the position to decide whether the fragment RB was first generated and then trapped by the appropriate unsaturated hydrocarbon or whether compounds **9–11** were formed by chloroboration of An^H, An$^{R'}$, or COT and subsequent dedichlorination; the latter is the more plausible.

There are precedents for the reactions of the type shown in Scheme 15.3 involving dichlorides of other elements, for example eqs. (15.1) [12] and (15.2) [13].

$$\text{Li}_2(\text{COT}) + \text{PhPCl}_2 \xrightarrow{0°\text{ C}} \text{[structure]} \xrightarrow{70°\text{ C}} \text{[structure]} \qquad (15.1)$$

$$\text{Mg(AnR}'_2\text{-9,10)} + \text{Et}_2\text{AlCl(thf)} \xrightarrow{\quad} \text{[structure]} \qquad (15.2)$$

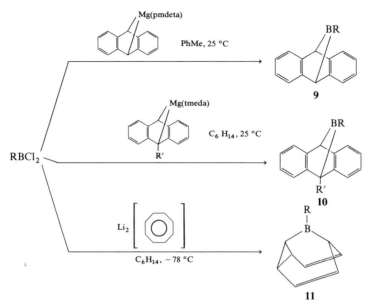

Scheme 15.3 ■ Elimination–trapping (dedichlorination) reactions of RBCl$_2$. *Abbreviations*: R = CH(SiMe$_3$)$_2$; R' = SiMe$_3$; pmdeta = [Me$_2$NCH$_2$CH$_2$]$_2$NMe; tmeda = [Me$_2$NCH$_2$]$_2$; **9** = RBAn^H; **10** = RBAn$^{R'}$; **11** = RB(COT').

15.2.4. Nucleophilic Addition and Properties of the New Compounds

Chloroboranes are well known to be strong Lewis acids, although their capacity to act in this fashion is often determined by steric effects. In the event, RBCl$_2$ formed a stable 1 : 1 adduct (**8**) with pyridine.

The new compounds **1–12** have been satisfactorily characterized by microanalysis. For the most part, they were found to be crystalline; selected data are in Table 15.1.

15.3. Physical Properties of the New Compounds

Each of the boron compounds **1–12** has been identified by spectroscopic and mass-spectrometric data, some of which are also summarized in Table 15.1.

15.3.1. The X-Ray Structure of the Boronic Anhydride [RB(μ-O)]$_3$ [R = CH(SiMe$_3$)$_2$] (7)

The asymmetric unit of the anhydride **7** was shown to comprise a single molecule consisting of a planar, six-membered B$_3$O$_3$ ring, with all the B—O bonds essentially equal. Electron-diffraction data on the gaseous molecules [XB(μ-O)]$_3$ (X = H [14] and X = Me [15]) are available. A low-temperature (-160 °C) X-ray structure has recently

Table 15.1 ▪ Yields and Melting (or Boiling) Points, and Selected Mass-Spectral and NMR Data for the New Boron Compounds 1–12

Compound[a]	No.	Yield (%)	Melting point (°C) [boiling point (°C/Torr^{-1})]	Highest m/1 in MS	$\delta[^{11}B\{^1H\}]$ (ppm)[c]
$RBCl_2$	1	74	[61–62/40]	$[P-Me]^+$	57.6
$RB(Cl)Bu^t$	2	68	[95/10^{-2}]	$[P-Me]^+$	73.5
RB(Cl)NHAr	3	72	131–133	P^+	b
$RB(Cl)NR'_2$	4	69	44–46	$[P-Me]^+$	47.3
$[RB(\mu\text{-}NR')]_2$	5	79	186–187	P^+	45.0
$RB=NAr$	6	63	100–102	P^+	b
$[RB(\mu\text{-}O)]_3$	7	96	197–198	$[P-Me]^+$	57.6
$RBCl_2(py)$	8	92	108	$[P-R]^+$	31.8
$RBAn^H$	9	61	135–137	P^+	56.2
$RBAn^{R'}$	10	41	110–112	P^+	57.2
RB(COT')	11	70	[150/10^{-2}]	P^+	78.3
$ArBCl_2$	12	35	111 [55/10^{-2}]	P^+	61.5

[a]*Abbreviations*: R = CH(SiMe$_3$)$_2$; Ar = C$_6$H$_2$Bu$^t_3$-2,4,6; R' = SiMe$_3$; An^H = anthracene;
An$^{R'}$ = 9-trimethylsilylanthracene; COT' = an isomer of cyclooctatetraene; P^+ = parent ion.
[b]Not recorded.
[c]Referenced to external BF$_3$ · OEt$_2$.

been carried out on [EtB(μ-O)]$_3$ (a liquid at ambient temperature) [16]. Significant structural features for **7** are summarized and compared with earlier data in Table 15.2, and the X-ray structure is given in Figure 15.1.

The boroxines have been the focus of considerable attention with regard to the extent of π charge delocalization from oxygen to boron within the six-membered ring (Scheme 15.4). This delocalization of charge, or aromaticity, is consistent with the ring being planar and with the B—O bond distances being considerably shorter than the

Table 15.2 ▪ Some Structural Data for the Boronic Anhydride [RB(μ-O)]$_3$ [R = CH(SiMe$_3$)$_2$] (7) and Some Related (XBO)$_3$ Compounds

X in [XB(μ-O)]$_3$	Mean B—O bond length (Å)	B—O—B angle (degrees)	O—B—O angle (degrees)	Reference
R[a]	1.39(1)	124(1)	116(1)	This work[b]
H[c]	1.375$_8$(2)	120(3)	120(3)	14
Me[c]	1.38(2)	112 ± 4		15
Et[a]	1.384(5)	121.6(1)	118.4(1)	16

[a]X-ray.
[b]$R = 0.049$ for 333 diffractometer reflections with $I > 3\sigma(I)$ in space group $P6_3/m$, $a = 11.330(5)$, $c = 17.004(8)$ Å, $Z = 2$ trimers; atom coordinates are available from the Cambridge Crystallographic Data Centre.
[c]Electron diffraction.

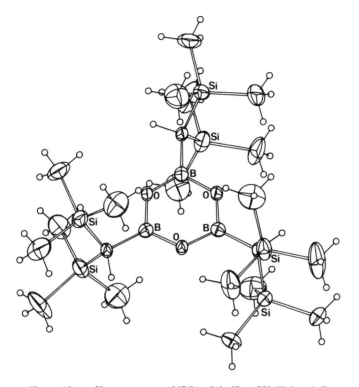

Figure 15.1 ■ X-ray structure of [RB(μ-O)]$_3$ [R = CH(SiMe$_3$)$_2$] (**7**).

sum (1.54 Å) of the B and O covalent radii [15]. In this respect the boroxines are similar to borazines [XB(μ-NY)]$_3$ [9]. The recently reported boron–phosphorus analogue of a borazine [MesB(μ-PC$_6$H$_{11}$)]$_3$ (Mes = C$_6$H$_2$Me$_3$-2,4,6) was also found to exhibit similar characteristics of ring planarity and equal but short B—P bonds [17].

15.3.2. The Molecular Structure of RB(COT') (11) in Toluene Solution

We believe RB(COT') (**11**) (see Scheme 15.3) to be the RB derivative of the C$_8$H$_{10}$ hydrocarbon homotropilidene (**I**) [18]. Various derivatives (**13**) of **I** are known and showed fluxional behavior, as evident from both their ^{1}H and ^{13}C NMR spectra; this

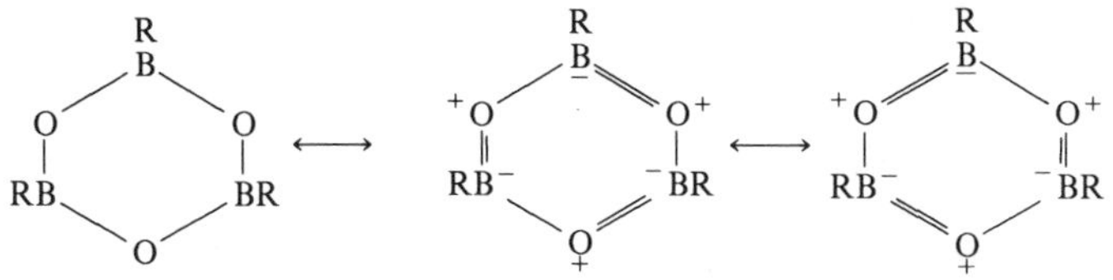

Scheme 15.4 ■ Valence bond representation of the structure of the boroxine [RB(μ-O)]$_3$.

can be explained in terms of a degenerate Cope rearrangement, which for **11** is illustrated in eq. (15.3), $\Delta G^{\ddagger}_{T_c} = 41.8 \pm 1.3$ kJ mol^{-1} (derived from the coalescence temperature, $T_c = 238$ K).

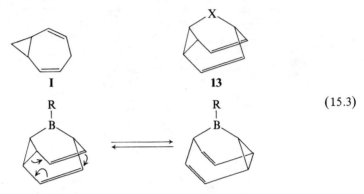

(15.3)

Compounds **13** include those with X = CH$_2$ (barbaralane) (**13a**) [19a], X = CO (barbaralone) (**13b**) [19b], X = CH = CH (bullvalene) (**13c**) [19c], X is absent (semi-bullvalene) (**13d**) [19d], and X = $^+$PEt$_2$ or $^+$PPh$_2$ (**13e**) [19e]. It appears that the tighter the "basket" structure in **13**, the faster the exchange rate, because **13d** had the lowest $\Delta G^{\ddagger}_{T_c}$ (23 kJ mol^{-1}) and coalescence temperature ($T_c = 106$ K) [19d].

Assignments of the low (−90 °C) and high (+30 °C) temperature ^{1}H NMR spectra for RB(COT') (**11**) are shown in Tables 15.3 and 15.4. As for Table 15.3, the assignments are based on the following considerations:

1. for Me and H-9, the integrals, shifts (δ), and the singlet signals;
2. for H-2 and H-8, δ;
3. for H-2 and H-8, H-9, H-1, and H-3, and H-7, selective decoupling experiments (see **II** and Table 15.4 footnote c).

Table 15.3 ■ The ^{1}H NMR Spectrum of RB(COT') (11) in Toluene-d$_8$ at −90 °C at 360 MHz, with Assignments of Signals to Particular Protons of the Major or Minor Isomer (11a/11b)

| Proton^a | Chemical shift (δ ppm) | | Integral | Multiplicity^c |
	Major^b	Minor^b		
Me	0.11	0.11	18	s
H-9	1.04	1.35	1	s
H-1	2.41	2.55	1	t
H-5	1.19	1.27	1	t
H-2, H-8	5.81	5.65	2	t
H-3, H-7	5.56	5.56	2	m
H-4, H-6	2.20	2.17	2	m

^{a} For details of numbering, see eq. (15.4).
^{b} Structures **11a/11b**.
^{c} *Abbreviations*: s = singlet; t = triplet; m = multiplet.

Table 15.4 ■ **The ^{1}H NMR Spectrum of RB(COT') (11) in Toluene-d_8 at $+30$ °C at 360 MHz, with Assignments of Signals to Particular Protons**

Proton[a]	Chemical shift (δ ppm)	Integral	Multiplicity[b]
Me	0.11	18	s
H-9	1.23	1	s
H-1, H-5	1.93	2	t
H-2, H-8, H-4, H-6	3.98	4	br
H-3, H-7	5.51	2	cx[c]

[a]For details of numbering, see eq. (15.4).
[b]*Abbreviations*: s = singlet; t = triplet; br = broad; cx = complex.
[c]This complex multiplet, when observed with selective decoupling of the signal at 1.93 ppm, was identified as the AA' component of an AA'XX'X''X''' spin system (**II**) with $J(AX) = 7.1$ Hz and $J(XX') = 3.8$ Hz, taking all other coupling constants as being small.

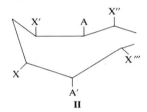

II

At the lowest temperature (-90 °C), two isomers were observed. From the integrals (Table 15.3), the minor isomer is estimated to constitute $32 \pm 3\%$ of the total, which translates into a free energy difference between the two isomers **11a** and **11b** of $\Delta G = 1.7$ kJ mol^{-1} at -90 °C. They showed similar shifts and coupling patterns (Table 15.3), and hence we propose that **11a** and **11b** are conformers arising from a restriction to rotation about the C—B bond; their interconversion is represented by eq. (15.4).

$$\text{(structures 11a and 11b)} \tag{15.4}$$

11a   **11b**

The assignment of structures **11a** and **11b** is further supported by the observation that symmetry appears to be maintained about the C_5BC_1 plane; that is, H-2 is always chemically equivalent to H-8, and so on. Results of nuclear Overhauser effect (nOe) experiments are awaited to determine which of **11a** and **11b** is the major isomer. It is not possible to make an estimate of $\Delta G^{\ddagger}$ for the process of eq. (15.4) without a full line-shape analysis or a two-dimensional exchange spectroscopy experiment (EXSY) at low temperature.

15.4. Some Exploratory Reactions of the Anthracene Adducts 9 and 10

The original objective in making the compounds 9,10-RBAn^H (9) and 9,10-RBAn$^{R'}$ (10) [R = CH(SiMe$_3$)$_2$, R' = SiMe$_3$, An^H = anthracene, An$^{R'}$ = 9-trimethylsilylanthracene] was to use them as thermal or photochemical sources of the boron analogue of a carbene RB̈. Experiments to this end proved unsuccessful.

Both of the compounds 9 and 10 exhibited high thermal stability with respect to elimination of anthracene or trimethylsilylanthracene, respectively. Indeed, both 9 and 10 sublimed with relative ease at about 90–100 °C at 10^{-2} Torr. Similarly, each was recovered in quantitative yield after photolysis for 5 h in n-hexane using a 1-kW Hg–Xe lamp at ambient temperature. It is conceivable that B—C homolysis did occur, but that recombination of the fragments was faster than associative RB coupling to yield RB=BR or an (RB)$_n$ cluster.

Photolysis of 9 and 10 in the presence of a 10 molar excess of diphenylethyne in THF gave similar results, whence we conclude that such B—C homolysis was unlikely to have occurred.

15.5. Synthesis and Structures of Some 1,3-Diphospha-2-metallapentanes [20]

Reactions of L$_n$MCl$_2$ with 1,2-C$_6$H$_4$(PPh)$_2$[Li(tmeda)]$_2$ yielded the metallacycles [L$_n$M{P(Ph)C$_6$H$_4$PPh-1,2}] [ML$_n$ = ZrCp''$_2$ (14), rac-SnR$_2$ (15), $meso$-SnMe$_2$, (16a) and rac-SnMe$_2$ (16b), and BAr (17); tmeda = (Me$_2$NCH$_2$)$_2$, Cp'' = η-C$_5$H$_3$(SiMe$_3$)$_2$-1,3; R = CH(SiMe$_3$)$_2$, Ar = C$_6$H$_2$Bu$^t_3$-2,4,6]. Complex 14 (unlike 15) had distinct P environments in the crystal, P(1) being planar and P(2) pyramidal and Zr — P(1) shorter than Zr—P(2). In toluene-d$_8$ solution, the low-temperature PMPCC skeletal arrangement in 14 and 17 corresponded (NMR) to that of crystalline 14, with exchange processes being evident at higher temperatures (also noted for 15). It is noteworthy that the formation of the R$_2$Sn (15) rather than the Me$_2$Sn (16) complex X$_2$Sn[P(Ph)C$_6$H$_4$PPh-1,2] was shown to be diastereoselective.

15.6. Lipophilic Lithium Dialkylboroxides [21]

Treatment of R$_2$BOH [R = CH(SiMe$_3$)$_2$] with LiBu^n yielded the colorless crystalline [Li(μ-OBR$_2$)]$_2$ (18). The dialkylboroxide 18 reacted with tmeda, MeCOCl, or [Ti(η-C$_5$H$_5$)$_2$Cl$_2$] to yield Li(OBR$_2$)(tmeda) (19) [tmeda = (Me$_2$NCH$_2$)$_2$], R$_2$BOH, or [Ti(η-C$_5$H$_5$)$_2$(Cl)(OBR$_2$)], respectively. X-ray data (18 and 19) showed the following significant features: a rare two-coordinate lithium environment in 18 and exceedingly short Li—O [1.677(5) Å] and O—B [1.308(8) Å] distances in 19.

15.7. Transition-Metal Complexes Having an M=P Bond

Recently, we reported the synthesis and characterization, including the X-ray structure, of the first thermally stable transition-metal analogue of an arylimide, [Mo(η-C$_5$H$_5$)$_2$(=PAr)] [22]. In Lappert's lecture [1], further data were presented relating to its chemistry [23], as well as to the X-ray structure of the tungsten analogue [24].

Acknowledgment. Thanks are due to S.E.R.C. and the Leverhulme Trust (M. F. L.) and the Australian Research Council (C. L. R. and A. H. W.) for support.

References

1. Lappert, M. F. Lecture given at the Second International Meeting on Heteroatom Chemistry, Albany, N.Y., July, 1989.
2. Collier, M. R.; Kingston, B. M.; Lappert, M. F.; Truelock, M. M. British Patent 36021 (1969); Collier, M. R.; Lappert, M. F.; Truelock, M. M. *J. Organomet. Chem.*, **1970**, *25*, C36; Yagupsky, G.; Mowat, M.; Shortland, A.; Wilkinson, G. *J. Chem. Soc. Chem. Commun.*, **1970**, 1369.
3. Barker, G.K.; Lappert, M. F.; Howard, J. A. K. *J. Chem. Soc., Dalton Trans.*, **1978**, 734.
4. Hitchcock, P. B.; Lappert, M. F.; Smith, R. G.; Bartlett, R. A.; Power, P. P. *J. Chem. Soc., Chem. Commun.*, **1988**, 1007.
5. Davidson, P. J.; Harris, D. H.; Lappert, M. F. *J. Chem. Soc., Dalton Trans.*, **1976**, 2268; Goldberg, D. E.; Hitchcock, P. B.; Lappert, M. F.; Thomas, K. M.; Thorne, A. J.; Fjeldberg, T.; Haaland, A.; Schilling, B. E. R. *J. Chem. Soc., Dalton Trans.*, **1986**, 2387.
6. (a) Hudson, A.; Lappert, M. F.; Lednor, P. W. *J. Chem. Soc., Dalton Trans.*, **1976**, 2369; (b) Gynane, M. J. S.; Hudson, A.; Lappert, M. F.; Power, P. P.; Goldwhite, H. *J. Chem. Soc. Dalton Trans.*, **1980**, 2428.
7. Davidson, P. J.; Lappert, M. F.; Pearce, R. *Acc. Chem. Res.*, **1974**, *7*, 209; Davidson, P. J.; Lappert, M. F.; Pearce, R. *Chem. Rev.*, **1976**, *76*, 219.
8. Al-Hashimi, S.; Smith, J. D. *J. Organomet. Chem.*, **1978**, *153*, 253.
9. See Paetzold, P. *Adv. Inorg. Chem.*, **1987**, *31*, 123.
10. Lappert, M. F.; Majumdar, M. K. *Proc. Chem. Soc.*, **1963**, 88.
11. Nöth, H.; Weber, S. *Z. Naturforsch., Teil B*, **1983**, *38a*, 1460.
12. Katz, T. J.; Nicholson, C. R.; Reilly, C. A. *J. Am. Chem. Soc.*, **1966**, *88*, 3832; Turnblom, E. W.; Katz, T. J. *J. Am. Chem. Soc.*, **1973**, *95*, 4292; Märkl, G.; Alig, B. *Tetrahedron Lett.*, **1982**, 4915; reactions related to those of the first part of eq. (15.1) include those between $M(COT)$ and $RPCl_2$ [M = Mg, Ca, Sr, or Ba; R = $Ph_2C(C_5H_4N — 2)$], see Hutchings, D. S.; Junk, P. C.; Patalinghug, W. C.; Raston, C. L.; White, A. H. *J. Chem. Soc., Chem. Commun.*, **1989**, 973.
13. Lehmkuhl, H.; Mehler, K.; Shakoor, A.; Krüger, C.; Tsay, Y.-H.; Benn, R.; Rufinska, A.; Schroth, G. *Chem. Ber.*, **1985**, *118*, 4248; Lehmkuhl, H.; Shakoor, A.; Mehler, K.; Krüger, C.; Tsay, Y.-H. *Z. Naturforsch., Teil B*, **1985**, *40*, 1504.
14. Chang, C. H.; Porter, R. F.; Bauer, S. H. *Inorg. Chem.*, **1969**, *8*, 1689.
15. Bauer, S. H.; Beach, J. Y. *J. Am. Chem. Soc.*, **1941**, *63*, 1394.
16. Boese, R.; Polk, M.; Bläser, D. *Angew. Chem., Int. Ed. Engl.*, **1987**, *26*, 245.
17. Dias, H. V. R.; Power, P. P. *Angew. Chem., Int. Ed. Engl.*, **1987**, *26*, 1270.
18. Bicker, R.; Kessler, H.; Ott, W. *Chem. Ber.*, **1975**, *108*, 3151.
19. (a) Günther, H.; Runskink, J.; Schmickler, H.; Schmitt, P. *J. Org. Chem.*, **1985**, *50*, 289. (b) Lambert, J. B. *Tetrahedron Lett.*, **1963**, 1901. (c) Günther, H.; Ulmen, I. *Tetrahedron*, **1974**, *30*, 3781. (d) Cheng, A. K.; Anet, F. A. L.; Mioduski, J.; Meinwald, J. *J. Am. Chem. Soc.*, **1974**, *96*, 2887. (e) Märkl, G.; Alig, B. *Tetrahedron Lett.*, **1983**, 3981.
20. Bohra, R.; Hitchcock, P. B.; Lappert, M. F.; Leung, W.-P. *J. Chem. Soc., Chem. Commun.*, **1989**, 728.
21. Beck, G.; Hitchcock, P. B.; Lappert, M. F.; MacKinnon, I. A. *J. Chem. Soc. Chem. Commun.*, **1989**, 1312.
22. Hitchcock, P. B.; Lappert, M. F.; Leung, W.-P. *J. Chem. Soc., Chem. Commun.*, **1987**, 1282.
23. Leung, W.-P. unpublished work.
24. Hitchcock, P. B.; Leung, W.-P., unpublished work.

Stereochemistry of Sulfoxide and Its Application to Synthesis

Atsuyoshi Ohno

Institute for Chemical Research
Kyoto University
Uji, Kyoto 611
Japan

Two protons in a methylene group adjacent to a sulfinyl group are diastereotopic due to the chirality of the sulfinyl sulfur. The *pro-R* and *pro-S* protons thus have different reactivity/selectivity toward base-catalyzed electrophilic substitution. Biellman and co-workers [1] reported that the reaction of the solid lithium salt from methyl phenylmethyl sulfoxide (1) with gaseous methyl iodide to afford the $R,S/S,R$-type product shows much greater stereoselectivity than the corresponding reactions in solution. In solution chemistry, solvents of higher dielectric constant tend to afford the $R,S/S,R$-type products [2–5]. A subsequent series of papers by many authors has made it clear that the stereospecificity of base-catalyzed reactions of diastereotopic protons α to the sulfinyl group is quite general both in acyclic and in relatively rigid cyclic sulfoxides [6–10].

Molecular orbital calculations have also been done to predict the stable conformation for the α-sulfinyl carbanion [11,12]. However, calculations that ignore the effects of the reaction medium and the countercation are of limited value in explaining the variation of stereochemistry observed in the reaction of a sulfinyl carbanion with an electrophile. A calculation on $LiCH_2S(O)H$ indicates the importance of the oxygen–lithium interaction in the stable conformation [13]. An X-ray crystallographic analysis supports the importance of the oxygen–lithium interaction [14]. At the same time, NMR studies revealed that no coupling between the α-carbon and lithium cation could be observed [9,15,16]. Stereochemical aspects of the reactions of cyclic sulfoxides suggest the importance of chelation [17,18].

The stereochemical course of the reactions of 1 and *t*-butyl phenylmethyl sulfoxide (2) have been studied in the greatest detail. Durst and co-workers showed that the *pro-S* hydrogen is abstracted preferentially from (S)-1 by a kinetic factor of $1.7:1$ with

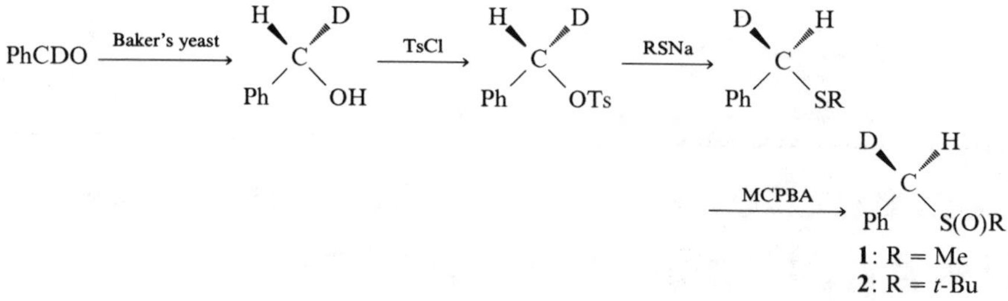

Scheme 16.1

n-BuLi in THF at $-70\ °C$ [19], which is in contrast with the factor (1 : 1.5) observed by Nishihata and Nishio [20]. Later, Iitaka and co-workers pointed out that the stereochemistry of the reaction of α-lithiated **2** was erroneously assigned [21], which casts doubt not only on the stereochemical assignment for the prochiral benzylic protons but also on the stereochemistry for electrophilic attack on α-sulfinyl carbanion.

Without confirmation of the configuration of the deuterated benzyl group, no prediction on the stability and/or reactivity of an α-sulfinyl carbanion is meaningful. Therefore, in order to interpret correctly the stereochemistry of α-sulfinyl carbanions, we first reexamined the configuration of the deuterated benzyl group and reinterpreted [22] the results reported by Durst and co-workers.

(S)-Benzyl alcohol-α-d was obtained from benzaldehyde-α-d by reduction with baker's yeast [23,24]. The alcohol was converted into the desired sulfoxides. Because the alkylthiolation proceeds with inversion of configuration [24], **1** and **2** so obtained have the R-configuration at the benzylic position, as shown in Scheme 16.1.

The ^1H NMR spectrum of S_R-**2** prepared according to the literature procedure [25] and contaminated by a small amount of S_{rac}-**2** is shown in Figure 16.1a. Figure 16.1b shows the spectrum of $C_R S_{rac}$-**2**-d contaminated by a small amount of $C_{rac} S_{rac}$-**2**-d. From parts a and b of Figure 16.1, it is obvious that the signals from the benzylic protons in the S_S isomer appear at lower fields than those of the S_R isomer regardless of the configuration of the benzylic carbon. Thus, the signals shown in Figure 16.1b are assigned to the benzylic protons in the $C_R S_S$, $C_S S_R$, $C_S S_S$, and $C_R S_R$ isomers, respectively, from lower to higher field.

The carbanion from S_R-**2** in THF was deuterated at $-78\ °C$, and the product was subjected to ^1H NMR spectroscopy. The spectrum is shown in Figure 16.1c. There is no doubt that the product is the isomer of $C_S S_R$ configuration, in agreement with the result from the recent crystallographic study [21].

The same procedure applied to the deuterated product from S_S-**1*** revealed that this is the $C_S S_S$ isomer, in agreement with the result reported by Durst and co-workers [19]. Consequently, the reaction may be summarized as shown in Scheme 16.2.

It is now obvious that deuteration and methylation of **1** in THF occur with opposite stereospecificity, whereas the corresponding reactions of **2** in THF proceed with the same stereospecificity.

---

*Note that S_S-**1** and S_R-**2** have the *same* configuration at sulfur.

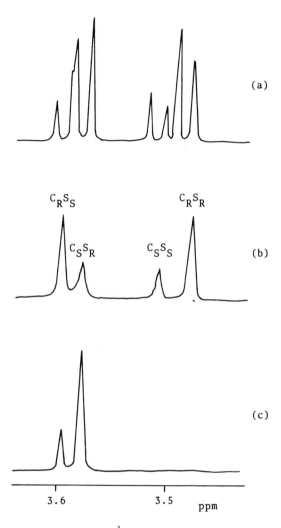

Figure 16.1 ■ ^{1}H NMR spectra of **2**.

The stereochemistry of the reaction products depends on three factors associated with the α-sulfinyl carbanion:

1. kinetic acidity, which controls the stereochemistry of the carbanion initially formed;
2. thermodynamic acidity, which defines the stereochemistry of the conformation of the intermediate carbanion;
3. reactivity of the carbanion, which may be important to control the stereochemistry of the products.

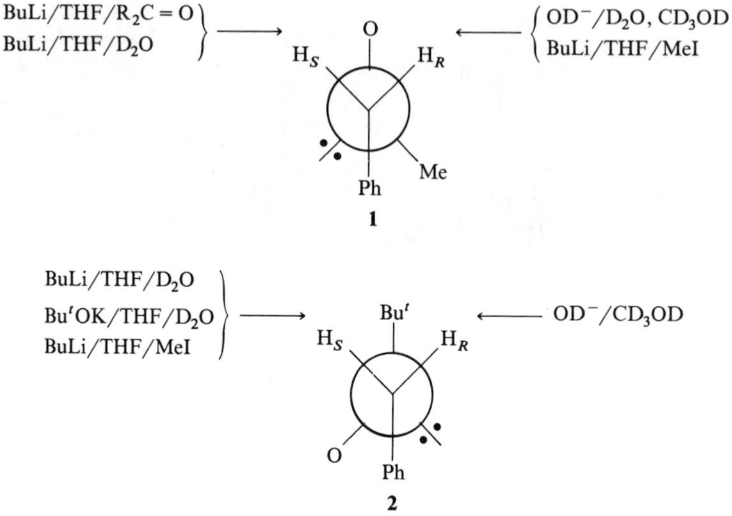
BuLi/THF/R₂C = O
BuLi/THF/D₂O
O
H_S
H_R
OD⁻/D₂O, CD₃OD
BuLi/THF/MeI
Me
Ph
1
BuLi/THF/D₂O
Bu'OK/THF/D₂O
BuLi/THF/MeI
Bu'
H_S
H_R
OD⁻/CD₃OD
O
Ph
2

Scheme 16.2

It should also be noted that the carbanion, as well as the base used to form it, is not *free* but is always accompanied by a counterion. Thus, the stereochemistry of the products does not necessarily reflect the stable conformation of the *free* carbanion.

For the present reaction system, the contribution of kinetic acidity can be neglected because the carbanion in THF has enough time to reorganize into its most stable conformation before it reacts with an electrophile.

Nuclear Overhauser effect (NOE), optical rotatory dispersion (ORD), circular dichroism (CD), and NMR studies predict that the carbanions from S_R-**2** and S_S-**1** have equilibrium conformations shown in Scheme 16.3. A polar electrophile such as water

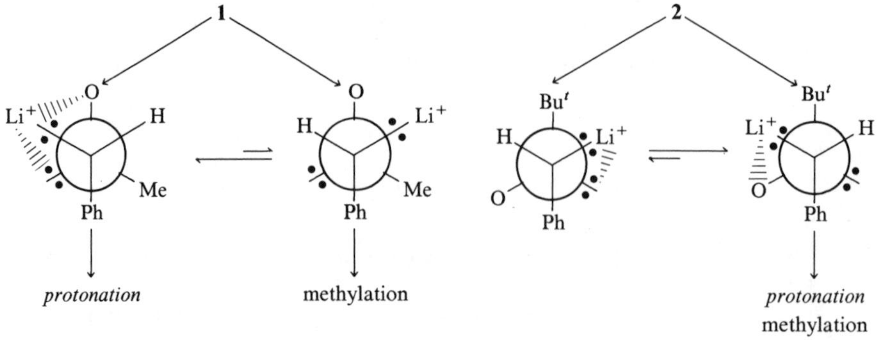
1
2
Li⁺
O
H
Me
Ph
protonation
O
H
Li⁺
Me
Ph
methylation
Bu'
H
Li⁺
O
Ph
Bu'
Li⁺
H
O
Ph
protonation
methylation

Scheme 16.3

(deuterium oxide) comes from the lithiated side of the carbanions because its polarization causes it to interact initially with the countercation. A nonpolar electrophile such as methyl iodide prefers to react on the more nucleophilic side, which is anti to the sulfur lone pair. In other words, the si and re faces of the carbanion from S_S-**1** are hard and soft reaction centers, respectively; hard reagents such as proton and carbonyl compounds react on the si face, whereas a soft reagent such as methyl iodide reacts on the re face. Because the carbanion S_R-**2** has both hard and soft reaction centers on the same si face, both hard and soft reagents react on this same face.

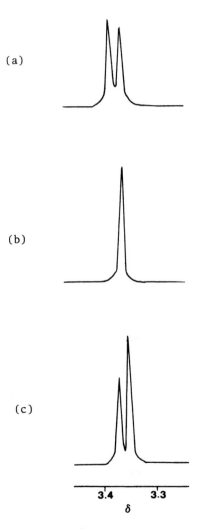

Figure 16.2 ■ ^{1}H NMR spectra of **1**.

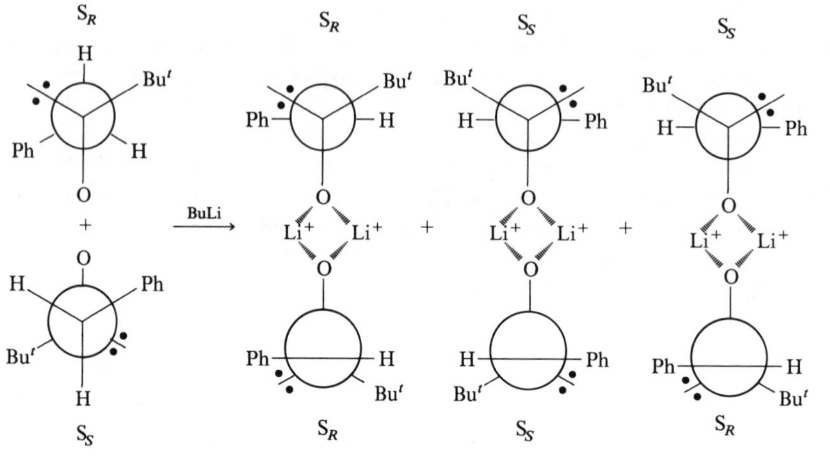

Scheme 6.4

In a polar solvent, however, the carbanion is not tightly paired with a countercation and the carbanionic lobe tends to keep away from the anionic face of the sulfoxide. Thus, H_R in both S_R-2 and S_S-1 is the reacting hydrogen in a polar solvent.

The ^{1}H NMR spectrum of the carbanion from racemic 2 in THF at -100 °C shows two singlets of equal intensity for the benzylic protons at $\delta = 3.34$ and 3.36 (Figure 16.2a) [26]. When the carbanion from optically pure S_R-2 is subjected to ^{1}H NMR spectroscopy, the benzylic proton exhibits a singlet at $\delta = 3.34$ (Figure 16.2b). Because it has been confirmed that the abstraction of a benzylic proton in 2 by a base results in highly diastereospecific carbanion thermodynamically, there is no doubt that the two singlets observed for racemic 2 are not due to diastereomerism of benzylic protons with respect to the chirality on a single sulfur. That is, the spectra reveal that the lithiated carbanion from 2 in THF is in dimeric forms of dl- and meso-isomers (Scheme 16.4). Although a dimer of the carbanion from a sulfoxide has been known in crystalline state [27] and the possibility of its existence in a solution was suggested [15,16], the present observation is the first one to confirm its presence in solution.

The dimeric forms shown in Scheme 16.4 indicate that the carbanion generated from 2 is rather in an oxy-anion form. This is confirmed by ^{7}Li NMR spectroscopy with 2-^{17}O by the appearance of a signal from the ^{7}Li–^{17}O bonding at $\delta = 28.1$ in THF-d_8 [28]. Thus, it may be said that the lithium salt of the anion formed from 2 is best represented by the oxylate anion structure with $C=S$ double bond (3).

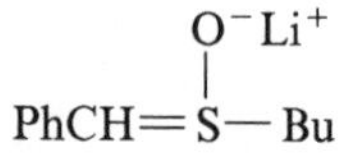

$$\begin{array}{c} O^-Li^+ \\ | \\ PhCH{=}S{-}Bu \end{array}$$

3

The stereochemistry described previously has been applied to the synthesis of compounds that have a chiral substituted-benzyl group. That is, the sulfinyl group was converted into the corresponding sulfonium salt, which was then subjected to nucle-

Scheme 16.5 ▪ *Abbreviations*: (a) BuLi/MeI in THF; (b) washed with hexane; (c) Ph_3P-I_2/NaI in CH_3CN; (d) $AgBF_4$/MeI in CH_2Cl_2; (e) AcOH (1.1 eq.)/K_2CO_3 (0.5 eq.) in $CHCl_3$ at room temperature for 24 h; (ee, enantiomer excess; de, diastereomer excess; op, optical purity; cy, chemical yield.

ophilic substitution. It is found that, under certain reaction conditions, the S_N2 reaction takes place efficiently without racemization (Scheme 16.5).

References

1. Biellmann, J. F.; Blanzat, J. F.; Vicens, J. J. *J. Am. Chem. Soc.*, **1980**, *102*, 2460.
2. Rauk, A.; Buncel, R. Y.; Moir, R. Y.; Wolfe, S. *J. Am. Chem. Soc.*, **1965**, *87*, 5498.
3. Wolfe, S.; Rauk, A. *J. Chem. Soc., Chem. Commun.*, **1966**, 778. It should be noted that an incorrect conclusion was deduced in this paper based on the erroneous assignment of *pro-R* and *pro-S* benzylic protons.
4. Baldwin, J. E.; Hackler, R. E.; Scott, R. M. *J. Chem. Soc., Chem. Commun.*, **1969**, 1415.
5. Durst, T.; Fraser, R. R.; McClory, M. R.; Swingle, R. B.; Viau, R.; Wigfield, Y. Y. *Can. J. Chem.*, **1970**, *48*, 2148.
6. Fraser, R. R.; Schuber, F. J. *J. Chem. Soc., Chem. Commun.*, **1970**, 633.
7. Fraser, R. R.; Schuber, F. J.; Wigfield, Y. Y. *J. Am. Chem. Soc.*, **1972**, *94*, 8795.
8. Hutchinson, B. J.; Andersen, K. K.; Katritzky, A. R. *J. Am. Chem. Soc.*, **1969**, *91*, 3839.
9. Lett, R.; Marquet, A. *Tetrahedron Lett.*, **1971**, 2851, 2855, 3255.
10. Nishio, M. *J. Chem. Soc., Chem. Commun.*, **1968**, 562.
11. Rauk, A.; Wolfe, S.; Csizmadia, I. G. *Can. J. Chem.*, **1968**, *47*, 113.
12. Wolfe, S.; Stolow, A.; LaJohn, L. A. *Tetrahedron Lett.*, **1983**, *24*, 4071.
13. Wolfe, S.; LaJohn, L. L.; Weaver, D. F. *Tetrahedron Lett.*, **1984**, 2863.
14. Marsch, M.; Massa, W.; Harms, K.; Baum, G.; Boche, G. *Angew. Chem., Int. Ed. Engl.*, **1986**, *25*, 1011.
15. Chassaing, G.; Marquet, A. *Tetrahedron*, **1978**, *34*, 1399.
16. Chassaing, G.; Marquet, A. *J. Organomet. Chem.*, **1982**, *232*, 293.
17. Chassaing, G.; Lett, R.; Marquet, A. *Tetrahedron Lett.*, **1978**, 471.
18. Biellmann, J. F.; Vicens, J. J. *Tetrahedron Lett.*, **1978**, 471.
19. Viau, R.; Durst, T.; McClory, M. R. *J. Am. Chem. Soc.*, **1971**, *93*, 3077.
20. Nishihata, K.; Nishio, M. *J. Chem. Soc., Perkin Trans. II*, **1972**, 1730.

21. Iitaka, Y.; Itai, A.; Tomioka, N.; Ichikaw, K.; Nishihata, K.; Nishio, M.; Izumi, M.; Doi, K. *Bull. Chem. Soc. Jpn.*, **1986**, *59*, 2801.
22. Nakamura, K.; Higaki, M.; Adachi, S.; Oka, S.; Ohno, A. *J. Org. Chem.*, **1987**, *52*, 1414.
23. Streitwieser, A., Jr.; Wolfe, J. R., Jr.; Schaeffer, W. D. *Tetrahedron*, **1959**, *6*, 338.
24. Althouse, V. E.; Feigl, D. M.; Sanderson, W. A.; Mosher, H. S. *J. Am. Chem. Soc.*, **1966**, 3595.
25. Axelrod, M.; Bickart, P.; Jacobus, J.; Green, M. M.; Mislow, K. *J. Am. Chem. Soc.*, **1968**, *90*, 4835.
26. Ohno, A.; Higaki, M.; Oka, S. *Bull. Chem. Soc. Jpn.*, **1988**, *61*, 1721.
27. Marsch, M.; Massa, W.; Harms, K.; Baum, G.; Boche, G. *Angew. Chem.*, *Int. Ed. Engl.*, **1986**, *25*, 1011.
28. Higaki, M.; Goto, M.; Ohno, A. *Heteroatom Chem.* **1990**, *1*, 181.

Phosphaalkynes—New Building Blocks in Heteroatom Chemistry*

Manfred Regitz

Fachbereich Chemie der Universität
Erwin-Schrödinger-Strasse
D-6750 Kaiserslautern
Federal Republic of Germany

17.1. Introduction

Alkynes and nitriles (compounds containing carbon–carbon and carbon–nitrogen triple bonds, respectively), are now well established and indispensable building blocks in synthetic organic chemistry. In contrast, the phosphaalkynes **3** [1], which are characterized by analogous carbon–phosphorus triple bonds, were of absolutely no synthetic relevance before the beginning of the 1980's. Even though the parent compound of this substance class—phosphaacetylene (**1**)—has been known since 1961 [2] (see below), the so-called double bond rule, in particular, has severely hindered the development of the chemistry of low-coordinated phosphorus compounds containing $P–X$ double bonds. Thus, for example, this rule "forbids" the formation of $(p–p)\pi$-multiple bonds between phosphorus and the elements in the first period of the periodic table. The rule was finally and completely discredited by the syntheses of the phosphamethine-cyanine cation [3] and the series of phosphabenzenes [4], which both contain $\lambda^3\sigma^2$-phosphorus atoms.

Phosphaacetylene (**1**) represents the beginning of the short history of the phosphaalkynes; it is a short-lived species, was first prepared by striking an electric arc between graphite electrodes in an atmosphere of PH_3, and was identified by IR spectroscopy after chromatographic purification [2] (see Scheme 17.1). The fact that it could also be used profitably in cycloaddition reactions with 1,3-dipoles for the

---

*Unusually coordinated phosphorus compounds, part 38; for part 37, see Wettling, T.; Schneider, J.; Wagner, O.; Kreiter, C. G.; Regitz, M. *Angew. Chem.*, **1989** *101*, 1035; *Angew. Chem., Int. Ed. Engl.*, **1989**, *28*, 1013.

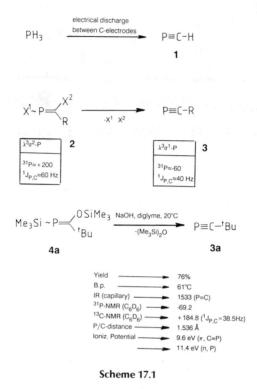

Scheme 17.1

synthesis of phospholes was only recognized very much later [5]. In general, β-elimination reactions of suitably substituted phosphaalkenes have proved to be valuable for the syntheses of kinetically stabilized phosphaalkynes ($2 \to 3$) [1]. The transformation of the trivalent, dicoordinated phosphorus atom of **2** (^{31}P NMR, δ approximately $+200$) to the trivalent, but only monocoordinated phosphorus atom of **3** (^{31}P NMR, δ approximately -60) is accompanied by pronounced high field shifts of the signals in the ^{31}P NMR spectra. A decisive factor for further progress was the report of the NaOH-catalyzed β-elimination of hexamethyldisiloxane from the phosphaalkene **4a** to furnish the thermally stable *tert*-butylphosphaacetylene (**3a**) [6]. This compound has been unequivocally characterized by IR, NMR, microwave, and photoelectron spectroscopy [4c, 4d]. Although the electronically unadulterated phosphaalkyne moiety of **3a** is shielded somewhat by the presence of the bulky *tert*-butyl group, this has only a limited influence on the reactivity. It is just for this reason that *tert*-butylphosphaacetylene (**3a**) has been employed preferentially in almost all of the model studies peformed so far.

17.2. General Synthetic Principles

The requirements for larger amounts of *tert*-butylphosphaacetylene (**3a**) as well as the question of the range of variability of the group R in compounds of the type **3** have led to a generalization of the previously mentioned synthesis in which, above all, the key

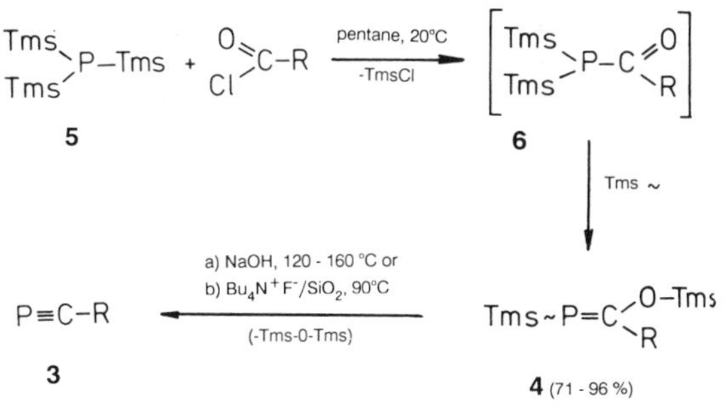

3,4	R	yield [%]	^{31}P (C$_6$D$_6$)[ppm]	^{13}C (C$_6$D$_6$)[ppm]
a	^{t}Bu	96	-69.2	184.8 ($^1J_{P,C}$=38.5 Hz)
b	CH$_2$^{t}Bu	75	-51.4	173.7 ($^1J_{P,C}$=45.5 Hz)
c	^{i}Pr	63	-64.3	183.4 ($^1J_{P,C}$=41.3 Hz)
d	1-Ad	71	-66.4	184.7 ($^1J_{P,C}$=39.0 Hz)
e	⬡⟨Me	76	-57.0	184.9 ($^1J_{P,C}$=37.2 Hz)

Scheme 17.2

β-elimination step was optimized. In every case, tris(trimethylsilyl)phosphane (5) is used as the starting material. When appropriate safety precautions are observed, this compound can be prepared relatively easily from white phosphorus, sodium–potassium alloy, and chlorotrimethylsilane [7]. Reactions of 5 with carboxylic acid chlorides in pentane at room temperature (Scheme 17.2) furnish the carboxylic acid phosphides 6 that can be detected [8, 9] but, in general, cannot be isolated. [The more reactive lithium bis(trimethylsilyl)phosphide · THF complex [10] can be used to the same effect in place of 5 and the reactions then take place at considerably lower temperatures.]

Compounds 6, in turn, undergo rapid [1,3]-trimethylsilyl shifts from phosphorus to oxygen that most certainly profit from the silicon–oxygen bond energy and are at the same time responsible for the formation of the phosphaalkenes 4. The question as to the E–Z geometry at the phosphorus–carbon double bond is apparently of no significance for the subsequent cleavage of hexamethyldisiloxane. In contrast to the original procedure for 3a in which this step is performed in solution [6], it has since been found to be extremely advantageous with respect to the scope of the phosphaalkyne synthesis to perform the elimination step in the absence of a solvent at a higher temperature. The phosphaalkynes are then rapidly removed from the reaction vessel and subsequently separated by distillation, whenever possible [11–13]. Compounds 3a [13] and 3e [11], which both contain tertiary carbon substituents, were prepared in this way. For the synthesis of 1-adamantylphosphaacetylene (3d), tetrabutylammonium fluoride on silica gel was used in place of sodium hydroxide to bring about the hexamethyldisiloxane cleavage [9]. Even the phosphaalkynes 3c and 3d bearing sterically less-demanding substituents are relatively easily accessible using this "solvent-free" distillation technique [11]. In addition to the typical high-field ^{31}P NMR resonances for the $\lambda^3\sigma^1$-phosphorus atoms of **3**, the diagnostic value of the ^{31}C NMR signals of the *sp*-hybridized carbon atoms (^{13}C NMR, $\delta = 173.7$–184.9) for characterization purposes should be emphasized. These signals, furthermore, are split by $^1J_{C,P}$ couplings of 37.2–45.5 Hz.

17.3. Phospholes

In analogy to the cycloaddition potential of the acetylenes, phospholes with widely varying further heteroatoms should be accessible by [3 + 2]-cycloaddition reactions of 1,3-dipoles with phosphaalkynes. The class of diazoalkane dipoles is the most thoroughly investigated to date [14,15]. Thus, diazomethane, diazoethane, 1-diazo-2,2-dimethylpropane, diazoacetophenone, methyl diazoacetate, and diazomethyl(diphenyl)-phosphine oxide do indeed undergo smooth addition reactions with the phosphaalkyne **3a** in diethyl ether at 20 °C to furnish the heteroaromatic compounds **8** as the final products [13]. (See Scheme 17.3.)

The initial step in the reaction of **3a** with R—CH=N$_2$ is regiospecific and presumably proceeds under electronic control (the unsubstituted compound **1** reacts with diazomethane with the same dipole orientation [5]). The 3H-1,2,4-diazaphospholes **7** are the initial products that can be detected [13,14] but not isolated because they undergo transformation to the energetically more favored 1H-1,2,4-diazaphospholes **8** via sigmatropic [1,5]-H migrations. The phosphorus heterocycles **8** are formed in high yields (76–93%) and exhibit ^{31}P NMR resonances in the range $\delta = 75.3$–110.8 [13]. By-products with the opposite dipole orientation have not been detected within the limits of the ^{31}P NMR monitoring method but their resonances would be expected to

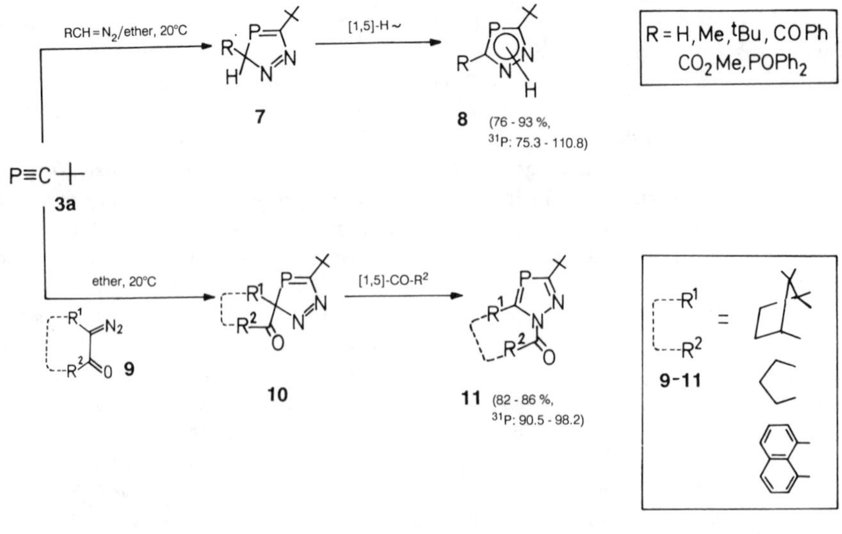

Scheme 17.3

appear at considerably lower field regions (δ approximately 205) [16]. The phosphaalkynes **3b**, **3c**, **3e** [11], and **3d** [9] react analogously with diazomethane.

α-Diazoketones **9** in which the functional groups are incorporated in ring systems, such as diazocamphor, 2-diazacyclopentan-1-one, or diazoacenaphthenone, also readily take part in 1,3-dipolar cycloaddition reactions with **3a**. The primarily formed spirocyclenes **10** then also undergo similar isomerizations to furnish the thermodynamically stable and, in these cases, condensed 1,2,4-diazaphospholes **11**. Here [1,5]-acyl shifts [13] occur in place of the previously mentioned H-migrations; the yields and ^{31}P NMR signals of the products are comparable with those of compounds **8**. ^{31}P NMR spectroscopic monitoring of the reaction of **3a** with **9** provides convincing evidence that the isomerization of **10** to **11** is preceded by the [1,5]-sigmatropic acyl shift to phosphorus, which is apparently kinetically controlled [13].

When the 3*H*-1,2,4-diazaphospholes (of the types **7** and **10**) need to be isolated—this can be of major significance for the subsequent reactions (see, for example, the section on phosphirenes)—the reactions following the cycloaddition process must be suppressed in these cases. This means that one has to start from diazoalkanes with substituents that are less susceptible to rearrangements. Alkyl-substituted diazomethanes, such as 3-diazo-2,2,4,4-tetramethylpentane, 3-diazo-2,2-dimethylbutane, and 6-diazo-1,1,5,5-tetramethylcyclohexane (**12a–12c**) fulfill these prerequisites and do indeed yield the target molecules **13a–13f** [17,18] in quantitative yields when allowed to react with **3a–3c** and **3e** at temperatures of $-40\ °C$ or less. (See Scheme 17.4.)

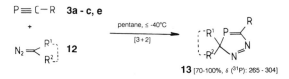

13 [70-100%, δ (^{31}P): 265 - 304]

12	a	b	c
R¹	ᵗBu	ᵗBu	⌐⟨
R²	ᵗBu	Me	

25°C, CHCl₃
[1,5]-R¹

14 [82-86%, δ (^{31}P): 50 - 58]

13,14	R	R¹	R²
a	ᵗBu	ᵗBu	ᵗBu
b	ᵗBu	ᵗBu	Me
c	ᵗBu	*	
d	CH₂ᵗBu	ᵗBu	ᵗBu
e	CHMe₂	ᵗBu	ᵗBu
f	⟨⟩×Me	ᵗBu	ᵗBu

*Same substituent as in **12c**.

Scheme 17.4

These products are characterized by low-field signals in their ^{31}P NMR spectra at $\delta = 265$–304, which clearly demonstrate the olefinic character of 13. The $3H$-1,2,4-diazaphospholes undergo isomerization to the $4H$-isomers 14a, 14c–14f (82–100% yields) on standing in deuterochloroform at room temperature, with the exception that 13b is stable under these conditions. The conversion from $\lambda^3\sigma^2$- to $\lambda^3\sigma^3$-phosphorus is apparent from, among others, the enormous shifts of the ^{31}P NMR signals of compounds 14 to high field ($\delta = 50$–58) [17,18]. When, for example, we want to study the photochemistry of 13—and the possibility to obtain $2H$-phosphirenes by nitrogen cleavage for the first time in this way is immediately apparent—we must avoid the isomerization process 13 $\rightarrow$ 14 by using lower reaction temperatures (see Section 17.4).

Azides, which are isoelectronic with diazo compounds, all take part in orientation-specific [3 + 2]-cycloaddition reactions with phosphaalkynes and thus render the previously unknown $3H$-1,2,3,4-triazaphospholes readily accessible [5,9,11,14,19]. Hence, the reactions of 3a with methyl azide, methyl azidocarbonate, and phenyl azide give rise to the corresponding adducts with the constitution 15; only in the cases of the reactions of hydrazoic acid or trimethylsilyl azide with 3a does the H- or trimethylsilyl-shift producing the $2H$-1,2,3,4-triazaphosphole isomers follow the cycloaddition step yielding 15 [19]. The ^{31}P NMR resonances of the azide adducts 15 appear at relatively high fields ($\delta = 160.8$–183.2) [19] in comparison to those of compounds 8 or 11; this is a consequence of the nitrogen atom bonded directly to the phosphorus atom. (See Scheme 17.5.) The products 15 are thermally extremely stable heterocycles that only cleave nitrogen under flash vacuum pyrolysis conditions. In the case of 15 with R = Ph, pyrolysis results in the formation of a benzo-annelated azaphosphole [19].

On account of the well-known affinity of phosphorus for oxygen, we would expect the reactions of nitrile oxides with phosphaalkynes to produce 1,2,5-oxazaphospholes. However, the experimental result is quite different: The regioisomeric 1,2,4-oxazaphospholes 16 (X = O) are in fact formed [14,20]. This is without doubt mainly a consequence of the steric requirements of the two reaction partners. Other phosphaalkynes can be used as the dipolarophile in this reaction instead of 3a [5,9,11,20]. Sometimes the formation of 1 : 2 adducts is also observed, that is, the initially formed 1,2,4-oxazaphosphole of the type 16 (X = O) can undergo addition of a further molecule of nitrile oxide in the same orientation to give a heterobicyclic product [5,20,21].

Benzonitrile sulfide also undergoes addition to 3a with the same dipole orientation as observed with nitrile oxides and thus gives rise to the 1,2,4-thiazaphosphole 16 (X = S) [20], the first representative of this class of compounds. The $1H$-1,2,4-diazaphosphole 16 (X = NPh) is the sole product obtainable from the reaction of benzonitrile imine with 3a, as has been shown by ^{31}P NMR spectroscopic analysis of the crude reaction product [20]. The two isomers benzonitrile N-methylimine and acetonitrile N-phenylimine, in contrast, only take part in regioselective [3 + 2]-cycloaddition reactions with 3a to give product mixtures in which adducts of the type 16, however, predominate (approximately 98%) [20].

The reactions of mesoionic compounds with phosphaalkynes are, in comparison, relatively sluggish and require thermal activation. Thus, the 1,3-thiaphosphole 21 can only be obtained from 3a and the dithioliumolate 17 after heating of the reaction partners in toluene at 130 °C (Schlenk tube, under pressure) [22]. The bicyclic primary

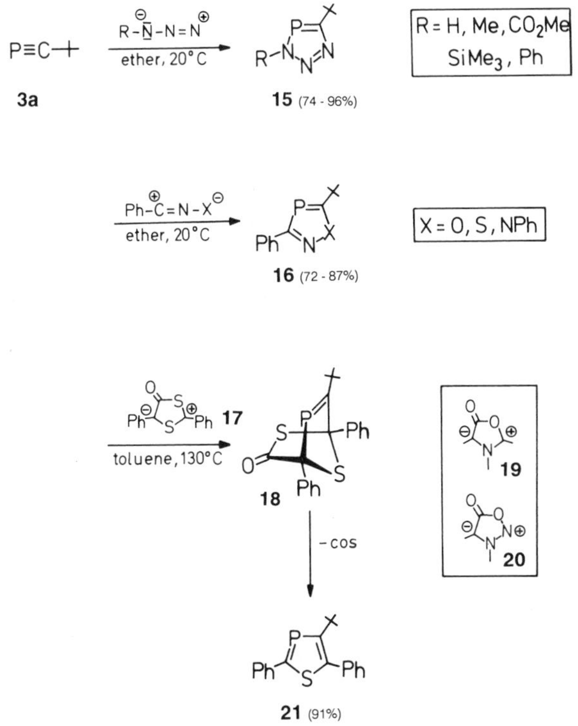

Scheme 17.5

product **18** cannot be detected because it undergoes rapid cycloelimination of carbon oxide sulfide to furnish **21** under the drastic reaction conditions required [22]. Münchnones **19** and sydnones **20** react analogously with **3a**, giving rise to $1H$-1,3-azaphospholes and $2H$-1,2,4-diazaphospholes, respectively [22].

17.4. 1-Phospha-1-cycloalkenes

$3H$-1,2,4-Diazaphospholes **14** offer particular possibilities for their synthesis because they are suitable precursors for generating the previously unknown phosphavinylcarbenes **22**. The latter, in turn, play a decisive role in the syntheses of the practically unknown class of cyclophosphaalkenes. To begin with, the cyclic hetero-1,3-dienes **13a**, **13d**, and **13e** represented the central point of this strategy; they were prepared from the phosphaalkynes **3a**–**3c** and the diazoalkane **12a**. When compounds **13** were subjected to photolysis at low temperatures, essential in order to suppress sigmatropic alkyl shifts according to **13 → 14**, the previously mentioned carbenes **22** are generated via ring-opened diazo isomers (see the following paragraph). These in turn, within the limits of the ^{31}P NMR spectroscopic detection method, do not undergo the desired 1,3-ring closure that furnishes the previously unknown $2H$-phosphirenes **24**, that is, the parent

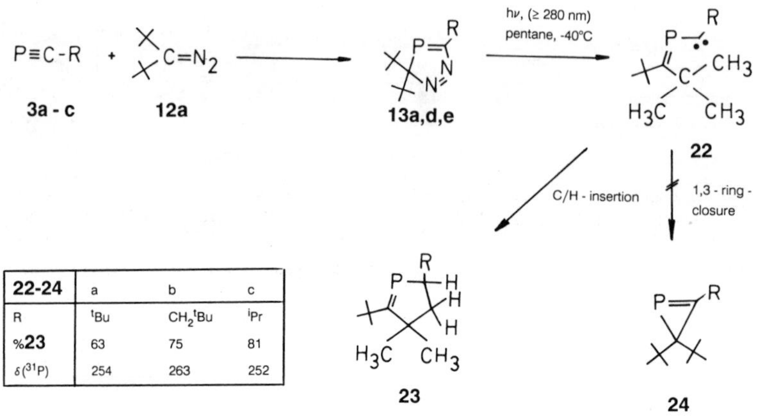

22-24	a	b	c
R	^tBu	$CH_2{}^tBu$	^iPr
%**23**	63	75	81
$\delta(^{31}P)$	254	263	252

Scheme 17.6

compounds of the 1-phospha-1-cycloalkene series. Instead, at this stage insertion of the carbene carbon atom into a carbon–hydrogen bond of a *tert*-butyl group takes place, yielding the 1-phospha-1-cyclopentenes **23a**–**23e** [18]. (See Scheme 17.6.) The products **23** are formed in high yields, are relatively stable in the air because their phosphorus–carbon double bonds are well shielded by the presence of bulky substituents, and are characterized by the low-field ^{31}P NMR resonances ($\delta = 252–263$) of the $\lambda^3\sigma^2$-phosphorus atoms [18].

The desired synthetic target was, however, finally achieved by low-temperature photolysis of the spirocyclic 3*H*-1,2,4-diazaphosphole **13c**. (See Scheme 17.7.) Irradiation initially induces ring opening to give rise to the phosphavinyldiazoalkane **25**; in the case of the photolysis of **13b**, the diazo isomer can even be isolated as a crude, red oil [$\nu = 2030\ cm^{-1}$ ($C{=}N_2$ stretch)] [17]. Conversion of **25** into the carbene **26** is followed by two reaction pathways, giving rise to a mixture of products, one of which is

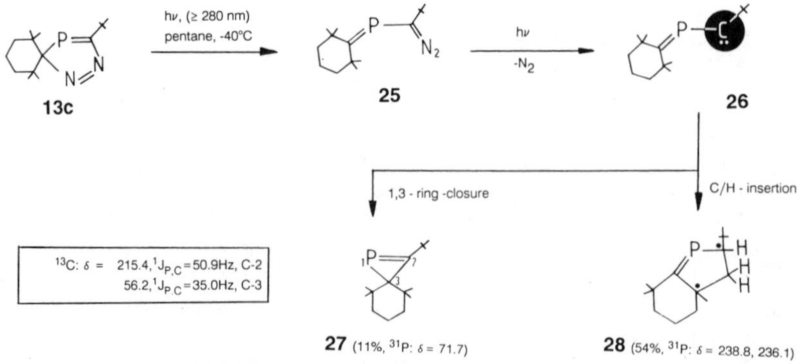

Scheme 17.7

the spirocyclic $2H$-phosphirene **27** and the other is the annelated 1-phospha-1-cyclo-pentene **28**. The former pathway comprises the desired [1,3]–ring closure to **27** and the latter, carbon–hydrogen-insertion to **28**. Bulb-to-bulb distillation and subsequent chro-matography on silica gel results in a 1 : 5 ratio of the two products in a combined yield of 65% [17]. Comparable results are also obtained when the tetramethyl-substituted six-membered ring of **13c** is replaced by a five- or seven-membered ring unit [18].

Four stereoisomers comprising two diastereotopic pairs of the bicyclic phos-phaalkene **28** are feasible because it contains two chiral carbon atoms. And, in fact, two closely adjacent signals are observed in the ^{31}P NMR spectrum of **28** that can be assigned to the two diastereomeric pairs of enantiomers [$\delta = 236.1$ and 238.8 (major signal)]. Although the spirocyclic phosphaalkene **27** exhibits the expected ^{13}C NMR data (see formula, Scheme 17.7), the high-field signal of the phosphorus atom ($\delta = 71.7$) is at first surprising [17]. Because the product is obtained as an oil that has not yet been induced to crystallize, a structure determination by X-ray crystallography cannot be performed and, therefore, attempts were made to resolve the problem by means of a metal complex.

When the $2H$-phosphirene **27** is allowed to react with a metal carbonyl complex of the type $M(CO)_5 \cdot THF$ (M = Cr, W) containing a loosely coordinated molecule of tetrahydrofuran [23], the latter is displaced by the spirocyclic product **27**. The "end-on" complexes **29a** and **29b** are then obtained as yellow, readily crystallizable products. (See Scheme 17.8.) In the chromium complex **29a**, the presence of the metal coordination brings about a paramagnetic shift of the ^{31}P NMR signal by 37 ppm relative to the noncoordinated form **27**. In the tungsten complex **29b**, the opposite effect, namely, a shift to higher field by 23 ppm, is observed [17]; similar phenomena have been reported previously [24]. The phosphorus resonance in **29b** is, furthermore, split by coupling with tungsten, $^1J_{P,W} = 223$ Hz.

The X-ray crystal structure analysis of **29b** unequivocally confirmed the presence of the three-membered ring moiety with a phosphorus–carbon double bond. As expected, the phosphirene ligand takes up one of the octahedral coordination positions of

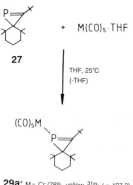

27

+ M(CO)₅·THF

THF, 25°C
(-THF)

(CO)₅M

29a: M = Cr (78%, yellow, ^{31}P: $\delta = 107.2$)

b: M = W (45%, yellow, ^{31}P: $\delta = 48.7$)

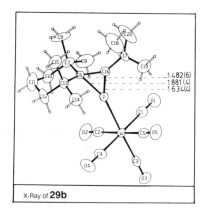

X-Ray of **29b**

Scheme 17.8

tungsten. As in other metal complexes with $\lambda^3\sigma^2$-phosphorus atoms, a practically planar environment at the phosphorus atom was observed. In comparison to other free and complex-bonded phosphaalkenes, the phosphorus–carbon double-bond length [1.634(4) Å] is remarkably short [25, 26] whereas, in contrast, the phosphorus–carbon single bonds in the three-membered ring are approximately 0.03 Å longer than comparable bonds in phosphaalkenes.

17.5. 1H-Phosphirenes

In the preceding section, we have seen the $2H$-phosphirenes are, to a limited extent, accessible via photolysis of $3H$-1,2,4-diazaphospholes, which, in turn, were obtained by 1,3-dipolar cycloadditions of diazoalkanes to phosphaalkynes. The complete reaction sequence ($3a + 12c \rightarrow 13c \rightarrow 25 \rightarrow 26 \rightarrow 27$) raises the question whether carbenes would undergo direct addition to phosphaalkynes and thus open a further access to the $2H$-phosphirenes. Diazo compounds themselves are disqualified as carbene sources because of their own inherent cycloaddition behavior towards phosphaalkynes; hence, we have opted for their cyclic isomers, that is, the diazirines, which are inert towards the phosphaalkynes.

When the diazirines **30a–30e** are subjected to thermolysis in the presence of a large excess of the phosphaalkyne, the 1-chloro-1H-phosphirenes **33a–33e** can be isolated in 24–63% yields after work-up by distillation as stable, colorless oils that are, however, sensitive towards hydrolysis [27]. (See Scheme 17.9.) The pyrolysis temperature in these cases is governed by the decomposition temperature of the respective diazirine. The above results can best be explained as follows: The carbenes **31**, initially generated from **30** by elimination of nitrogen, undergo [2 + 1]-cycloaddition to the phosphaalkyne **3**, yielding the $2H$-phosphirenes **32**. These products, however, cannot be isolated because they rearrange by rapid [1,3]-chlorine shifts to the 1H-isomers **33**. The driving force for this rearrangement is most certainly the degradation of the energy-rich phosphorus–carbon double bond in favor of the formation of a carbon–carbon double bond. Antiaromatic destabilization, such as could be assumed to be significant in the case of **33**, is not operative because the phosphorus atom has a pyramidal configuration (see the discussion on the crystal structure analysis of **33a**) [27]. Exactly the opposite isomerization behavior has been observed for the analogous nitrogen system [28]. Phosphasilirenes [29] and phosphagermirenes [30] (**32**, SiR$_2$ or GeR$_2$ in place of CClR^1, respectively) are accessible by analogous routes but the subsequent substituent shifts were not observed. The first report on a 1-chloro-1H-phosphirene was its detection by Deschamps and Mathey [31] with the aid of ^{31}P NMR spectroscopy.

The ^{31}P NMR resonances of 1-chloro-1H-phosphirenes **33** ($\delta = -29$ to -77) are paramagnetically shifted by more than 100 ppm when compared with the corresponding signals of the 1-alkyl- or 1-aryl-1H-phosphirenes [32]. A noteworthy result from the crystal structure analysis of **33a** [27] is the significant lengthening of the phosphorus–chlorine bond [2.166(2) Å] by about 0.1 Å in comparison to normal phosphorus–chlorine single bond lengths [33]. This observation provides a feasible explanation for the ready nucleophilic exchangeability of the halogen atom, which will be discussed later. Within the limits of the standard deviations, the phosphorus–carbon bonds (1.784 Å) are identical and shorter than those in triphenyl-1H-phosphirene [32a].

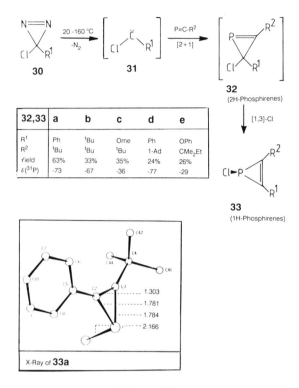

Scheme 17.9

It has been shown for the example of the 1-chloro-1H-phosphirene **33a** that a wide variety of nucleophilic substitution reactions at phosphorus is possible and that two procedures have proven to be valuable. (See Scheme 17.10.)

In the first variant, **33a** is allowed to come into contact with, for example, lithium bis(trimethylsily)phosphide ($\to$ **34**), lithium *tert*-butylacetylide ($\to$ **36**), or lithium di-azo(trimethylsilyl)methane ($\to$ **38**) at -78 °C and the reaction mixtures are allowed to warm up, whereupon quantitative reactions have occurred [27]. In the second variant, the nucleophiles are used in the form of their trimethylsilyl derivatives, for example, (2,2-dimethyl-1-trimethylsioloxypropylidene)-trimethylsilylphosphane ($\to$ **35**, as an E/Z mixture from which one of the species can be separated) and trimethylsilyl azide ($\to$ **37**). Both substitution processes take place smoothly at room temperature and profit from the high silicon–chlorine bond energy. The resonances of the phosphorus atoms in the three-membered ring element ($\delta = -106.7$ to -228.1), of course, depend very strongly on the newly introduced substituents whereby **36** and **37** exhibit ex-tremely high and low limiting values, respectively. The $\lambda^3\sigma^2$- and $\lambda^3\sigma^3$-phosphorus atoms in **35** show the expected large differences in their chemical shift values. The chemistry of these novel, functionalized phosphirenes is still practically unknown.

One as yet unanswered question in the chemistry of low-coordinated phosphorus compounds is that of the existance of the phosphirenylium cations (**39a** $\leftrightarrow$ **39b** $\leftrightarrow$ **39c**,

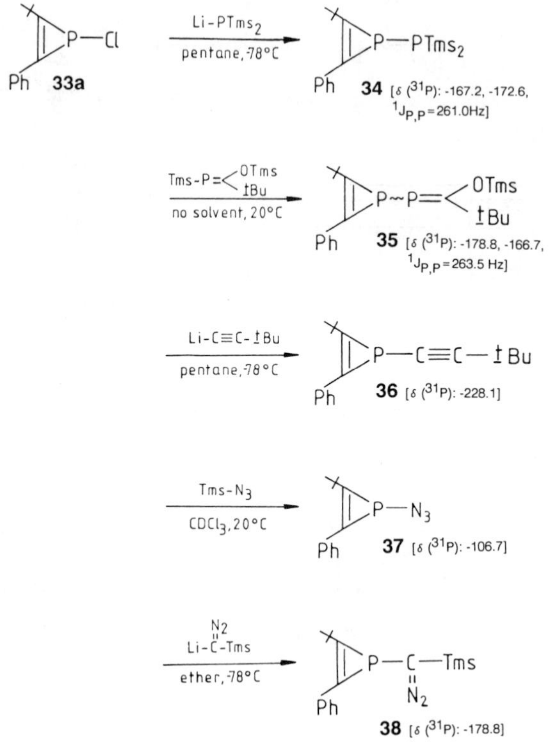

Scheme 17.10

or **39**). This should be considered under the aspect of Hückel aromatic compounds with two π electrons and in the light of the well-investigated cyclopropenylium compounds. The previously described 1-chloro-1H-phosphirenes could contribute to the solution of this question should it ever become possible to "detach" the halogen ionically from the phosphorus atom.

Up to now, model reactions have been carried out only with **33a** [18]. (See Scheme 17.11.) When this 1H-phosphirene is allowed to react with silver tetrafluoroborate in diethyl ether/pentane, the 1-fluoro-1H-phosphirene **41** is finally isolated in quantitative yield; the ^{31}P NMR signal of this product is split by a $^{1}J_{P,F}$ coupling constant of 1062 Hz as is, of course, the ^{19}F NMR signal ($\delta = 18.5$) [18]. A plausible explanation for this result is that the phosphirenylium tetrafluoroborate **40** is formed initially and that its cation extracts a fluorine atom from the BF$_4^-$ counterion to furnish **41** and boron trifluoride. This unexpected reaction, however, does have a parallel in the chemistry of benzenediazonium tetrafluoroborate: thermolysis of the latter results in evolution of nitrogen to give fluorobenzene and BF$_3$ [34].

In another experiment, **33a** was allowed to react with silver triflate; the product of this reaction was **43**, which was obtained in quantitative yield and could be purified by distillation. This fact alone is sufficient to suggest that the product does not exist in the ionic form **42** even though the ^{31}P NMR signal appears at relatively very low field

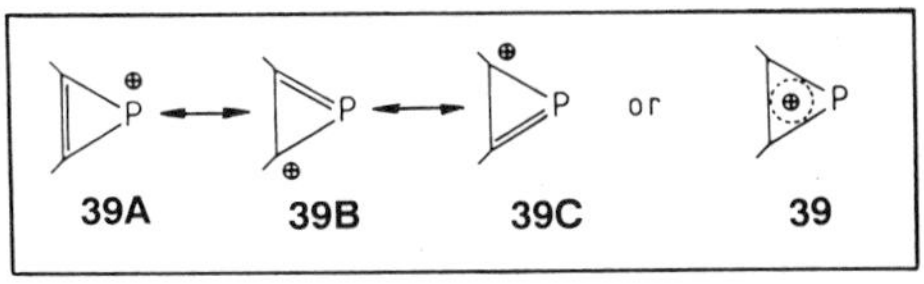

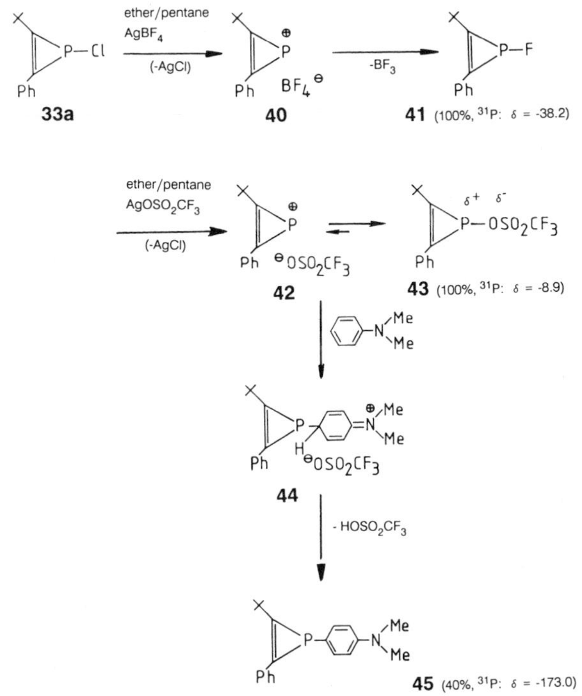

Scheme 17.11

($\delta = -8.9$). On the other hand, this value does indicate that the phosphorus atom in **43** does at least carry a partial positive charge, that is, that the phosphorus–oxygen is strongly polarized. This situation is also indicated by the observed solvent dependency of the ^{31}P NMR signal of **43** [18]. Further evidence in support of this assumption is provided by the electrophilic substitution reaction of N,N-dimethylaniline with **43**. The σ-intermediate in the process leading to the 1-aryl-1H-phosphirene **45** is **44**. We can speculate that the substitution reaction either starts from the highly polarized **43** or that a low equilibrium proportion of the phosphirenium triflate **42** is successively trapped by the aniline derivative [18].

17.6. Valence Isomers of Phosphabenzenes

Phosphabenzenes (phosphinenes) represent a class of thoroughly investigated heteroarenes [35] and have played a decisive role in the development of the chemistry of low-coordinated phosphorus compounds [4]. The classical valence isomers of the phosphabenzenes recently became accessible for the first time and their development was further stimulated by results obtained with the valency isomers of benzene. The reaction partners cyclobutadiene and acetylene have provided a comprehensive and versatile route to the benzene derivatives [36] and so it was to be expected that reactions of the same cyclic 1,3-dienes with phosphaalkynes would render the title compounds accessible.

When the kinetically stabilized cyclobutadienes **46** (R′ = Me, *t*-Bu) are allowed to react with the phosphaalkynes **3a–3e** at temperatures between − 78 and + 20 °C in pentane, the corresponding adducts are formed in quantitative yields. (See Scheme 17.12.) Only the reaction of **46** (R′ = Me) with **3d** is regiospecific, as demonstrated by the formation of the 2-Dewar phosphabenzene **47** (R′ = Me, R = 1-Ad); all other reactions of the type **46** + **3**, which may be formally classified as Diels–Alder reactions, are regioselective and products of the type **47** (≥ 85%) clearly predominate over those of the type **48** (≤ 15%) [37]. The major reaction (→ **47**) is sterically controlled, as can be deduced from the fact that the proportion of **48** noticeably increases on going from the methyl to the *tert*-butyl ester **46**; products **48** must then be assumed to be the result of an electronically controlled reaction. Both isomers exhibit the typical ^{31}P NMR signals for $\lambda^3\sigma^2$-phosphorus atoms in a narrow range (δ = 312–317). A complete set of ^{13}C NMR data for the skeletal carbon atoms of the 2-Dewar phosphabenzene **47**

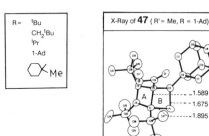

Scheme 17.12

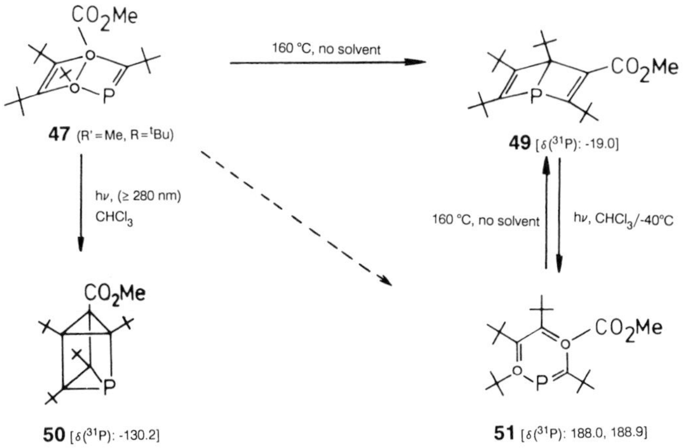

Scheme 17.13

(R' = Me, R = i-Pr) is listed in the formula scheme; in addition, the structure of the 2-Dewar phosphabenzene **47** (R' = Me, R = 1-Ad) has been confirmed by a crystal structure analysis [37]. With a length of 1.675 Å, the phosphorus–carbon double bond of this compound is somewhat longer than that of **29b**; the folding angle of the rings A and B amounts to 113.0°. These bicyclic products **47** are surprisingly stable towards oxygen and moisture; they can also be subjected to considerable thermal stress before isomerization processes take place (see the following paragraphs).

When the 2-Dewar phosphabenzene **47** (R' = Me, R = t-Bu) is heated in the absence of a solvent at 160 °C, it undergoes isomerization to furnish the 1-isomer **49** (see Scheme 17.13); Dewar benzenes of this type were also previously unknown [38]. It is logical to assume that thermodynamic control is responsible for the skeletal rearrangement because it is associated with a transformation of $\lambda^3\sigma^2$- to $\lambda^3\sigma^3$-phosphorus. This is also apparent from the enormous diamagnetic shift of the ^{31}P NMR signal from $\delta = 315.0$ to $\delta = -19.0$.

An obvious but, at first glance, difficultly conceivable assumption is that the phosphabenzene **51** is an intermediate in the transformation of **47** (R' = Me, R = t-Bu) to **49**; this would involve cleavage of the zero atom bridge in the 2-Dewar phosphabenzene and subsequent ring closure between the phosphorus atom and the carbon atom at position 4 in the final product (**51 → 49**); that means that, of this set of three isomers, not the heteroarene but rather the bicyclic member is the energetically more stable. In the first place, the phosphabenzene **51** can be prepared photochemically from **49**. Compound **51** is thermally rather stable but does isomerize, as already mentioned, under authentic thermal conditions (160 °C in the absence of a solvent) to reform **49** [38]. This behavior can be understood when the particular steric situation of **51**, which will be discussed later, is taken into consideration.

It is known that Dewar benzenes with highly sterically demanding substitution patterns can be converted very selectively via intramolecular [2 + 2]-cycloaddition processes into prismanes under photochemical conditions [36]. Thus, when the 2-Dewar phosphabenzene **47** (R' = Me, R = t-Bu) is appropriately excited, a very rapid reaction takes place to give **50**, the first representative of the previously unknown phosphaprismanes [38]. As expected for compounds containing a phosphirane unit, the ^{31}P NMR signal appears at surprisingly high field ($\delta = -130.2$). It is interesting to note that, when it is interrupted prematurely after an appropriate period of time, the photoreaction comes to a standstill at the stage of the tetracyclic product; when this is not the case, a further, subsequent skeletal rearrangement takes place to furnish the isomeric phosphabenzvalenes [38], which will not be discussed here.

The crystal structure analysis of the phosphabenzene 51 [39] (see Scheme 17.14) provides some decisive information for an understanding of its key role in the 2-Dewar

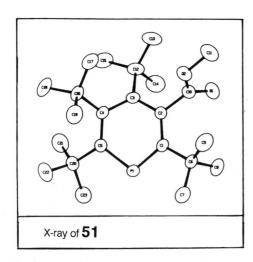

X-ray of **51**

Bond distances [Å] :

P1-C1	1.730(2)
P1-C5	1.758(2)
C1-C2	1.404(3)
C2-C3	1.411(3)
C3-C4	1.437(3)
C4-C5	1.402(3)

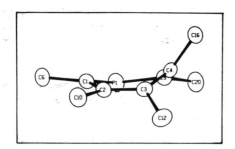

Side profile of the phosphinine ring (including neighbouring substituent atoms)

Deviation from planarity

$\alpha =$ 33.7°
$\beta =$ 13.9°
$\gamma =$ 17.9°
$\delta =$ 8.2°
$\Sigma \alpha + \gamma = $ 51.6°

Scheme 17.14

phosphabenzene/1-Dewar phosphabenzene isomerization process shown in Scheme 17.13. This heteroaromatic compound is no longer planar as a consequence of the accumulation of sterically demanding substituents. It actually exists in a boat conformation with bow and stern angles of $\alpha = 33.7°$ and $\gamma = 17.9°$. Even the plane of atoms C-2, C-3, C-5, and P is twisted and, furthermore, the *tert*-butyl groups at C-1 and C-4 are also inclined ($\beta = 13.9°$, $\delta = 8.2°$), as can be clearly seen from the side profile view of the phosphabenzene ring (Scheme 17.14).

In other words, **51** constitutes the most deformed, nonannelated and nonbridged 6π-system known to date and, of course, the aromatic stabilization suffers from this fact.

17.7. Phosphaaromatic Compounds

The classical synthesis of phosphabenzenes is based on the condensation reaction of pyrylium salts with a synthetic equivalent of PH_3 [35]. The Diels–Alder behavior of the phosphaalkynes towards cyclic 1,3-dienes in which the 1 and 4 positions are joined by a good leaving group [$C(=O)-O$, $C=O$, or $Ph-P(=S)$] opens up new perspectives in phosphabenzene chemistry. Because phosphaalkynes with bulky substituents are thermally very stable, they are well able to withstand the drastic reaction conditions frequently necessary to bring about the cycloaddition processes (temperatures of approximately 110–140 °C, glass pressure vessels). Only a few significant examples of the numerous known processes [40, 41] will be discussed here.

Thus, the reaction of **3a** with α-pyrone (**52**) gives rise to 2-*tert*-butylphosphabenzene (**54**) in high yield [40]. (See Scheme 17.15.) Because the cycloelimination of CO_2 that follows the [4 + 2]-cycloaddition is very rapid, there is no chance of detecting the heterobicyclic intermediate **53**. Even more complicated phosphaarenes such as **57** are accessible by this route. In the latter case, the formation of the intermediate **56** from **3a** and the condensed α-pyrone **55** is most certainly sterically controlled; this can be deduced from the positions of the two alkyl groups on the phosphabenzene ring, as

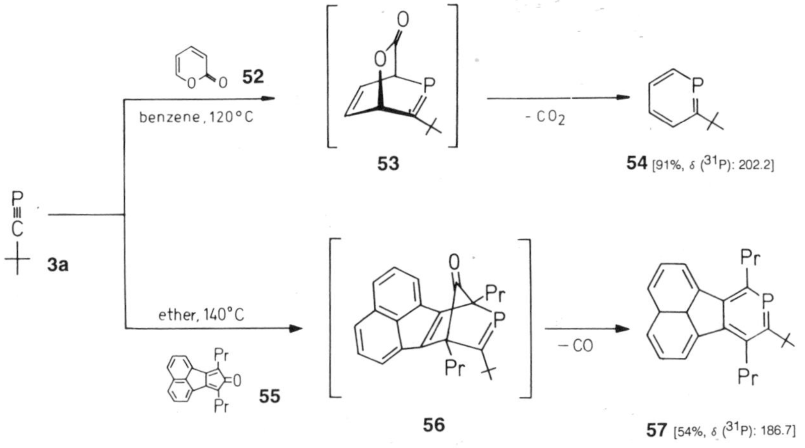

Scheme 17.15

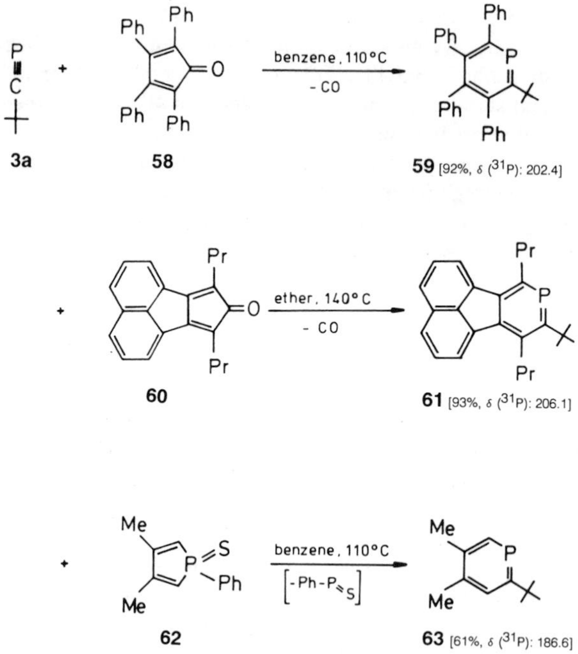

Scheme 17.16

determined by NMR spectroscopic analysis [41]. Typical characteristics for the structure of the two heteroarenes are the low-field positions ($\delta = 202.2$ and 186.7, respectively) of the phosphorus signals in their ^{31}P NMR spectra.

The reactions of 3a with the cyclopentadienes 58 and 60 (Scheme 17.16) also proceed without significant side-product formation to give, after elimination of carbon monoxide, the highly substituted phosphabenzenes 59 [40] and 61 [41].

The leaving group in the cyclic 1,3-diene 62 is of special interest; phenyl(thioxo)-phosphane (Ph—P=S) constitutes a compound containing a $\lambda^3\sigma^2$-phosphorus atom, as does also the product of the reaction, 63 [40]. However, in contrast to the heteroarene, Ph—P=S has only a short lifetime but it can be trapped by various means [42]. The phosphorus atoms in the products 59, 61, and 63 all show the expected low-field signals ($\delta = 186.6$–206.1) in their ^{31}P NMR spectra.

17.8. Polycyclic Phosphaalkenes

Diels–Alder reactions of phosphaalkynes with cyclic 1,3-dienes that, in contrast to 52, 55, 58, 60, or 62, do not possess an easily cleavable bridging unit open up a new possibility to prepare bicyclic phosphaalkenes that, until recently, were also unknown. Thus, the reaction of 3a with cyclohexa-1,3-diene at 120 °C in benzene (Schlenk tube, under pressure) gives rise to 2-phosphabicyclo[2.2.2]octa-2,5-diene (64) exclusively [40]

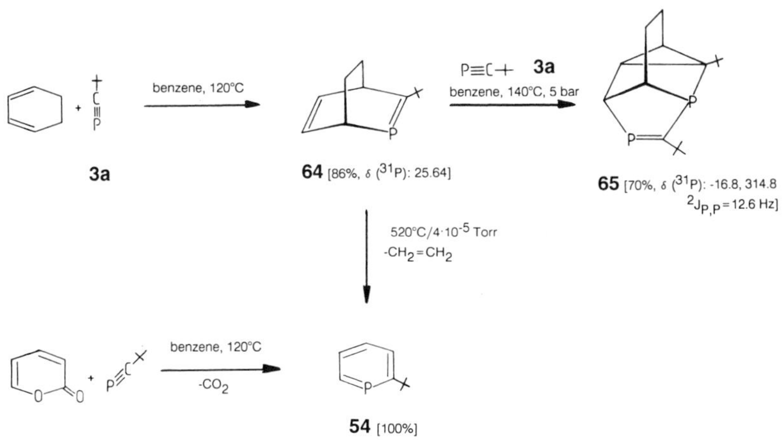

Scheme 17.17

as a thermally very stable product. (See Scheme 17.17.) Under flash vacuum pyrolysis conditions, this product undergoes the retro-Diels–Alder reaction, furnishing ethylene and 2-*tert*-butylphosphabenzene (**54**, previously obtained from the reaction of **3a** with α-pyrone) [40]; neither the retro reaction (to **3a** and cyclohexadiene) nor the formation of acetylene and 3-*tert*-butyl-2-phosphacyclohexa-1,3-diene take place. Anthracene also reacts analogously with **3a** and thus provides an access to the series of phosphabarrelenes [43].

The phosphabicyclic product **64** fulfills the constitutional prerequisites for a homo-Diels–Alder reaction—a very rare type of reaction in carbon chemistry. In organophosphorus chemistry, this reaction type in combination with the actual Diels–Alder reaction makes it possible to incorporate both $\lambda^3\sigma^2$- and $\lambda^3\sigma^3$-phosphorus atoms in the same polycyclic system.

Thus, under the "usual" conditions, the reaction of **3a** with **64** gives rise to the diphosphatetracyclodecene **65** [44]. The signals for the dicoordinated and tricoordinated phosphorus atoms are observed in the expected extreme positions ($\delta = 314.8$ and -16.8, respectively) in the ^{31}P NMR spectrum. The phosphorus–phosphorus coupling constant of only 12.6 Hz indicates that the two heteroatoms are not in directly adjacent positions. The orientation of the two reaction partners in the cycloaddition process is most certainly electronically controlled.

Cyclopentadiene reacts in principle with **3a** analogously to cyclohexa-1,3-diene except that the reaction takes place already at room temperature. However, the 2-phosphabicyclo[2.2.1]hepta-2,5-diene (^{31}P NMR, $\delta = 243.7$) cannot be isolated and this is apparently a result of an "oligomerization" process [41].

A further, unmistakeable indication for the smooth [4 + 2]-cycloaddition process between cyclopentadiene and **3a** is the formation of the product **66** from the 1:2 molar reaction of these substrates at 180 °C. (See Scheme 17.18.) The homo-Diels–Alder reaction is only conceivable when it is preceded by a normal Diels–Alder reaction [44]. The 1:2 reactions of dimethyl 5,5-dimethylcyclopentadiene-2,3-dicarboxylate and 2,4-di-*tert*-butylcyclopentadiene with **3a** that finally give rise to the tetracyclic products **67**

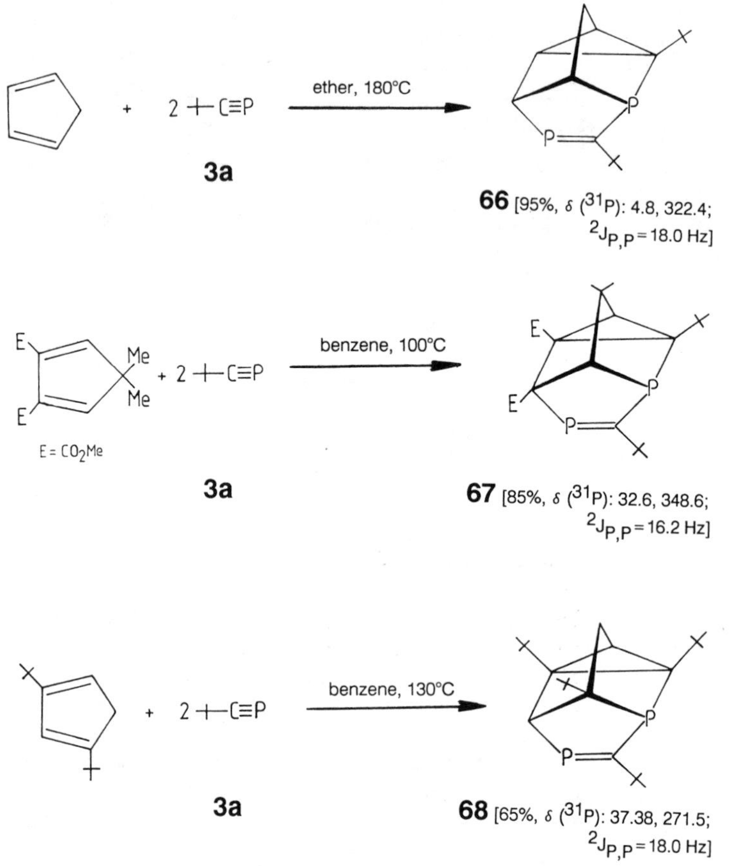
ether, 180°C
3a
66 [95%, δ (31P): 4.8, 322.4; 2JP,P = 18.0 Hz]
E
Me
Me
E = CO2Me
benzene, 100°C
3a
67 [85%, δ (31P): 32.6, 348.6; 2JP,P = 16.2 Hz]
benzene, 130°C
3a
68 [65%, δ (31P): 37.38, 271.5; 2JP,P = 18.0 Hz]

Scheme 17.18

and **68**, respectively [41], show that this sequence is capable of generalization. The ^{31}P NMR spectroscopic behavior of compounds **66**–**68** is very similar to that of **65** (described previously).

One limitation to the synthesis of polycyclic compounds from cyclic 1,3-dienes and phosphaalkynes is illustrated by the reaction of 1,2,3,4,5-pentamethylcyclopentadiene with **3a**. (See Scheme 17.19.) The expected product **71** is not formed, either as a result of general steric reasons or because of the "electron-richness" of the diene. Instead, an ene reaction to yield the phosphavinylcyclopentadiene **70** [44] occurred. Up to this point, ene reactions of this type were unknown in phosphaalkyne chemistry. On the basis of the generally accepted ideas concerning the course of ene reactions, the transition state **69** should play a decisive role in the formation of the product **70** obtained. The ^{1}H NMR of **70** at room temperature is at first surprising: It shows only one signal ($\delta = 1.69$, $^{3}J_{H,P} = 1.8$ Hz) for 15 methyl protons. This phenomenon can only

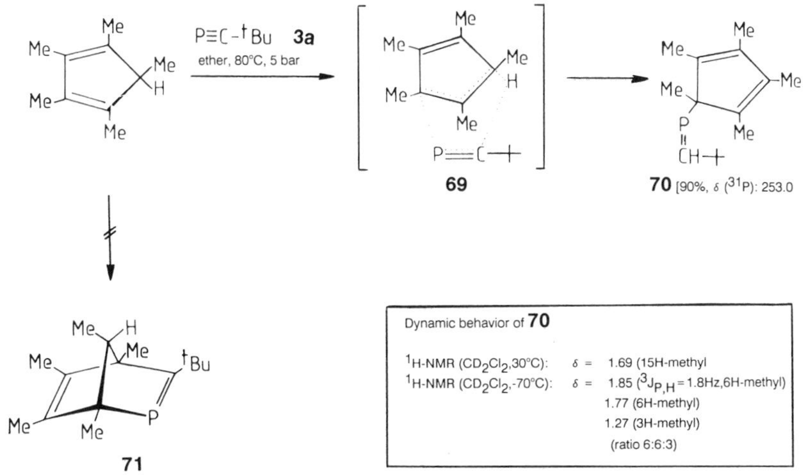

Scheme 17.19

be explained by the assumption that the phosphaalkene moiety is undergoing a rapid, degenerate sigmatropic [1,5]-shift.

On varying the measurement temperature, it is seen that the spectrum exhibits a coalescence point at about -40 °C and then, at -70 °C, three methyl signals are observed ($\delta = 1.27, 1.77$, and 1.85; ratio $3:6:6$), as would be expected for compound **70** in the absence of dynamic behavior. The enthalpy of activation for this process amounts to 45.7 kJ mol^{-1} [41,44].

17.9. Diphosphatricycloalkenes

Phosphabenzenes possessing a methyl group in an α-position can, at least formally, be considered as enophiles. Whether or not this is actually the case in reactions with the phosphaalkyne **3a** can only be determined by experiment. The phosphabenzenes need not be employed as such, it is sufficient to allow an appropriate α-pyrone to react with two equivalents of **3a**. The phosphabenzenes, for example, **72a–72c**, are formed as intermediates (for comparison, see the reactions of **3a** with **52** to **54**) and are thus available for the possible ene reaction.

The reaction with the second equivalent of **3a** takes place as shown by the arrows in Scheme 17.20 to bring about the specific formation of a phosphorus–phosphorus bond (**73**). Because this product fulfills the structural prerequisites for an intramolecular Diels–Alder reaction, it cannot be isolated. The reaction comes to a stop only after an apparently rapid cycloaddition process with formation of the diphosphatricyclooctenes **74a–74c** [41]. At this point, it is sufficient to state that the NMR spectroscopic data obtained are in full accord with the proposed structure. This applies in particular to the ^{31}P NMR data: The chemical shifts ($\delta = -169.6$ to -192.5) and the mutual coupling constants ($^{1}J_{P,P} = 144.0$–147.0 Hz) unambiguously demonstrated the presence of the

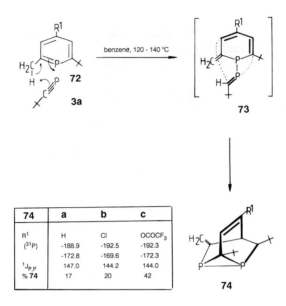

Scheme 17.20

diphosphirane structural unit [41]. Unexpectedly, compounds with the same skeleton are formed when acyclic 1,3-dienes are allowed to react with **3a**. Thus, the diphospha-tricyclooctene **77** is obtained in an optimized yield of 90% when the reaction partners are employed in a diene : phosphaalkyne ratio of 1 : 2 [45]. (See Scheme 17.21.)

The mechanism for the formation of the tricyclic product can be interpreted as follows: The reaction sequence commences with a Diels–Alder reaction to furnish the

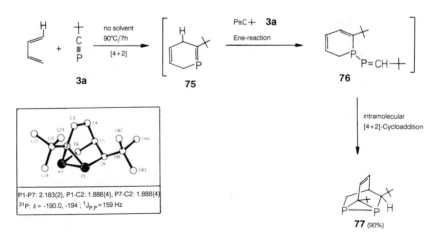

Scheme 17.21

intermediate **75**. Although this step is obvious, it was not recognized for a long time because the following steps could not be elucidated. The sequence now proceeds in the same way as the reactions of **72a–c** with **3a**, namely, an ene reaction (**75** + **3a** → **76**) and the concluding intramolecular [4 + 2]-cycloaddition process (**76** → **77**). When substituted 1,3-butadienes are allowed to react with **3a**, the initial Diels–Alder reaction is generally not regiospecific but this is not the case for the subsequent reactions. The end result is the formation of two isomeric diphosphatricyclooctenes with exchanged substituents on the carbon atoms originating from the 1,3-butadiene [45].

As in the case of products **74a–74c**, the ^{31}P NMR signals of **77** ($\delta = -190.0$ and -194.5) with a $^1J_{P,P}$ coupling constant of 159 Hz unambiguously demonstrate that the two phosphorus atoms originating from the two phosphaalkyne molecules now form part of a three-membered ring. An X-ray crystallographic analysis [45] was necessary to clarify completely the situation in the reaction product **77**. This provided the following values for the bond lengths of the three-membered ring: P-1—P-7 = 2.183(2) Å; P-1—C-2 = 1.886(4) Å; and P-7—C-2 = 1.888(4) Å.

17.10. Cyclooligomerization

The aforementioned results have clearly shown that the reactivity of the phosphaalkynes is very similar to that of the acetylenes. Furthermore, the numerous reaction possibilities of the phosphaalkynes have provided new aspects in the chemistry of organophosphorus compounds. The same is valid for their ligand behavior, which has also given rise to new and exciting incentives in complex chemistry [1c, 1d, 46]. As a typical example, we will here only mention the cyclodimerization in the coordination sphere of cobalt, which, at the same time, raises the question whether phosphaalkynes can also be induced thermally to undergo cyclooligomerization.

In analogy to the results obtained from the cyclooligomerization of alkynes with the bis(ethene)cobalt complex **78** [47], it was also possible to bring about cyclodimerization of the phosphaalkyne **3a** [48–50]. (See Scheme 17.22.) The reaction takes place regiospecifically in diethyl ether/pentane of -30 °C and gives rise to the η^5-cyclopentadienyl(η^4-1,3-diphosphacyclobutadiene)cobalt complex **79** in 67% yield. When a pronounced molar excess of **78** is employed, binuclear and trinuclear complexes are formed in which, finally, ethylene is expelled and the phosphorus atom of the diphosphacyclobutadiene system becomes involved in the coordination [49, 50].

As expected, only one signal for the two magnetically equivalent *tert*-butyl groups ($\delta = 0.91$) is observed in the ^{1}H NMR spectrum of **79**. This is also the case for the ring carbon atoms in the ^{13}C NMR spectrum of the hetero-antiaromatic compound ($\delta = 107.5$). The X-ray crystal structure analysis of **79** shows a phosphorus–carbon bond length in the η^4 ligand of 1.797 Å and that the two *tert*-butyl groups are inclined downwards with respect to the cyclopentadienyl ring plane [48, 49].

From a mechanistic point of view, the cyclodimerization presumably proceeds in such a way that, initially, the two "side-on" coordinated ethylene molecules of **78** are each replaced by a phosphaalkyne ligand (→ **80**). This process is followed by a coupling of the first phosphorus–carbon bonds to give the diphosphacobaltacene **81** from which the 1,3-diphosphacyclobutadiene **79** in the energetically more favorable complexed form results. This reaction sequence is supported by a series of experimentally sound arguments [51].

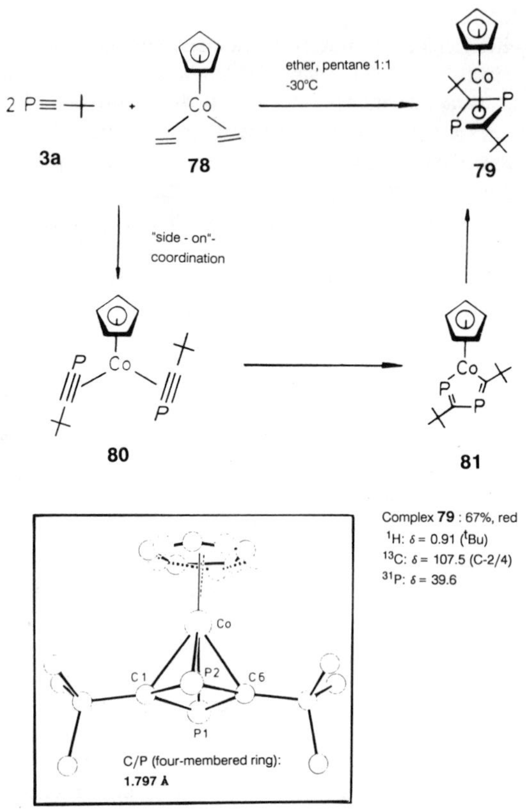

Scheme 17.22

Of course, the aforementioned results in conjunction with those obtained from another cyclodimerization type [52], which will not be discussed here, and from cyclotrimerization reactions [53, 54] in the coordination sphere of metals open the question of whether such reactions can also be initiated thermally. Preliminary results will be mentioned later.

When the phosphaalkyne 3a is heated in a glass pressure vessel at 180 °C for about 3 h and subsequently worked up by bulb-to-bulb distillation, a crystalline, pale yellow substance that can be sublimed is obtained. On the basis of its spectroscopic and analytical data, this product has been unequivocally identified as a cyclotetramer, that is, as tetra-*tert*-butyltetraphosphacubane (84) [55]. (See Scheme 17.23.) The structure of a further tetramer, presumed to be tricyclic, has not yet been elucidated completely [55].

The mass spectrum of 84 exhibits not only the molecular ion peak but also shows that the cyclotetramer is systematically degraded down to the monomeric units 3a. Disregarding the signals for the *tert*-butyl groups in the ^1H and ^{13}C NMR spectra, the resonances for the skeletal atoms are extremely interesting. The ^{31}P NMR signal appears at an almost unbelievably low-field position ($\delta = +257.4$; completely atypical

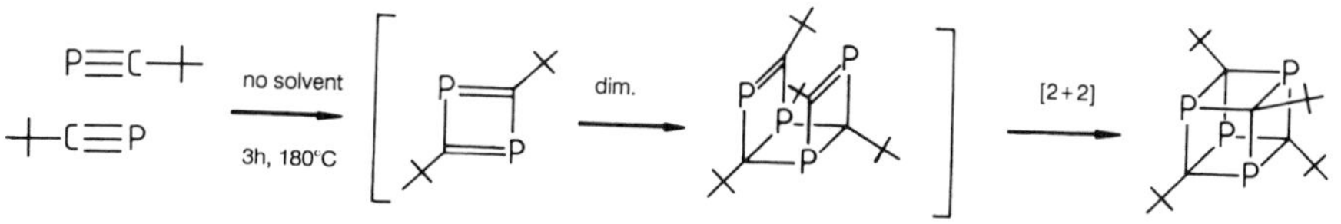

$$P\equiv C \!-\!\!< \quad + \quad \!\!>\!\!-C\equiv P \qquad \xrightarrow[\text{3h, 180°C}]{\text{no solvent}} \qquad \mathbf{82} \qquad \xrightarrow{\text{dim.}} \qquad \mathbf{83} \qquad \xrightarrow{[2+2]} \qquad \mathbf{84}$$

3a   **82**   **83**   **84**

84   mp ≈ 241°C (pale yellow), can be sublimed
MS: m / z = 400 (M +); degradation to **3a**
^{1}H: δ = 1.10; ^{13}C: δ = −29.07 ^{31}P: δ = 257.4

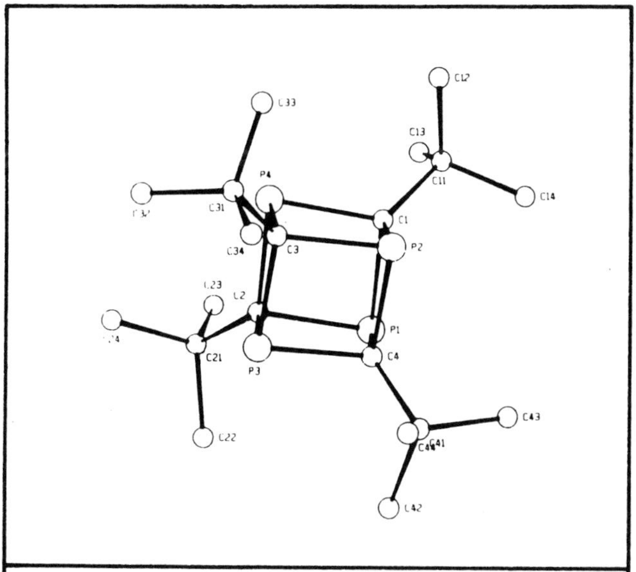

P-C: 1.881 Å; P-C-P: 94.4; C-P-C: 85.6

Scheme 17.23

for a $\lambda^3\sigma^3$-phosphorus atom) and the ^{13}C NMR signal is at $\delta = -29.7$ (certainly also a rarity). Both nuclei form part of an AX_3Y spin system and the A part of this can be simulated most satisfactorily. The X-ray crystal structure analysis of **84** shows a distorted cube structure with interior angles that are in part larger (P—C—P = 94.4°) and in part smaller (C—P—C = 85.6°) than those of cubane [56]. The average phosphorus–carbon bond length amounts to 1.881 Å and must be considered as normal.

From a mechanistic point of view, the formation of the tetraphosphacubane can only be explained in terms of the assumption of an initial head-to-tail dimerization step of **3a** to give **82**. The question whether this dimerizes to form the tetracyclic product **83** or, rather, undergoes cycloaddition with two further molecules of **3a** to produce **83** is still unanswered and, at present, irrelevant. Subsequent intramolecular phosphalkene dimerization (intermolecular parallels of which are known) finally leads to the end product **84** [55].

17.11. Summary

The phosphaalkynes **3** were only recognized as useful and versatile synthetic building blocks just a few years ago and have now become indispensable for studies on the chemistry of low-coordinated phosphorus compounds (Scheme 17.24). Numerous novel substances and new routes to other known classes of compounds have been achieved with their help. The 2*H*- and 1*H*-phosphirenes (**I** and **II**, respectively) are accessible from phosphaalkynes by the carbene route. Phospholes (**III**) with an extremely wide variety of other heteroatoms can be prepared through 1,3-dipolar cycloaddition reactions. The synthetic potential of the phosphaalkynes as dienophiles is practically

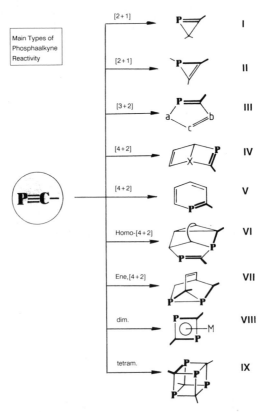

Scheme 17.24

inexhaustible. The 2-Dewar phosphabenzenes **IV** can be transformed to the other classical valency isomers of phosphabenzene. Furthermore, a direct route to the heteroarenes **V** consists of the reaction of cyclic 1,3-dienes containing easily removable ring units with phosphaalkynes. Combinations of Diels–Alder and homo-Diels–Alder reactions (→ **VI**), as well as of Diels–Alder, ene, and intramolecular [4 + 2]-cycloaddition reactions (→ **VII**) result in the formation of diphosphapolycyclic compounds. Dimerizations can be achieved within the coordination sphere of cobalt (→ **VIII**). Finally, the purely thermal cyclotetramerization of **3** to furnish the tetraphosphacubane system **IX** has opened up a new field of phosphaalkyne chemistry that will probably provide many new and surprising results in the future.

References

1. Reviews: (a) Appel, R.; Knoll, F.; Ruppert, I. *Angew. Chem.*, **1981**, *93*, 771; *Angew. Chem., Int. Ed. Engl.*, **1981**, *20*, 731. (b) Regitz, M. in *Houben–Weyl, Methoden der Organischen Chemie*, Regitz, M., ed., Vol. E1, Thieme, Stuttgart, 1982, p. 28. (c) Regitz, M.; Binger, P. *Angew. Chem.*, **1988**, *100*, 1541; *Angew. Chem., Int. Ed. Engl.*, **1988**, *27*, 1484. (d) Regitz, M.; Binger, P. in *Multiple Bonds and Low Coordination in Phosphorus Chemistry*, Regitz, M., and Scherer, O.J., eds., Thieme, Stuttgart, 1990, in press.
2. Gier, T. E. *J. Am. Chem. Soc.*, **1961**, *83*, 1769.
3. Dimroth, K.; Hoffmann, P. *Angew. Chem.*, **1964**, *76*, 433; *Angew. Chem., Int. Ed. Engl.*, **1964**, *3*, 384.
4. Märkl, G. *Angew. Chem.*, **1966**, *78*, 907; *Angew. Chem., Int. Ed. Engl.*, **1966**, *5*, 846.
5. Fuchs, E. P. O.; Hermesdorf, M.; Schnurr, W.; Rösch, W.; Heydt, H.; Regitz, M. *J. Organomet. Chem.*, **1987**, *338*, 329.
6. Becker, G.; Gresser, G.; Uhl, W. *Z. Naturforsch. B*, **1981**, *36*, 16.
7. Becker, G.; Hölderich, W. *Chem. Ber.*, **1975**, *108*, 2484.
8. Becker, G. *Z. Anorg. Allg. Chem.*, **1977**, *430*, 66: **6a**, ^{31}P NMR, $\delta = -107.0$.
9. Allspach, T.; Regitz, M.; Becker, G.; Becker, W. *Synthesis*, **1986**, 31: **6d**, ^{31}P NMR, $\delta = -110.6$.
10. Fritz, G. *Z. Anorg. Allg. Chem.*, **1977**, *422*, 104.
11. Rösch, W.; Vogelbacher, U.; Allspach, T.; Regitz, M. *J. Organomet. Chem.*, **1986**, *306*, 39.
12. Regitz, M.; Rösch, W.; Allspach, T.; Annen, U.; Blatter, K.; Fink, J.; Hermesdorf, M.; Heydt, H.; Vogelbacher, U.; Wagner, O. *Phosphorus and Sulfur*, **1987**, *30*, 479.
13. Rösch, W.; Hees, U.; Regitz, M. *Chem. Ber.*, **1987**, *120*, 1645.
14. Rösch, W.; Regitz, M. *Angew. Chem.*, **1984**, *96*, 898; *Angew. Chem., Int. Ed. Engl.*, **1984**, *23*, 900.
15. Yeng Lam Ko, Y. Y. C.; Carrié, R.; Münch, A.; Becker, G. *J. Chem. Soc., Chem. Commun.*, **1984**, 1634.
16. Weinmaier, J. H.; Brunnhuber, G.; Schmidpeter, A. *Chem. Ber.*, **1980**, *112*, 2278.
17. Wagner, O.; Maas, G.; Regitz, M. *Angew. Chem.*, **1987**, *99*, 1328; *Angew. Chem., Int. Ed. Engl.*, **1987**, *26*, 1257.
18. Regitz, M.; Wagner, O., unpublished results, Kaiserslautern, 1988.
19. Rösch, W.; Facklam, T.; Regitz, M. *Tetrahedron*, **1987**, *43*, 3247.
20. Rösch, W.; Regitz, M. *Synthesis*, **1987**, 689.
21. Hermesdorf, M.; Birkel, M.; Heydt, H.; Regitz, M.; Binger, P. *Phosphorus and Sulfur*, **1989**, *29*, 31.
22. Rösch, W.; Richter, H.; Regitz, M. *Chem. Ber.*, **1987**, *120*, 1809.
23. Strohmeier, W.; Müller, F. J. *Chem. Ber.*, **1969**, *102*, 3608.
24. Weinmaier, J. H.; Tautz, H.; Schmidpeter, A.; Pohl, S. *J. Organomet. Chem.*, **1980**, *185*, 53.

25. Becker, G.; Mundt, O. *Z. Anorg. Allg. Chem.*, **1978**, *443*, 53.
26. Klebach, T. C.; Lourens, R.; Bickelhaupt, F.; Stam, C. H.; van Herk, A. *J. Organomet. Chem.*, **1981**, *210*, 211.
27. Wagner, O.; Ehle, M.; Regitz, M. *Angew. Chem.*, **1989**, *101*, 227; *Angew Chem.*, *Int. Ed. Engl.*, **1989**, *28*, 225.
28. Anderson, D. J.; Gilchrist, T. L.; Rees, C. W. *J. Chem. Soc., Chem. Commun.*, **1969**, 147.
29. Schäfer, A.; Weidenbruch, M.; Saak, W.; Pohl, S. *Angew. Chem.*, **1987**, *99*, 806; *Angew. Chem.*, *Int. Ed. Engl.*, **1987**, *26*, 776.
30. Cowley, A. H.; Hall, S. W.; Nunn, C. M.; Power, J. M. *J. Chem. Soc., Chem. Commun.*, **1988**, 753.
31. Deschamps, B.; Mathey, F. *New J. Chem.*, **1988**, *12*, 755.
32. (a) Marinetti, A.; Mathey, F.; Fischer, J.; Mitschler, A. *J. Chem. Soc., Chem. Commun.*, **1984**, 45. (b) Lochschmidt, S.; Mathey, F.; Schmidpeter, A. *Tetrahedron Lett.*, **1986**, *27*, 2635.
33. *International Tables of X-Ray Crystallography*, Vol. 3, Kynach Press, Birmingham, 1968, p. 266.
34. Balz, G.; Schiemann, G. *Ber. Dtsch. Chem. Ges.*, **1927**, *60B*, 1186.
35. Märkl, G. in *Houben–Weyl, Methoden der Organischen Chemie*, Regitz, M., ed., Vol. E1, Thieme, Stuttgart, 1982, p. 72.
36. Wingert, H.; Regitz, M. *Chem. Ber.*, **1986**, *119*, 244; Wingert, H.; Regitz, M. *Z. Naturforsch. B*, **1986**, *41*, 1306; Wingert, H.; Maas, G.; Regitz, M. *Tetrahedron*, **1986**, *42*, 5341.
37. Fink, J.; Rösch, W.; Vogelbacher, U.-J.; Regitz, M. *Angew. Chem.*, **1986**, *98*, 265; *Angew. Chem.*, *Int. Ed. Engl.*, **1986**, *25*, 280.
38. Blatter, K.; Rösch, W.; Vogelbacher, U.-J. Fink, J.; Regitz, M. *Angew. Chem.*, **1987**, *99*, 67; *Angew. Chem.*, *Int. Ed. Engl.*, **1987**, *26*, 85.
39. Maas, G.; Fink, J. Wingert, H.; Blatter, K.; Regitz, M. *Chem. Ber.*, **1987**, *120*, 819.
40. Rösch, W.; Regitz, M. *Z. Naturforsch. B*, **1986**, *41*, 931.
41. Regitz, M.; Annen, U., unpublished results, Kaiserslautern, 1988.
42. Hussong, R.; Heydt, H.; Regitz, M. *Phosphorus and Sulfur*, **1985**, *25*, 20 (addition of alcohols); Hussong, R.; Heydt, H.; Regitz, M. *Z. Naturforsch. B*, **1986**, *41*, 915 ([4 + 1] cycloaddition); Hussong, R.; Heydt, H.; Maas, G.; Regitz, M. *Chem. Ber.*, **1987**, *120*, 1263 [trapping reaction with hexacarbonyl bis(cyclopentadienyl)dimolybdenum].
43. Annen, U.; Regitz, M. *Tetrahedron Lett.*, **1987**, *28*, 5141.
44. Annen, U.; Regitz, M. *Tetrahedron Lett.*, **1988**, *29*, 1681.
45. Fuchs, E. P. O.; Rösch, W.; Regitz, M. *Angew. Chem.*, **1987**, *99*, 1058; *Angew. Chem.*, *Int. Ed. Engl.*, **1987**, *26*, 1011.
46. Review: Nixon, J. F. *Chem. Rev.*, **1988**, *88*, 637.
47. Jonas, K. *Angew. Chem.*, **1985**, *97*, 292; *Angew. Chem.*, *Int. Ed. Engl.*, **1985**, *24*, 295.
48. Hitchcock, P. B.; Maah, M. J.; Nixon, J. F. *J. Chem. Soc., Chem. Commun.*, **1986**, 737.
49. Binger, P.; Milczarek, R.; Mynott, R.; Regitz, M.; Rösch, W. *Angew. Chem.*, **1986**, *98*, 645; *Angew. Chem.*, *Int. Ed. Engl.*, **1986**, *25*, 644.
50. Binger, P.; Milczarek, R.; Mynott, R.; Tsai, Y.-H. Raabe, E.; Krüger, C.; Regitz, M. *Chem. Ber.*, **1988**, *121*, 637.
51. Binger, P.; Biedenbach, B.; Mynott, R.; Krüger, C.; Betz, P.; Regitz, M. *Angew. Chem.*, **1988**, *100*, 1219; *Angew. Chem.*, *Int. Ed. Engl.*, **1988**, *27*, 1157.
52. Binger, P.; Biedenbach, B.; Krüger, C.; Regitz, M. *Angew. Chem.*, **1987**, *99*, 798; *Angew. Chem.*, *Int. Ed. Engl.*, **1987**, *26*, 764.
53. Barron, A. R.; Cowley, A. H. *Angew. Chem.*, **1987**, *99*, 956; *Angew. Chem.*, *Int. Ed. Engl.*, **1987**, *26*, 907.
54. Milczarek, R.; Rüsseler, W.; Binger, P.; Jonas, K.; Angermund, K.; Krüger, C.; Regitz, M. *Angew. Chem.*, **1987**, *99*, 957; *Angew. Chem.*, *Int. Ed. Engl.*, **1987**, *26*, 907.
55. Wettling, T.; Schneider, J.; Wagner, O.; Kreiter, C.; Regitz, M. *Angew. Chem.*, **1989**, *101*, 1035; *Angew. Chem.*, *Int. Ed. Engl.*, **1989**, *28*, 1013.
56. Fleischer, E. B. *J. Am. Chem. Soc.*, **1964**, *86*, 3889.

Recent Advances in Silicon-Containing Cyclic Polyacetylenes

Hideki Sakurai

Department of Chemistry
Faculty of Science
Tohoku University
Aoba-ku, Sendai 980
Japan

18.1. Introduction

Macrocyclic compounds containing both carbon–carbon triple bonds and silicon are an interesting class of compounds that have received much attention recently. When we initiated research on these compounds, there was only one example—reported by Voronkov and Pavlov [1], who prepared a 12-membered tetrasilacyclododecatetrayne in low yield.

In 1982 we prepared octamethyl-3,4,7,8-tetrasilacycloocta-1,5-diyne (**1**), as the first compound which contains both $C≡C$ and $Si—Si$ bonds [2, 3]. Our motivation in the project was the expectation that, as shown in a qualitative molecular-orbital diagram (Figure 18.1), two silicon–silicon σ orbitals can overlap with one of two π orbitals of each $C≡C$ bond to make up new molecular orbitals with through-bond conjugation. Consequently, interesting physical and chemical properties were expected. This expectation was fully realized in various spectroscopic and chemical studies [4, 5]. Because the development of the chemistry of the field has been summarized recently [6], the author will describe herein recent progress in the field, especially with emphasis on transition-metal–mediated cyclooligomerization of macrocyclic siloxane tethered polyynes, which can be derived by selective oxidation of the $Si—Si$ to $Si—O—Si$ bonds.

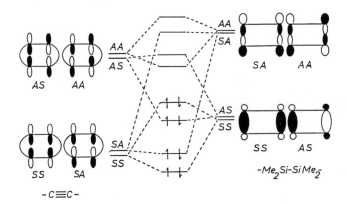

Figure 18.1 ■ Qualitative molecular-orbital diagram of octamethyl-3,4,7,8-tetrasilacycloocta-1,5-diyne (**1**).

18.2. Iron Carbonyl–Catalyzed Intramolecular Dimerization of Two Acetylene Units to Trimethylenemethane Complex

Selective oxidation of two tetramethyldisilanylene units of **1** with trimethylamine N-oxide gave a 10-membered macrocyclic dioxatetrasiladiyne, **2**, in 89% yield.

The reaction of the diyne **2** with diiron nonacarbonyl in refluxing benzene for 21 h resulted in the formation of iron dinuclear complex **3** in 35% yield [7]. The compound **3** (red crystals, mp 102–104 °C) was isolated and purified from the reaction mixture by passage through a silica gel short column followed by recrystallization from ethanol.

The structure of **3** was established by X-ray diffraction (Figure 18.2): The unit cell contains four molecules, which make up two very similar but independent molecules; one of the iron atoms forms a ferracycle that is almost planar. Interestingly, an exomethylene group forms a trimethylenemethane structure with the ferracycle; the exomethylene double bond of **3** is bent to iron by 33.0° from the planar ferracycle.

The transition-metal–catalyzed intermolecular and intramolecular dimerization of alkynes to η^4-cyclobutadiene complexes is well documented.

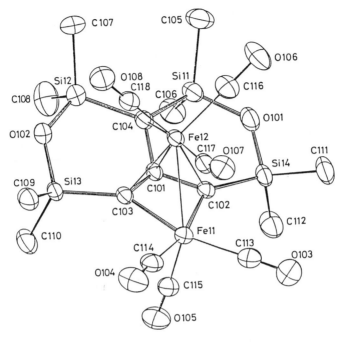

Figure 18.2 ▪ ORTEP diagram of **3**.

In particular, King and co-workers reported the formation of (cyclobutadiene)(cyclo-pentadienyl)cobalt complexes [8] and (cyclobutadiene)iron tricarbonyl and dinuclear ferracyclopentadiene complexes from macrocyclic alkadiynes [9]. However, there is no previous example of the formation of trimethylenemethane by dimerization of two acetylenes and **3** is the first example. The mechanism of formation of this interesting complex will be discussed later.

18.3. Iron Carbonyl–Catalyzed Intramolecular Dimerization of Two Acetylene Units to Novel Methylenecyclopropene Complex

Because the reaction of a 10-membered macrocyclic polyacetylenic siloxane compound resulted in the formation of a very interesting complex, we next examined the reaction of a macrocyclic diyne of a larger ring size, 3,3,5,5,7,7,10,10,12,12,14,14-dodecamethyl-4,6,11,13-tetraoxa-3,5,7,10,12,14-hexasilacyclotetradeca-1,8-diyne (4) with diiron non-acarbonyl. Compound 4 was prepared from 3,3,4,4,5,5,8,8,9,9,10,10-dodecamethyl-3,4,5,8,9,10-hexasilacyclodeca-1,6-diyne (5) by oxidation with trimethylamine N-oxide.

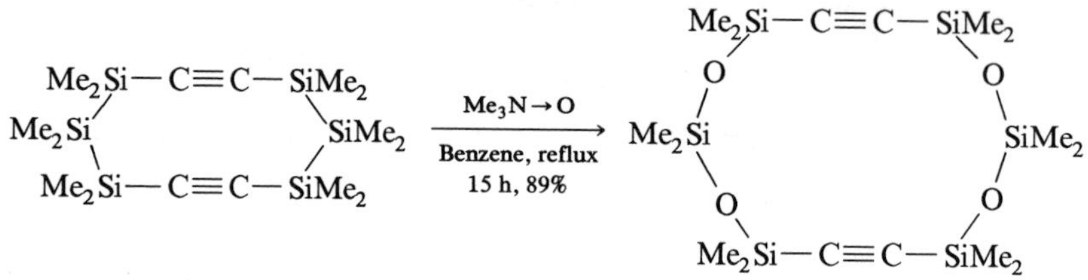

5

4

The reaction of the diyne 4 with 1.5 eq of diiron nonacarbonyl in refluxing benzene afforded the novel (methylenecyclopropene)iron tricarbonyl derivative 6 in 23% yield, as orange crystals, mp 122–123 °C$_{dec}$, isolated and purified from the reaction mixture by silica-gel thin-layer chromatography [10].

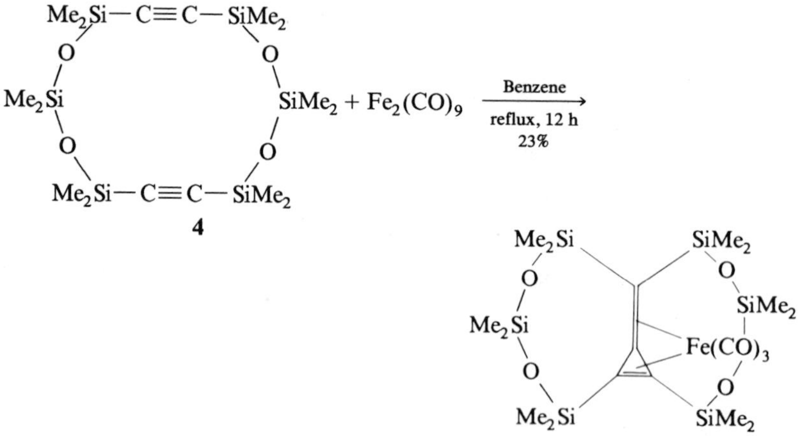

4

6

Methylenecyclopropene (7) is one of the most interesting target molecules for experimental and theoretical studies in organic chemistry. However, the molecule 7 is very unstable, having been synthesized and characterized quite recently only at low temperature. The compound is highly dipolar and, therefore, some derivatives, stabilized by both electron-donating and -withdrawing substituents, such as 1,2-diphenyl-3-(1,1-dicyanomethylene)cyclopropene (8), have been prepared as stable compounds. Compound 7 is the simplest member of the class of crossconjugated nonalternant hydrocarbons. Because it is isoelectronic with cyclobutadiene, complexes of which are well known, it would also be expected to function as an η^4 ligand.

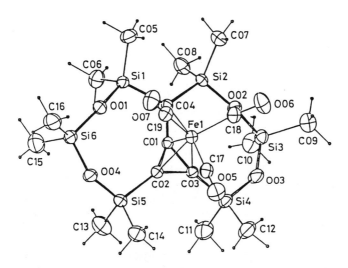

Another reactive $C_4\pi$ system, trimethylenemethane, is also known to serve as an η^4 ligand (one of which is already shown in the previous section). However, transition-metal complexes of **7** or its derivatives are unknown, perhaps because of the quite high reactivity of the strained C—C bonds toward ring opening. Actually, the reaction of **8** with (ethylene)bis(triphenylphosphine)platinum led to cyclopropene ring opening to give a complex **9**. Similar reactions are known to occur with cyclopropenes, methylenecyclopropanes, and cyclopropanones. Therefore, the formation of the methylenecyclopropene complex has not been reported to date.

The structure of the new complex was established unequivocally by X-ray diffraction, as shown in Figure 18.3. The iron atom is bonded to four carbons of the methylenecyclopropene ligand and three carbonyl groups. It is noteworthy that both r(C-2 C-3) ($= 1.479$ Å) and to a lesser extent r(C-1 C-4) ($= 1.397$ Å) are elongated by comparison with the corresponding bonds of **7** ($= 1.323$ and 1.332 Å, respectively) as determined by microwave spectroscopy. In contrast, the analogous bond distances in r(C-1 C-2) ($= 1.427$ Å) and r(C-1 C-3) ($= 1.417$ Å) are shorter than **7** ($= 1.441$ Å). Unusually wide bond angles were observed for $\angle$SiOSi ($= 163.9°$ and $158.6°$) in comparison with the normal values of disiloxanes ($130 \pm 10°$). The C-1–C-4 double bond of **6** is bent to iron by $27.9°$ from the planar cyclopropene ring.

As shown by the previous two examples, the variation of the reaction mode of intramolecular dimerization of two acetylenes due to ring size is quite remarkable. Apparently, the capability of silicon to undergo 1,2-shifts in disilylalkynes is responsi-

Figure 18.3 ■ ORTEP diagram of **6**.

ble for the formation of the unusual product reported here. We propose a reaction mechanism, shown later, to account for the formation of these complexes. The 1,2-silyl shift from an initially formed acetylene complex to a vinylidene complex is a key step which is then followed by cyclization. The methylenecyclopropene complex should be the initial complex. However, in the case of the smaller ring size, further insertion of an $Fe(CO)_4$ unit into the very strained double bond occurs, resulting in formation of trimethylenemethane complex. In these cases, we could not isolate the key intermediates, such as acetylene and vinylidene complexes. Isolation and characterization of similar complexes will be demonstrated in the next section.

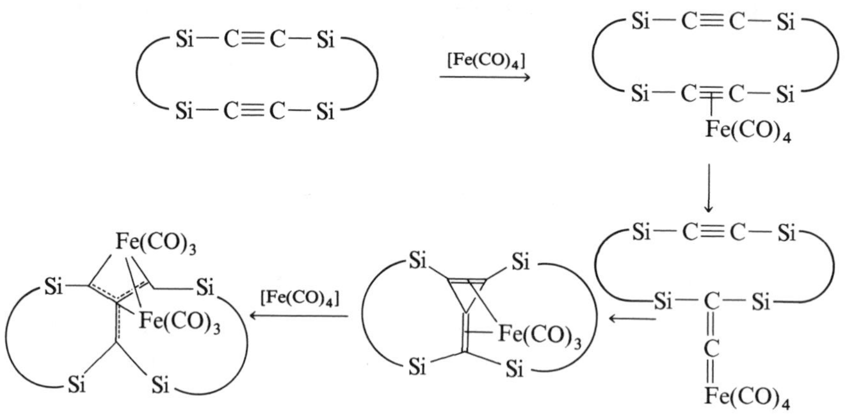

18.4. Iron Carbonyl–Catalyzed Intramolecular Trimerization of Three Acetylene Units to Benzene and Fulvene Complexes

The interesting chemistry of octamethyl-3,4,7,8-tetrasilacyclo-1,5-octadiyne (1), has been extended successfully to the second member, dodecamethylhexasilacyclododeca-1,5,9-triyne (10). Highly efficient through-bond conjugation between the Si—Si and acetylene bonds are also indicated by a photoelectron spectroscopy (PES) study [5].

Selective oxidation of three tetramethyldisilanylene units of 10 with trimethylamine N-oxide gave a 15-membered macrocyclic trioxahexasilatriyne (11) in almost quantitative yield.

Transition-metal–catalyzed trimerization of alkynes has long attracted organic chemists. Several mechanisms and intermediates have been proposed for formation of triacetylene complexes (M = Cr, Ni), metalacyclopentadienes (M = Co, Ir), and halogenohexatrienyl metal complexes (M = Pd). Dodecamethyl-4,9,14-trioxa-3,5,8,10,13,15-hexasilacyclopentadeca-1,6,11-triyne (11) undergoes thermal intramolecular cyclotrimerization to the corresponding benzene derivative 12 only in low yield, but when 1 was heated in the presence of group VI–transition-metal carbonyls [$M(CO)_6$: M = Cr, Mo, and W], both black-violet crystalline fulvene complexes 13 and an intramolecular cycloadduct (benzene–12) were obtained in addition to a trace amount of the corresponding arene–metal tricarbonyl complexes 14 [11]. The formation of 12 was indeed catalyzed by the metal complexes, because in the absence of metal complexes, **11** did not isomerize under such mild conditions.

Interestingly, the main products of the reaction accompanied by the intramolecular trimerization of acetylenes are not benzene complexes but fulvene complexes **13**. Although direct complexation of fulvene by the reaction with group VI metal carbonyls is a straightforward route to the metal complexes, early efforts to obtain these complexes by this route failed because of concomitant hydrogen abstraction or disproportionation of dimeric products. Several monomeric fulvene complexes have recently been isolated by the use of tripyridine metal tricarbonyl.

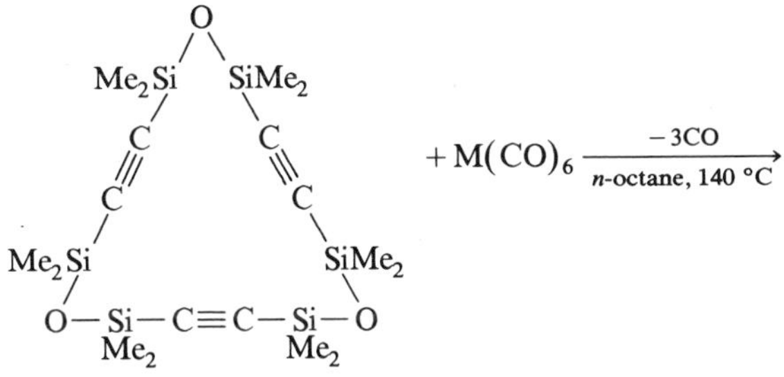

11

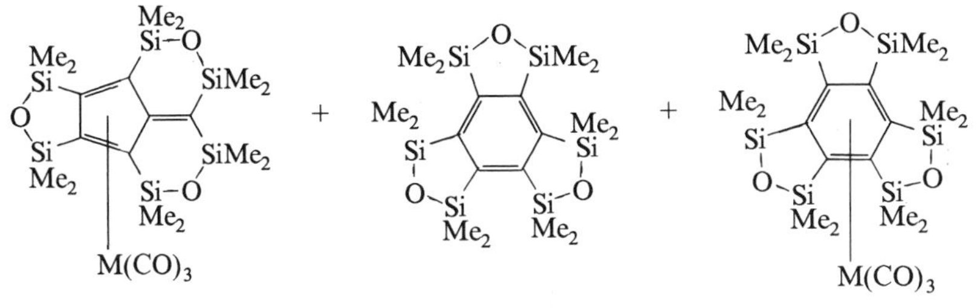

13   **12**   **14**

The molecular structure of **13b**, determined by X-ray diffraction analysis at room temperature, is given in Figure 18.4. The Mo complex **13b** was thermally stable, but the decomplexation with trimethylamine N-oxide or by irradiation with a high-pressure mercury lamp in the presence of triphenylphosphine, gave the liberated fulvene **15** in low yield. The observed low yield of fulvene is probably due to the instability of fulvene itself under both oxidative and photochemical conditions.

The formation of a fulvene complex indicates that a 1,2-silyl shift is a key step also in this reaction. Although we could not isolate the intermediates in the cases of diacetylenes, we could indeed isolate a new vinylidene complex **16** in the reaction of **11** and η^5-cyclopentadienyl(tricarbonyl)manganese. The structure of the complex was determined by X-ray diffraction. Thermal and photochemical reactions of **16** afforded the corresponding benzene and fulvene derivatives.

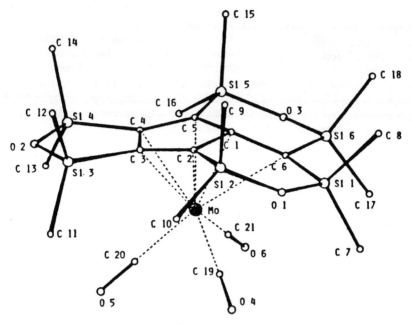

Figure 18.4 ▪ The structure of the fulvene complex **13b**.

18.5. Synthesis, Properties, and Molecular Structure of Highly Distorted Hexakis(trimethylsilyl)benzene

The author has described the formation of a hexasilylbenzene, **12**, in the previous section. The compound **12** also has been prepared in a different way [12]. The structure is, however, known to be not very much distorted. The molecular structures of overcrowded compounds are of interest in order to determine how the molecules relieve their inherent repulsive nonbonded interactions. Hexa-*t*-butylbenzene **17** and hexakis(trimethylsilyl)benzene **18** have not been synthesized yet and hence these sterically crowded molecules are interesting synthetic targets. Hexakis(trimethylgermyl)benzene **19** has recently been prepared by Mislow and co-workers [13] and its unique molecular structure has been demonstrated.

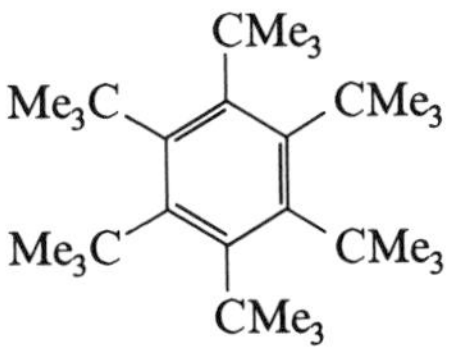

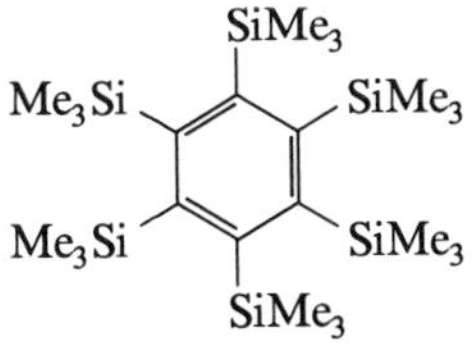

In connection with our studies on highly crowded silylethylenes [14] and transition-metal–catalyzed formation of **12**, it is quite interesting to investigate the physical and chemical properties of **18**.

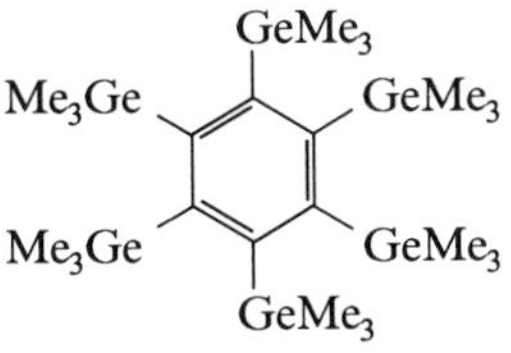

The preparation of **18** was previously attempted by treating hexabromobenzene with magnesium and chlorotrimethylsilane, but the reaction resulted in the formation of 1,1,3,4,6,6-hexakis(trimethylsilyl)-1,2,4,5-hexatetraene [15]. Our strategy for the synthesis of **18** is outlined in the following.

Hexakis(dimethylsilyl)benzene was subjected to bromination in carbon tetrachloride at 0 °C in the presence of pyridine, giving hexakis(bromodimethylsily)benzene as relatively labile yellow crystals. Without purification, treatment with an excess of methyllithium in THF followed by chromatography on silica gel with hexane afforded bright yellow crystals of **18**. The compound **18** is fairly stable to air and water, but readily decomposes by acids to give 1,2,4,5-tetrakis(trimethylsilyl)benzene [16].

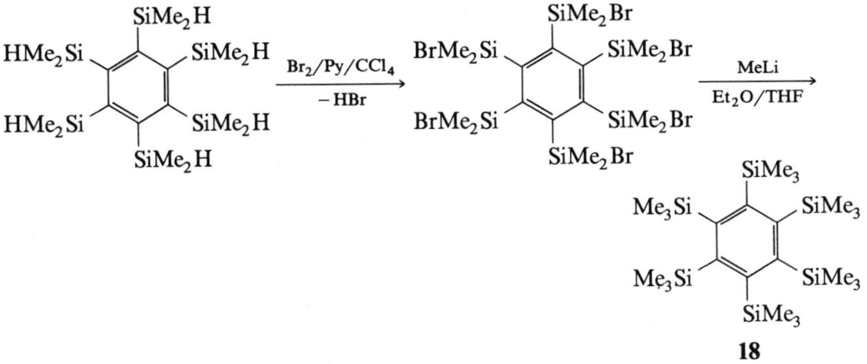

The compound **18** shows quite unique chemical features. Irradiation of **18** (λ > 300 nm) in hexane resulted in the formation of hexakis(trimethylsilyl)bicyclo[2.2.0]-

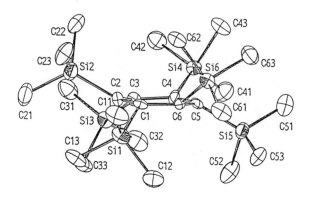

Figure 18.5 ■ ORTEP drawing of **18**.

hexa-2,5-diene **20** (Dewar benzene). The Dewar benzene **20** was isolated as deep red crystals. The photolysis of **18** is extremely clean, and the Dewar benzene is formed quantitatively.

In contrast to the photochemical reaction, thermolysis of **18** in octane at 200 °C cleanly produced 1,1,3,4,6,6-hexakis(trimethylsilyl)-1,2,4,5-hexatetraene **21** via the rupture of the benzene ring followed by 1,2-silyl migration.

The structure of **18** was determined by X-ray crystallography. Compound **18** crystallizes in two forms: orange and yellow crystals. The two crystals are isomorphous, consisting of similar structures. The orange ones belong to the space group *Pbca* and Si methyl groups are disordered. The yellow ones belong to the space group $P2_12_12_1$, with ordered Si methyl groups, and thus the data collection was carried out for the yellow ones at low temperature (-40 °C).

The four molecules in the unit cell sit in general positions but have approximate D_3 symmetry. An ORTEP drawing of **18** is shown in Figure 18.5.

The unique structure as well as interesting chemical properties of **18** are currently under active investigation and the results will be reported elsewhere.

References

1. Voronkov, M. G.; Pavlov, S. F. *Zh. Obsch. Khim.*, **1973**, *43*, 1408.
2. Sakurai, H.; Nakadaira, Y.; Hosomi, A.; Eriyama, Y., *Chem. Lett.*, **1982**, 1971.
3. Sakurai, H.; Nakadaira, Y.; Hosomi, A.; Eriyama, Y.; Kabuto, C. *J. Am. Chem. Soc.*, **1983**, *105*, 3359.

4. Sakurai, H.; Nakadaira, Y.; Hosomi, A.; Eriyama, Y.; Kabuto, C. *Chem. Lett.*, **1984**, 594.
5. Gleiter, R.; Schäfer, W.; Sakurai, H. *J. Am. Chem. Soc.*, **1985**, *107*, 3046.
6. Sakurai, H. *Pure Appl. Chem.*, **1987**, *59*, 1637.
7. Sakurai, H.; Hirama, K.; Nakadaira, Y.; Kabuto, C. *Chem. Lett.*, **1988**, 485.
8. King, R. B.; Efraty, A. *J. Am. Chem. Soc.*, **1972**, *94*, 3021.
9. King, R. B.; Haiduc, I.; Eaverson, C. W. *J. Am. Chem. Soc.*, **1973**, *95*, 2505.
10. Sakurai, H.; Hirama, K.; Nakadaira, Y.; Kabuto, C. *J. Am. Chem. Soc.*, **1987**, *109*, 6880.
11. Sakurai, H.; Nakadaira, Y.; Hosomi, A.; Eriyama, Y.; Hirama, K.; Kabuto, C. *J. Am. Chem. Soc.*, **1984**, *106*, 8315.
12. Fink, W. *Helv. Chim. Acta*, **1974**, *57*, 1010.
13. Weissensteiner, W.; Schuster, I. I.; Blount, J. F.; Mislow, K. *J. Am. Chem. Soc.*, **1986**, *108*, 6664.
14. (a) Sakurai, H.; Nakadaira, Y.; Kira, M.; Tobita, H. *Tetrahedron Lett.*, **1980**, *21*, 3077. (b) Sakurai, H.; Tobita, H.; Kira, M.; Nakadaira, Y. *Angew. Chem., Int. Ed. Engl.*, **1980**, *19*, 620. (c) Sakurai, H.; Nakadaira, Y.; Tobita, H.; Ito, T.; Toriumi, K.; Ito, H. *J. Am. Chem. Soc.*, **1982**, *104*, 300. (d) Sakurai, H.; Tobita, H.; Nakadaira, Y.; Kabuto, C. *J. Am. Chem. Soc.*, **1982**, *104*, 4288. (e) Sakurai, H.; Ebata, K.; Nakadaira, Y.; Kabuto, C. *Chem. Lett.*, **1987**, 301. (f) Sakurai, H.; Ebata, K.; Sakamoto, K.; Nakadaira, Y.; Kabuto, C. *Chem. Lett.*, **1988**, 965. (g) Sekiguchi, A.; Nakanishi, T.; Kabuto, C.; Sakurai, H. *J. Am. Chem. Soc.*, **1989**, *111*, 3748.
15. (a) Shiina, K.; Gilman, H. *J. Am. Chem. Soc.*, **1966**, *88*, 5367. (b) Fearon, F. W. G.; Gilman, H. *J. Chem. Soc., Chem. Commun.*, **1987**, 86. (c) Grennan, T.; Gilman, H. *J. Organomet. Chem.*, **1968**, *11*, 625. (d) Ballard, D.; Brennan, T.; Fearon, F. W. G.; Shiina, K.; Haiduc, I.; Gilman, H. *Pure Appl. Chem.*, **1969**, *19*, 449.
16. Sakurai, H.; Ebata, K.; Kabuto, C.; Sekiguchi, A. *J. Am. Chem. Soc.*, **1990**, *112*, 1799.

Unusual Bonding between Bismuth and Carbon in Bismuth Pentaaryls: The Color Problem of Bi(C$_6$H$_5$)$_5$

Konrad Seppelt

Freie Universität Berlin
Institut für Anorganische und Analytische Chemie
1000 Berlin 33
West Germany

19.1. Introduction

Organic bismuth compounds have been known for a long time: In 1850 and 1852, Bi(C$_2$H$_5$)$_3$ was described [1]; the first aromatic compound Bi(C$_6$H$_5$)$_3$ was also prepared in the last century [2]. Its crystal structure, however, was determined only recently [3] and is essentially identical to the structures of A(C$_6$H$_5$)$_3$ with A = N, P, As, and Sb.

On reacting (C$_6$H$_5$)$_3$BiCl$_2$ with phenylmagnesium bromide, Challenger [4] observed a purple color, and Gilman and Yablunsky [5] used this color as a qualitative test for phenyllithium and phenylmagnesium halides. In pursuing the preparation of phenyl compounds of metalloids in their highest oxidation states, Wittig finally isolated Bi(C$_6$H$_5$)$_5$ as a purple, crystalline material that is quite unstable in comparison to P(C$_6$H$_5$)$_5$, As(C$_6$H$_5$)$_5$, and Sb(C$_6$H$_5$)$_5$ [6]. In the course of the discussion on intramolecular rearrangements of five-coordinated compounds, the structures of P(C$_6$H$_5$)$_5$, As(C$_6$H$_5$)$_5$, and Sb(C$_6$H$_5$)$_5$ played an important role, because the first two are trigonal bipyramidal whereas Sb(C$_6$H$_5$)$_5$ is square pyramidal [7]. On the other hand, Sb(C$_6$H$_4$-4-CH$_3$)$_5$ and the cyclohexane solvate Sb(C$_6$H$_5$)$_5 \cdot \frac{1}{2}$C$_6$H$_{12}$ have trigonal bipyramidal structures [8].

For the terms trigonal bipyramidal and square pyramidal as used here, only the *ipso*-carbon atoms are regarded in describing the coordination (see Figure 19.1). The two largest angles are 180° and 120° in the trigonal bipyramid, and in the square

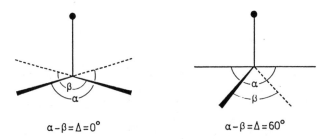

$$\alpha - \beta = \Delta = 0°$$ $$\alpha - \beta = \Delta = 60°$$

Figure 19.1 ■ Definition of angle difference $D = \alpha - \beta$ for distinguishing trigonal bipyramidal and square pyramidal structures [9]. α and β are the two largest angles between two ligands and the central atom. Ideal trigonal bipyramids have $D = 60°$, and ideal square pyramids $D = 0°$. In the molecular structures of bismuth pentaaryls, D often is close to but never exactly $0°$ or $60°$. Thus defined, D can be regarded as the reaction coordinate of the Berry mechanism [10]. This definition is only useful if C_{2v} symmetry is at least qualitatively retained, as is the case in most structures discussed here. The definition of the angles is similar but not identical to an earlier definition by Hoffmann, Howell, and Muetterties [15].

pyramid they are of equal size. The angle difference $D = 60°$ or $D = 0°$ is thus a characteristic for the ideal structures [9].

Structures with angle differences $D \geq 45°$ are regarded as distorted trigonal bipyramids, with $D \leq 15°$ as distorted square pyramid, and with $15° < D < 45°$ as irregular cases. In other words, D is the reaction coordinate of the Berry mechanism [10].

19.2. Structural Aspects of BiAr$_5$ Compounds

Structures of pentavalent pentacoordinated Bi compounds such as $(C_6H_5)_3BiCl_2$ [3] or $(C_6F_5)_3BiF_2$ [11] (see Figure 19.2) are trigonal bipyramidal with the halogen atoms in axial positions. This is in accord with other group V structures of this type.

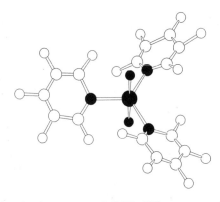

Figure 19.2 ■ Crystal and molecular structure of $(C_6F_5)_3BiF_2$ as an example for an ideal trigonal bipyramidal structure with the electronegative fluorine occupying axial positions, $D = 60°$ [11].

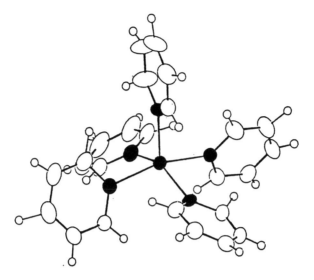

Figure 19.3 ▪ Crystal and molecular structure of $Bi(C_6H_5)_5$, ORTEP plot, including H atoms, $-96\ °C$ [12]. The structure is square pyramidal, $D = 13.2°$. The orientation of the propeller-like basal phenyl groups is very similar in $Bi(C_6H_5)_5$ and $Sb(C_6H_5)_5$, and all the hydrogen atoms are also normally bonded. So the knowledge of the molecular structure does not provide an explanation of the intense absorption. The color of the material is violet-blue.

It was assumed that the strong color of $Bi(C_6H_5)_5$ is an indication of an unusual bonding situation. It was quickly shown that the absorption is not due to impurities, that $Bi(C_6H_5)_5$ is not paramagnetic, and that the absorption is a molecular phenomenon, because the colors of the solutions are identical to those of the solid state.

At the beginning of this work nothing was known about the structure of $Bi(C_6H_5)_5$ because it is quite sensitive toward light, X-rays, temperature, and (less so) moisture. The crystal structure of $Bi(C_6H_5)_5$ was finally solved at low temperatures (see Figure 19.3) [12]. It is clearly square pyramidal, and in all respects very similar to the structure of $Sb(C_6H_5)_5$. The structure does not offer any explanation for the deep color. In large crystals a dichroism can be observed. This finding is important and will be discussed later on in detail.

At first we tried to obtain structures of as many $BiAr_5$ compounds as possible. However, the crystallographic work was often hampered by the tendency of the compounds to form oily materials or twinned crystals, so the successful crystallographic investigations look somewhat accidental. The overall picture is simple, however, because the majority of the structures are square pyramidal (see Figures 19.4–19.7) [13,11]. By use of the o,o'-diphenylene ligand, the structure is constrained and the structure becomes irregular, but it is still very close to the square pyramidal configuration (see Figure 19.8).

A more interesting case is the structure of $Bi(C_6H_4\text{-}4\text{-}CH_3)_5$, because the corresponding antimony material is trigonal bipyramidal and because it is the only other

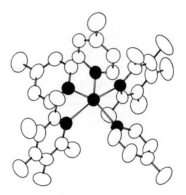

Figure 19.4 ■ Crystal and molecular structure of $Bi(C_6F_5)_3(C_6H_4\text{-}4\text{-}CH_3)_2$ [9]. *ipso*-Carbon atoms are blackened to facilitate the view on the coordination around Bi ($D = 16°$; distorted square pyramid; color, yellow).

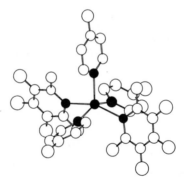

Figure 19.5 ■ Crystal and molecular structure of $Bi(C_6F_5)_3(C_6F_4\text{-}4\text{-}F)_2$, ($D = 5°$; square pyramid, color, yellow) [9].

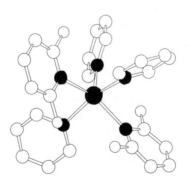

Figure 19.6 ■ Crystal and molecular structure of $Bi(C_6H_5)_3(C_6H_3\text{-}2,6\text{-}F_2)_2$ ($D = 11.5°$; square pyramid; color, red) [9].

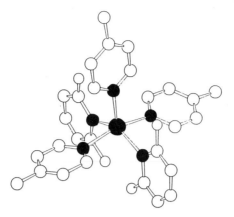

Figure 19.7 ▪ Crystal and molecular structure of $Bi(C_6H_4\text{-}4\text{-}CH_3)_3(C_6H_3\text{-}2,6\text{-}F_2)_2$, ($D = 17.5°$; square pyramid, color, red) [11]. Note: Introduction of *para*-methyl groups does not change the structure significantly.

compound with five equal ligands (homoleptic). Again, however, it is an irregular case, here closer to a trigonal bipyramid (see Figure 19.9) [11].

Based on the available data one can generalize that pentaaryl bismuth compounds prefer the square pyramidal environment quite in contrast to the other group V compounds. Only antimony exhibits some intermediate behavior.

In a simplified bonding model this can be explained as follows: As a result of the ineffective shielding by $4f$ electrons (or as a result of the relativistic behavior of this heavy atom), the $6s$ electrons participate only to a small degree in bonding, resulting in

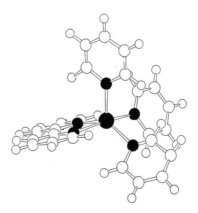

Figure 19.8 ▪ Crystal and molecular structure of $Bi(C_6H_5)_3(C_{12}H_8)$ [11]. The *o,o*-diphenylene ligand enforces planarity in some parts of the molecule. The structure is irregular, because of the constraints caused by the five-membered ring: $D = 21.7°$ and $28.8°$ (two molecules in the asymmetric unit). The color is orange, weakly dichroitic.

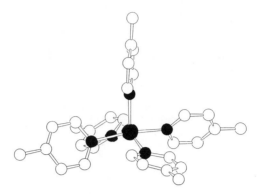

Figure 19.9 ▪ Crystal and molecular structure of dichroitic $Bi(C_6H_4\text{-}4\text{-}CH_3)_5$ ($D = 35°$; color, violet) [11]. For comparison, $Sb(C_6H_4\text{-}4\text{-}CH_3)_5$ is trigonal pyramidal, $D = 48°$. It is interesting to note that the five tolyl ligands force the molecule into a trigonal bipyramidal structure, but more successfully in the case of antimony than of bismuth.

an overwhelmingly *p*-bonding situation, which prefers the square pyramidal configuration. However, there are clearly exceptions to this generalization, because violet $Bi(C_6H_5)_3(C_6H_4\text{-}2\text{-}F)_2$ is square pyramidal (see Figure 19.10) whereas orange $Bi(C_6H_5\text{-}4\text{-}CH_3)_3(C_6H_4\text{-}2\text{-}F)_2$ is clearly trigonal bipyramidal (see Figure 19.11) [13]. It is inconceivable why the introduction of three *para*-methyl groups should change the structure so drastically, if it were not for minute changes in the lattice energy.

A more-detailed look into these pentacoordinated structures shows that the axial bonds in the tetragonal pyramid are always quite short (210 pm), whereas the four basal bonds are longer (220 pm). The basal Bi—C bond lengths are further lengthened if fluorine is introduced; in C_6F_5 ligands it may go up to 240 pm. This alone gives some insight into the nature of the bonding and suggests that the bonds are electron deficient. (Electron-rich bonds are normally shortened and strengthened by introduction of electron-withdrawing groups.) As Bi(V) itself is very electron withdrawing, the bond is likely to be polarized according to Bi($\delta +$)—C($\delta -$). This bond-length pattern is also retained in the trigonal bipyramidal case and in the irregular cases.

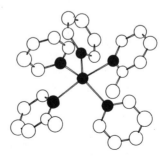

Figure 19.10 ▪ Crystal and molecular structure of $Bi(C_6H_5)_3(C_6H_4\text{-}2\text{-}F)_2$ (square pyramidal; $D = 16°$; color, violet, dichroitic) [13].

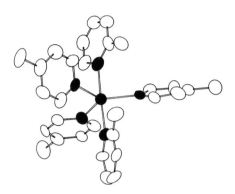

Figure 19.11 ▪ Crystal and molecular structure of $Bi(C_6H_4\text{-}4\text{-}CH_3)_3(C_6H_4\text{-}2\text{-}F)_2$ (trigonal bipyramidal; $D = 45.9°$; color, yellow, nondichroitic) [13].

19.2.1. Optical Phenomena in BiAr₅ Compounds

So far this discussion of structures does not give any insight into the nature of the chromophore. The color of $Bi(C_6H_5)_5$ is due to a weak ($\log \varepsilon = 2.4$) and broad absorption at $\lambda_{max} = 528$ nm [9]. Introduction of methyl substituents on the phenyl group as in $Bi(C_6H_4\text{-}4\text{-}CH_3)_5$ does not significantly shift the absorption maximum (see Table 19.1). Introduction of two C_6F_5 groups, however, transforms the color into yellow [9]. *Para*-substitution by fluorine, however, has obviously only a small effect, if at all [9]. Certainly the strongest influence is expected for *ortho*-F: $Bi(C_6H_5)_3(C_6H_3\text{-}2,6\text{-}F_2)_2$ and $Bi(C_6H_4\text{-}4\text{-}CH_3)_3(C_6H_3\text{-}2,6\text{-}F_2)_2$ are both square pyramidal, red, and not dichroitic [11].

19.2.2. The Dichroism of Bi(C₆H₅)₅

Crystals of $Bi(C_6H_5)_5$, $Bi(C_6H_4\text{-}4\text{-}CH_3)_5$, and many others are dichroitic. Because of the size of the crystals, this effect is most beautiful with $Bi(C_6H_4\text{-}4\text{-}CH_3)_3(C_6H_4\text{-}4\text{-}CF_3)_2$ (see the figure in [12]) (but because of twinning its structure could not be solved yet). Because the orientation of the molecules in space is known from crystal structures,

Table 19.1 ▪ **Absorption Maxima of BiAr₅ Compounds (in THF, at Room Temperature)**

Compound	λ_{max} (nm)
$Bi(C_6H_5)_5$	528 [12]
$Bi(C_6H_4\text{-}4\text{-}CH_3)_5$	521 [11]
$Bi(C_6H_5)_3(C_6H_3\text{-}2,6\text{-}F_2)_2$	454 [9]
$Bi(C_6H_4\text{-}4\text{-}CH_3)_3(C_6H_3\text{-}2,6\text{-}F_2)_2$	447 [11]
$Bi(C_6H_5)_3(C_6H_4\text{-}2\text{-}F)_2$	482 [13]
$Bi(C_6H_4\text{-}4\text{-}CH_3)_3(C_6H_4\text{-}2\text{-}F)_2$	478 [13]

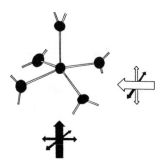

Figure 19.12 ▪ Dichroism of the Bi(C$_6$H$_5$)$_5$ molecule [12]. Fat arrows indicate the direction of light, thin arrows indicate the direction of the electric dipole. Light with an electric dipole only in the basal plane is totally absorbed (black arrows). Light with an electric dipole partially in the basal plane is partially absorbed and therefore polarized after transmission through the crystal.

the light absorption could be traced back to the plane formed by the four basal Bi — C bonds: Whenever the electric dipole of the light is in the plane of these four bonds, it is absorbed; light perpendicular to this plane is fully absorbed. Light in this plane is polarized, because only light with the electric vector in-plane is filtered out (see Figure 19.12.)

The most interesting case of color is that of the pair Bi(C$_6$H$_5$)$_3$(C$_6$H$_4$-2-F)$_2$ and Bi(C$_6$H$_4$-4-CH$_3$)$_3$(C$_6$H$_4$-2-F)$_2$ [13]. The first is violet, dichroitic, and square pyramidal, the latter is yellow, nondichroitic, and trigonal bipyramidal (see Figures 19.10 and 19.11). This is the only example of the latter type observed so far.

Both compounds are chemically identical with the exception of *para*-methyl substitution, which ought not make too much difference except minute changes in lattice energy. Indeed, both materials have virtually identical colors (and UV–visual spectra) in solution. On cooling we observe a sharpening of the broad bands, but no remarkable change in intensity.

Because the absorption at $\lambda = 520$ nm is attributed to the square pyramidal configuration, and the strong end absorption to the trigonal bipyramidal case, the energy change between them is close to zero. In other words, one can follow the Berry pseudorotation with one's own eyes in solution.

19.3. Preliminary Conclusions

Absorption, structure, bonding, and chemical reactivity are interconnected. Only square pyramidal molecules have the violet color and only if the aromatic rings do not carry too many electron-withdrawing groups. A very simple molecular-orbital (MO) calculation including relativistic effects on BiH$_5$ revealed the following qualitative picture (see Figure 19.13) [14].

The HOMO–LUMO gap is much smaller in the square pyramidal case. This is so already in PH$_5$ [15], but ineffective shielding by $4f$ electrons and relativistic effects lower the LUMO in BiH$_5$ further, because the HOMO has mainly basal ligand character, whereas the LUMO has mainly Bi $6p$-character. Therefore, the electron excitation is a charge transfer from the basal ligands to the central bismuth.

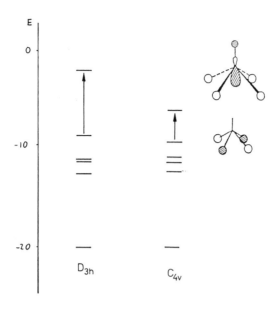

Figure 19.13 ■ Qualitative MO diagram of BiH$_5$ in trigonal bipyramidal and square pyramidal configurations [13,14]. The HOMO–LUMO gap is much smaller in the square pyramidal case. Qualitatively the HOMO has largely basal ligand character, the LUMO has largely Bi 6s,p-character, as indicated by the orbital representation on the right.

This is also consistent with bond-length arguments and the high electronegativity of the Bi(V). In other words, in the excited state bismuth behaves like Bi^{3+}. Simplistically argued, the excited state is pseudooctahedral with a nonbonding electron pair; therefore, only square pyramidal states have the probability of charge transfer (Frank Condon principle). Two chemical consequences are drawn. First, because the Bi — C bond is polarized towards bismuth, particularly if in the excited state, Bi(V)-aryl materials react as cationic phenylating agents, as is documented in many cases by Barton and Finet [16]. Second, and more speculative, is the following: Bismuth-containing superconductors have square pyramidal configurations with oxygen deficiencies, thus coupling the electron pair transportation with an octahedral–square-pyramidal or square-pyramidal–pyramidal transformation. This is only possible because Bi(V) and Bi(III) states are energetically close (in comparison to Sb, As, and P). In other words, the same physical principles that cause the color of BiAr$_5$ compounds are effective in bismuth-containing superconductors.

References

1. Leurig, C.; Schweizer, E. *Liebigs Ann. Chem.*, **1850**, *75*, 315; Breed, D. *Liebigs Ann. Chem.*, **1852**, *82*, 106.
2. Michaelis, A.; Polis, A. *Ber Dtsch. Chem. Ges.*, **1887**, *20*, 54; Michaelis, A.; Marquardt, A. *Liebigs Ann. Chem.*, **1889**, *251*, 323.

3. Hawley, D. M.; Ferguson, G.; Harris, G. S. *J. Chem. Soc., Chem. Comm.*, **1966**, 111; Hawley, D. M.; Ferguson, G. *J. Chem. Soc. A*, **1968**, 2059.
4. Challenger, F. *J. Chem. Soc.*, **1914**, *105*, 2210.
5. Gilman, H.; Yablunsky, H. L. *J. Am. Chem. Soc.*, **1941**, *63*, 839.
6. Wittig, G.; Clauß, K. *Liebis Ann. Chem.*, **1952**, *578*, 136.
7. Wheatley, P. J.; Wittig, G. *Proc. Chem. Soc.* (*London*), **1962**, 251; Wheatley, P. J. *J. Chem. Soc.*, **1964**, 2206; Brock, C. P.; Webster, D. F. *Acta Cryst. B*, **1976**, *32*, 2089; Wheatley, P. J. *J. Chem. Soc.*, **1964**, 3718; Beauchamp, A. C.; Bennett, M. J.; Cotton, F. A. *J. Am. Chem. Soc.*, **1968**, *90*, 6675.
8. Brabant, C.; Blanck, B.; Beauchamp, A. L. *J. Organomet. Chem.*, **1974**, *82*, 231; Brabant, C.; Hubert, J.; Beauchamp, A. J. *Can. J. Chem.*, **1973**, *51*, 2952.
9. Schmuck, A.; Leopold, D.; Seppelt, K. *Chem. Ber.*, **1989**, *122*, 803.
10. Berry, R. S. *J. Chem. Phys.*, **1960**, *32*, 933.
11. Schmuck, A.; Leopold, D.; Wallenhauer, S.; Seppelt, K. *Chem. Ber.*, **1990**, in press.
12. Schmuck, A.; Buschmann, J.; Fuchs, J.; Seppelt, K. *Angew. Chem.*, **1987**, *99*, 1206; *Angew. Chem., Int. Ed. Engl.*, **1987**, *26*, 1180.
13. Schmuck, A.; Pyykkö, P.; Seppelt, K. *Angew. Chem.*, **1990**, *102*, 211; *Angew. Chem., Int. Ed. · Engl.*, **1990**, *29*, 213.
14. We thank P. Pyykkö for sending us the program ITEREX-87.
15. Hoffmann, R.; Howell, J. M.; Muetterties, E. L. *J. Am. Chem. Soc.*, **1972**, *94*, 3047.
16. Barton, D. H. R.; Finet, J. P. *Pure Appl. Chem.*, **1987**, *59*, 937.

Syntheses and Structures of Copper and Silver Cluster Complexes with Sulfur-Containing Ligands*

Kaluo Tang and Youqi Tang

Institute of Physical Chemistry
Peking University, Beijing 100871,
China

20.1. Introduction

Interest in the study of the chemistry of copper and silver cluster complexes with sulfur-containing ligands has grown quite considerably over the last 20 years. Complexes of copper with sulfur-containing ligands have been implicated in a number of important chemical and physiological systems: as synergistic deactivators in plastics, hydrocarbon oils, and rubbers; as potential models for copper binding sites in blue copper proteins; and as metallotherapeuticals of demonstrated anticancer, antibacterial, or antiviral activity [1]. Furthermore, copper and silver complexes with sulfur-containing ligands possess great potential as catalysts and reagents in organic synthesis. In addition, the diversity of their structural pattern and coordination mode has greatly stimulated the interest in their structural chemistry.

Copper(I) and silver(I) complexes are usually stabilized by cluster formation. Copper(I) forms small clusters with between three and eight metal atoms. Silver(I) forms discrete clusters and has the tendency to form polymeric arrangements containing chains of metal atoms joined by Ag—Ag bonds. We limit our discussion to the chemistry of copper and silver in the formally monovalent oxidation state with sulfur-containing ligands.

In 1963, Hesse studied the structure of copper(I) diethyldithiocarbamate $[CuS_2CNEt_2]_4$ [2a], which was one of the earliest copper(I) sulfur clusters to be

*Supported by China's National Natural Science Foundation as an item in the major project "Structural Chemistry and Molecular Design."

studied. Diisopropyldithiophosphato copper(I) $[Cu(i\text{-}C_3H_7O)_2PS_2]_4$ [2a] has a similar tetrahedral Cu_4 configuration. The crystal structures of hexameric silver(I) clusters $[AgS_2CNPr_2]_6$ [2a] and $[AgS_2CNEt_2]_6$ [3] have been determined. Coucouvanis [2b] studied three octameric copper(I) clusters, $[Cu_8(i\text{-}MNT)_6]^{4-}$, $[Cu_8(DED)_6]^{4-}$, and $[Cu_8(DTS)_6]^{4-}$, which all contain a Cu_8S_{12} core composed of a Cu_8 cube inscribed in a distorted icosahedron of 12 sulfur atoms. In recent years, a number of copper(I) or silver(I) thiolate anionic complexes $[M_x(SR)_y]^{x-y}$ and copper or silver thiolate clusters with heteroligands PPh_3 have been structurally characterized by Dance [4]. To date there are nearly one hundred copper(I) sulfur cluster structures and fewer than fifty silver(I) sulfur cluster structures found in the literature, but the types and patterns of their structures and the factors influencing them have not been fully realized.

Since 1981 we have worked on the chemistry of copper(I) and silver(I) cluster complexes with sulfur-containing ligands, such as dithiocarboxylates, perthiocarboxylates, thiolates, thioxanthates, and so on. We have synthesized and structurally characterized by X-ray more than 20 compounds. Their syntheses, structural characters, and factors influencing structures have been systematically investigated. A novel type of insertion reaction of CS_2 into Ag—S bonds has been found. This chapter reviews this research.

20.2. Cu(I) and Ag(I) Cluster Complexes with Dithiocarboxylate or Perthiocarboxylate Ligands

20.2.1. Syntheses

The methods used for preparing the dithioacids RCSSH have been summarized by Coucouvanis [2a]. The most versatile method is addition of carbon disulfide to a Grignard reagent [5a]. Because the ligand acids are generally not very stable, particularly in air, most of the acids are converted to tetraalkylammonium salts [5a] or hexahydropyridinium salts [5b].

20.2.1a. Synthesis of Ag(I) Cluster Complexes with Dithiocarboxylate Ligands. In general, the title complexes were prepared by adding a solution of $AgNO_3$ in $EtOH\text{–}H_2O$ (1:1) to a solution of corresponding dithioacids in ether or of their tetraalkylammonium salts in methanol. The resulting precipitate was recrystallized from pyridine or pyridine–DMF [6–9]. For example:

$$AgNO_3 + \alpha\text{-}C_{10}H_7CS_2H \xrightarrow[\text{EtOH–H}_2\text{O}]{\text{Et}_2\text{O}} \xrightarrow[\text{(recrystallized)}]{\text{pyridine}} Ag_4(\alpha\text{-}C_{10}H_7CS_2)_4(Py)_4 \cdot 2Py \quad [6]$$

$$(20.1)$$

$$AgNO_3 + Ph_3CCS_2H \xrightarrow[\text{MeOH–H}_2\text{O}]{\text{Et}_2\text{O}} \xrightarrow[\text{(recrystallized)}]{\text{pyridine–DMF}} Ag_6(Ph_3CCS_2)_6 \cdot Py \cdot 6DMF \quad [8]$$

$$(20.2)$$

$$AgNO_3 + [C_6H_5CS_2]^-(NEt_4)^+ \xrightarrow[\text{EtOH–H}_2\text{O}]{\text{MeOH}} \xrightarrow[\text{(recrystallized)}]{\text{pyridine}} [Ag_2(C_6H_5CS_2)_2]_n \quad [9]$$

$$(20.3)$$

In these reactions, the formation of Ag(I) cluster complexes did not involve a redox reaction, which is different from the formation of Cu(I) cluster complexes by the reaction between Cu(II) and dithioacids.

20.2.1b. Syntheses of Cu(I) Cluster Complexes with Perthiocarboxylate and Mixed Dithio-perthiocarboxylate Ligands.

The title complexes were prepared by reaction of $CuCl_2$ with the corresponding dithioacids or their tetraalkylammonium salts, followed by recrystallization from CS_2–EtOH, pyridine, pyridine–EtOH, or pyridine–Et_2O [7, 10–12]. For example:

$$CuCl_2 + \alpha\text{-}C_{10}H_7CS_2H \xrightarrow[\text{EtOH}]{CS_2} \xrightarrow[\text{(recrystallized)}]{CS_2\text{-EtOH}}$$

$$Cu_4(\alpha\text{-}C_{10}H_7CSS_2)_4 \cdot \tfrac{1}{2}CS_2 \quad [10] \quad (20.4)$$

$$CuCl_2 + (o\text{-}CH_3C_6H_4CS_2)^-(NEt_4)^+ \xrightarrow{MeOH} \xrightarrow[\text{(recrystallized)}]{\text{pyridine}}$$

$$Cu_2(o\text{-}CH_3C_6H_4CSS_2)_2(Py)_2 \quad [7] \quad (20.5)$$

$$CuCl_2 + (C_6H_5CS_2)^-(NEt_4)^+ \xrightarrow{MeOH} \xrightarrow[\text{(recrystallized)}]{\text{pyridine–Et}_2O}$$

$$Cu_4(C_6H_5CSS_2)_2(C_6H_5CS_2)_2(Py)_2 \quad [12] \quad (20.6)$$

The reactions (20.4)–(20.6) between $CuCl_2$ and dithioacids or their salts involve a redox reaction: Cu(II) is reduced to Cu(I), while dithiocarboxylate is oxidized to perthiocarboxylate. The disulfide $(RCS_2)_2$ is probably the intermediate product. It is cleaved unsymmetrically and forms a stable complex. The reaction proceeds as follows:

$$Cu(II) + 2\alpha\text{-}C_{10}H_7CS_2^- \longrightarrow Cu(I) + (\alpha\text{-}C_{10}H_7CS_2)_2 \longrightarrow Cu_4(\alpha\text{-}C_{10}H_7CSS_2)_4$$
$$(20.7)$$

This mechanism of the reaction has been confirmed by the reaction between Cu(I) and the disulfide $(\alpha\text{-}C_{10}H_7CS_2)_2$ resulting in the same cluster compound [10b].

Recently, Selbin and co-workers [13] reported that they had synthesized two tetranuclear copper(I) clusters with perthiobenzoates $Cu_4[2,4,6\text{-}(CH_3)_3C_6H_2CSS_2]_4$ and $Cu_4(o\text{-}CH_3C_6H_4CSS_2)_4$ by using the analogous redox reaction between Cu(II) and corresponding dithiobenzoates.

Camus and co-workers synthesized Cu(I) cluster complex with o-perthiotoluate ligand $Cu_4[o\text{-}CH_3C_6H_4CSS_2]_4$ by treating the dithiotoluato copper derivative with sulfur [14a]. They prepared copper(I) dithio-perthiocarboxylates by reacting $(CuS_3CAr)_4$ and $(CuS_2CAr)_4$ in 1:1 ratio in pyridine, then adding petroleum ether to the solution to obtain single crystals of $Cu_2(p\text{-}CH_3C_6H_4CSS_2)(p\text{-}CH_3C_6H_4CS_2)(Py)_2$ or $Cu_2(C_6H_5CSS_2)(C_6H_5CS_2)(Py)_2$, the structures of which have been determined by us recently [14b].

20.2.2. Structures

20.2.2a. Structural Patterns

Dinuclear Complexes. Dimeric copper(I) α-perthionaphthoato-pyridine $Cu_2(\alpha\text{-}C_{10}H_7CSS_2)_2(Py)_2$ [11] and o-perthiotoluato-pyridine $Cu_2(o\text{-}CH_3C_6H_4CSS_2)_2(Py)_2$ [17] have a similar molecular configuration, shown in Figure 20.1. Perthiocarboxylate

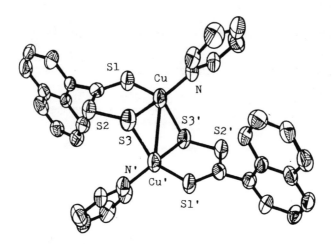

Figure 20.1 ■ Structure of $Cu_2(\alpha\text{-}C_{10}H_7CSS_2)_2(Py)_2$.

ligands coordinate to copper atoms through two sulfur atoms. One of them bridges two copper atoms whereas the other coordinates to only one of the two copper atoms. The pyridine ligands coordinate to two copper atoms. If the copper–copper interaction is ignored, the S and N atoms surrounding the Cu atoms are distributed in distorted tetrahedrons. The Cu—Cu distance is 2.616 Å in $Cu_2(\alpha\text{-}C_{10}H_7CSS_2)(Py)_2$ and 2.608 Å in $Cu_2(o\text{-}CH_3C_6H_4CSS_2)_2(Py)_2$, slightly longer than Cu—Cu single bond length 2.56 Å in metallic copper [16].

Dimeric copper(I) complexes with mixed dithiobenzoate and perthiobenzoate ligands $Cu_2(C_6H_5CSS_2)(C_6H_5CS_2)(Py)_2$ and $Cu_2(p\text{-}CH_3C_6H_5CSS_2)(p\text{-}CH_3C_6H_5CS_2)(Py)_2$ have a similar configuration, shown in Figure 20.2 [14b], that is different from that of the previously mentioned dimers. The dithiobenzoate ligand coordinates to two copper atoms. The coordination mode of perthiobenzoate is the same as that in other copper(I) perthiocarboxylate. Two pyridine ligands all coordinate to one copper atom so that the coordination geometries about the two copper atoms are different. One is a distorted tetrahedron formed by two sulfur atoms and two nitrogen atoms; the other is a trigonal plane formed by three sulfur atoms.

Tetranuclear Complexes. In the structure of tetranuclear copper(I) α-perthionaphthoate $Cu_4(\alpha\text{-}C_{10}H_7CSS_2)_4 \cdot \frac{1}{2}CS_2$ [10] or tetranuclear silver(I) α-dithionaphthoato-pyridine $Ag_4(\alpha\text{-}C_{10}H_7CS_2)_4(Py)_4 \cdot 2Py$, the nucleus possesses a distorted tetrahedral configuration that is similar to that of tetranuclear Cu(I) clusters $Cu_4[(C_2H_5)_2NCS_2]_4$ [2a] and $Cu_4[(i\text{-}C_3H_7O)_2PS_2]_4$ [2a].

In the structure of tetranuclear copper(I) α-perthionaphthoate $Cu_4(\alpha\text{-}C_{10}H_7CSS_2)_4 \cdot \frac{1}{2}CS_2$ (Figure 20.3), the coordination mode of α-perthionaphthoate ligands is the same as that in $Cu_2(\alpha\text{-}C_{10}H_7CSS_2)_2(Py)_2$. The coordination geometry about copper(I) atoms with sulfur ligands is trigonal planar. The copper–copper distances fall into two categories: the four bridged by sulfur average 2.789 Å; the other two, nonbridged, average 3.028 Å.

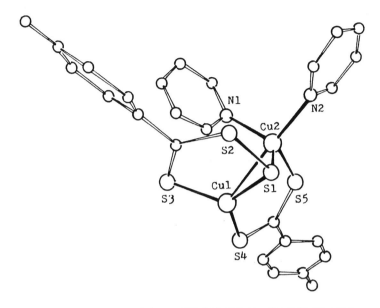

Figure 20.2 ▪ Structure of $Cu_2(p\text{-}CH_3C_6H_4CSS_2)(p\text{-}CH_3C_6H_4CS_2)(Py)_2$.

Comparison of the structure of $Cu_4(\alpha\text{-}C_{10}H_7CSS_2)_4 \cdot \frac{1}{2}CS$ to that of other two tetranuclear copper(I) clusters $Cu_4[2,4,6\text{-}(CH_3)_3C_6H_2CSS_2]_4$ and $Cu_4(o\text{-}CH_3C_6H_4CSS_2)_4$ reported by Selbin and co-workers [13] indicates that they have similar configurations and coordination modes.

Tetranuclear silver(I) clusters $Ag_4(\alpha\text{-}C_{10}H_7CS_2)_4(Py)_4$ [6] and $Ag_4(o\text{-}CH_3C_6H_4CS_2)_4(Py)_4$ [7] have similar structures, shown in Figure 20.4. Each dithiocarboxylate ligand coordinates to three silver atoms, as a tridentate ligand. One of the sulfur atoms bridges two silver atoms and the other coordinates to the third silver atom. The pyridine ligands coordinate to silver atoms through nitrogen atoms. If the silver–silver interaction is neglected, the silver atoms display distorted tetrahedral coordination.

The Cu_4 cluster with mixed dithio-perthiobenzoate ligands, $Cu_4(C_6H_5CSS_2)_2\text{-}(C_6H_5CS_2)_2(Py)_2$ [12], has a Cu_4 zigzag configuration (shown in Figure 20.5). The four copper atoms are coplanar, and the molecule has a symmetry center. Each perthiobenzoate ligand coordinates to two copper atoms and the dithiobenzoate acts as a tridentate ligand. The two copper atoms at the middle are also coordinated by pyridine ligands. The coordination geometry about the two middle copper atoms is a distorted tetrahedron, whereas that about the two terminal copper atoms is a trigonal plane. The cooper–copper distance between the middle atoms is slightly longer than the other two copper–copper distances: the former is 2.746 Å and the latter is 2.701 Å.

Hexanuclear Complex. The molecule of hexanuclear silver(I) dithiotriphenylacetate $Ag_6(Ph_3CSS_2)_6$ [8] exhibits threefold symmetry. The structure of the core is shown in Figure 20.6. Six surrounding dithiotriphenylacetate groups act as triple bridging

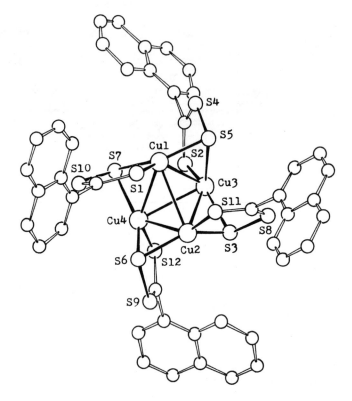

Figure 20.3 ■ Structure of $Cu_4(\alpha\text{-}C_{10}H_7CSS_2)_2$.

ligands. The Ag_6 configuration is an octahedron flattened in its axial direction, with six comparatively short and six longer edges. The silver–silver bridged distances (3.512 Å, avg.) are longer than the nonbridged distances (2.975 Å, avg.), which are different from those in tetranuclear silver dithiocarboxylates. The silver coordination is threefold but not planar, the metal atoms being situated "inside" the plane of the coordinating sulfur atoms 0.44 Å. The configuration and coordination mode of this cluster is similar to that of the neutral molecule $Ag_6[n\text{-}Pr_2Dtc]_6$ [2a] and the anion $[Ag_6(i\text{-}MNT)_6]^{6-}$ [17].

Polymeric Complex. The crystal of silver(I) dithiobenzoate consists of a nonmolecular polymeric structure $[Ag_2(C_6H_5CS_2)_2]_n$ [9], whose projection on plane (0 1 0) is shown in Figure 20.7. Two dithiobenzoate ligands and two silver atoms form a structural unit $Ag_2(C_6H_5CS_2)_2$ in which the distance between the unbridged silver atoms is only 2.890 Å, almost equal to the silver–silver distance (2.889 Å) in metallic silver [16]. Structural units are connected by Ag$\cdots$S interactions along the a and b axes to form polymeric structures, which contact each other by van der Waals force along the c axis. Each silver atom interacts with four sulfur atoms from four dithiobenzoate ligands and one silver atom nearest to it. If the silver–silver interaction is ignored, the coordination polyhedron around each silver atom is a distorted tetrahedron.

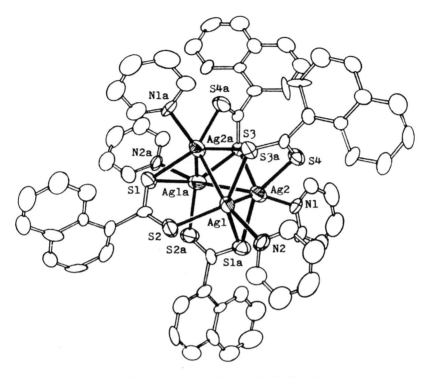

Figure 20.4 ▪ Structure of $Ag_4(\alpha\text{-}C_{10}H_7CS_2)_4(Py)_4$.

In order to compare the structures of Cu(I) and Ag(I) dithiocarboxylates or perthiocarboxylates with each other, we summarize the structural details of these complexes in Table 20.1.

20.2.2b. Coordination Numbers. In Cu(I) and Ag(I) clusters with 1,1-dithiolate ligands, the metals usually prefer trigonal planar coordination, for instance, in complexes $Cu_4(\alpha\text{-}C_{10}H_7CSS_2)_4$ [10], $[CuS_2CNEt_2]_4$ [2a], $[Cu(i\text{-}C_3H_7O)_2PS_2]_4$ [2a], $Ag_6(Ph_3CCS_2)_6$ [8], and so on. When the metals are coordinated by three sulfur atoms, they might be capable of further accepting an additional ligand, such as pyridine, triphenylphosphine, etc. Thus, when the crystals of tetranuclear copper cluster $Cu_4(\alpha\text{-}C_{10}H_7CSS_2)_4 \cdot \frac{1}{2}CS_2$ were recrystallized from pyridine, they were turned into the crystals of dimer $Cu_2(\alpha\text{-}C_{10}H_7CSS_2)_2(Py)_2$ [11]. The coordination geometry of copper atoms changed from trigonal plane to distorted tetrahedron. In the other complexes with pyridine as heteroligands $Ag_4(\alpha\text{-}C_{10}H_7CS_2)_4(Py)_4$ [6], $Ag_4(o\text{-}CH_3C_6H_4CS_2)_4(Py)_4$ [7], and $Cu_2(o\text{-}CH_3C_6H_4CSS_2)_2(Py)_2$ [7], which were formed by recrystallizing from pyridine, the metal atoms all exhibited a distorted tetrahedral coordination geometry as well. However, $[Ag_2(C_6H_5CS_2)_2]_n$ [9] was recrystallized from pyridine, but it had no pyridine molecule attached to silver. Each silver atom has already been coordinated by four sulfur atoms and is unable to further accept pyridine as a ligand.

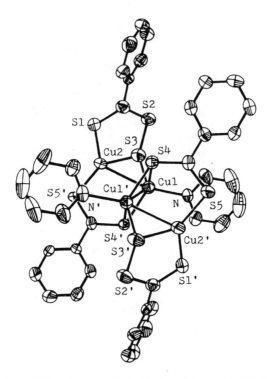

Figure 20.5 ▪ Structure of $Cu_4(C_6H_5CSS_2)_2(C_6H_5CS_2)_2(Py)_2$.

Copper(I) complexes with mixed dithio-perthiocarboxylate ligands $Cu_2(C_6H_5CSS_2)$-$(C_6H_5CS_2)(Py)_2$, $Cu_2(p\text{-}CH_3C_6H_5CSS_2)(p\text{-}CH_3C_6H_5CS_2)(Py)_2$ [14b], and $Cu_4(C_6H_5CSS_2)(C_6H_5CS_2)_2(Py)_2$ [12] formed by recrystallizing from pyridine have two kinds of copper atoms with different coordination numbers: One displays trigonal coordination; the other, coordinated also by pyridine, displays tetrahedral coordination.

Recently, Camus and co-workers [15] reported that when copper(I) perthiocarboxylate $(CuS_3C\text{-}p\text{-}Tolyl)_4$ reacted with triphenylphosphine (PPh_3) in $Cu:P = 1:1.5$ ratio, formation of PPh_3S occurred and $[CuS_2C\text{-}p\text{-}Tolyl)_4(PPh_3)_2]$ was isolated. Its structure shows a distorted trigonal-pyramidal arrangement of four copper atoms, in which two copper atoms display distorted trigonal coordination and the other two copper atoms, involved also in bonds with the PPh_3 ligands, show distorted tetrahedral coordination.

In the anionic complex $[Ag(Ph_3CCS_2)_2]^-$ [8] the silver coordination geometry is linear (shown in Figure 20.8), instead of the usually preferred trigonal-planar or tetrahedral coordination for Ag(I) 1,1-dithiolate. Only one of the two sulfur atoms in the ligand $Ph_3CCS_2^-$ is coordinated to the silver atom. Consequently, the two C—S distances of the ligand are different: C—S coordinated is 1.710 Å, slightly longer than C—S noncoordinated, 1.646 Å.

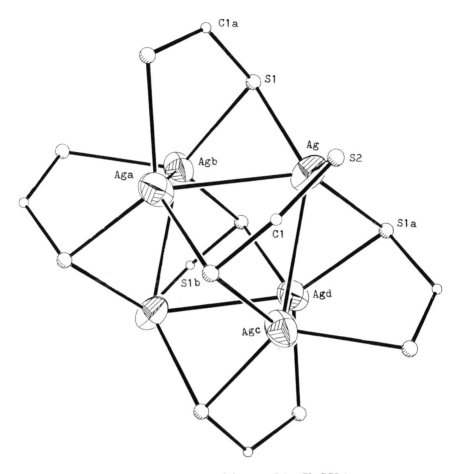

Figure 20.6 ▪ Structure of the core of $Ag_6(Ph_3CCS_2)_6$.

20.2.2c. Structural Influence of the Ligand Volume. In Cu(I) and Ag(I) clusters with 1,1-dithiolate ligands, it seems that the variation in the steric bulk of ligands influences the degree of cluster association and molecular configuration. In comparison with two tetranuclear silver dithiocarboxylates $Ag_4(\alpha\text{-}C_{10}H_7CS_2)(Py)_4$ and $Ag_4(o\text{-}CH_3C_6H_4CS_2)_4(Py)_4$, the complex $[Ag_2(C_6H_5CS_2)_2]_n$ with the ligand bulkiness reduced possesses a nonmolecular polymeric structure. In other words, as the ligands are not so bulky in silver dithiocarboxylates, the degree of association increases. The secondary $Ag\cdots S$ interactions will suppress the formation of discrete molecules, providing a driving force for further aggregation or polymerization.

A similar situation is observed in Ag(I) clusters with dithiocarbamate ligands. Silver dipropyldithiocarbamate consists of a discrete hexameric molecule $[(C_3H_7)_2NCS_2Ag]_6$ [2a]. However, the crystal structure of silver(I) diethyldithiocarbamate consists of

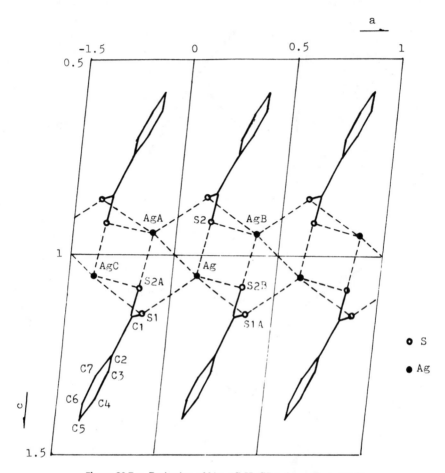

Figure 20.7 ■ Projection of $[Ag_2(C_6H_5CS_2)_2]_n$ on plane (0 1 0).

hexameric molecules $[AgS_2CN(C_2H_5)_2]_6$ [3], in which the terminal silver atom in one molecule is bridged by a sulfur atom to that in the adjacent hexamer, forming a chain structure along the X axis.

Compared with silver cluster complexes, copper cluster complexes with the same ligands seem to have a lower degree of association. Copper complexes $Cu_2(\alpha\text{-}C_{10}H_7CSS_2)_2(Py)_2$ and $Cu_2(o\text{-}CH_3C_6H_4CSS_2)_2(Py)_2$ are dimeric, whereas silver complexes $Ag_4(\alpha\text{-}C_{10}H_7CS_2)_4(Py)_4$ and $Ag_4(o\text{-}CH_3C_6H_4CS_2)_4(Py)_4$ are tetrameric. This may be due to the fact that the silver atom has a larger atomic radius than copper. The volume of the polyhedron formed by silver atoms is bigger than that formed by copper atoms, and bigger ligands can be accommodated around the polyhedron of silver atoms. For instance, when the tetranuclear silver cluster complex $Ag_4(\alpha\text{-}$

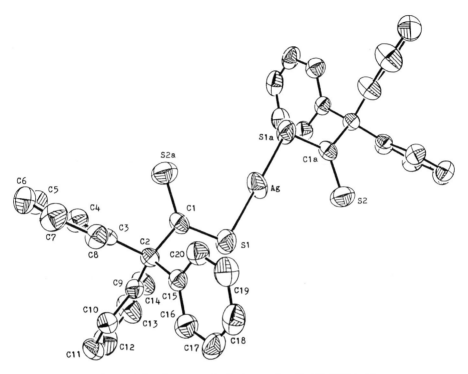

Figure 20.8 ■ Structure of anionic complex $(Ag(Ph_3CCS_2)_2)^-$.

$C_{10}H_7CS_2)_4$ accepts pyridine as a heteroligand, it remains a tetramer $Ag_4(\alpha\text{-}C_{10}H_7CS_2)_4(Py)_4$. However, when the copper cluster complex $Cu_4(\alpha\text{-}C_{10}H_7CSS_2)_4$ accepts pyridine as a heteroligand, the tetrameric complex dissociates into a dimeric complex $Cu_2(\alpha\text{-}C_{10}H_7CSS_2)_2(Py)_2$. The transition between the dimeric and tetrameric forms is reversible.

On the other hand, the molecular configuration of Cu(I) and Ag(I) 1,1-dithiolate complexes is affected by the ligand bulkiness. As discussed in Section 20.2.2a, the structure of Cu(I) α-perthionaphthoate is similar to that of Cu(I) o-perthiotoluate, and Ag(I) α-dithionaphthoate is similar to Ag(I) o-dithiotoluate. However, Cu(I) dithio-perthiobenzoate and p-dithio-perthiotoluate have similar configurations, which are different from that of Cu(I) o-perthiotoluate. These facts reflect the similar steric requirements of the o-methylphenyl and α-naphthoate groups and the similarity of p-methylphenyl and phenyl groups. The steric cone angle of p-toluate is smaller than that of o-toluate and similar to that of benzoate. However, the steric factor is only one factor influencing the configuration. Thus, Cu(I) with mixed dithio-perthiobenzoate and pyridine ligands forms two kinds of configuration under different conditions. One is a dimer, $Cu_2(C_6H_5CSS_2)(C_6H_5CSS_2)(Py)_2$, the other is a tetramer, $Cu_4(C_6H_5CSS_2)_2(C_6H_5CS_2)_2(Py)_2$, with a zigzag chain.

Table 20.1 ■ Structural Details of Cu(I) and Ag(I) Dithiocarboxylates or Perthiocarboxylates

Compound	M—M (Å) (bridged)	M—M (Å) (nonbridged)	M—S (Å) (bridged)	M—S (Å) (nonbridged)	M—N (Å)	Coordination geometry	Reference
$Cu_2(\alpha\text{-}C_{10}H_7CSS_2)_2(Py_2)$	2.616(2)		2.340(3) 2.475(2)	2.240(2)	2.026(6)	Tetrahedral	[11]
$Cu_2(o\text{-}CH_3C_6H_4CSS_2)_2(Py)_2$	2.608(3)		2.335(5) 2.429(4)	2.227(4)	2.049(7)	Tetrahedral	[7]
$Cu_2(C_6H_5CSS_2)(C_6H_5CS_2)(Py)_2$	2.660(2)		2.243(3) 2.307(3)	2.224 (avg.)	2.096 (avg.)	Tetrahedral Trigonal	[14b]
$Cu_2(p\text{-}CH_3C_6H_4CSS_2)(p\text{-}CH_3C_6H_4CS_2)(Py)_2$	2.642(2)		2.257(6) 2.347(6)	2.217 (avg.)	2.075 (avg.)	Tetrahedral Trigonal	[14b]
$Cu_4(\alpha\text{-}C_{10}H_7CSS_2)_4 \cdot \frac{1}{2}CS_2$	2.789 (avg.)	3.028 (avg.)	2.250 (avg.)	2.223 (avg.)		Trigonal	[10]
$Cu_4(o\text{-}CH_3C_6H_4CSS_2)_4$	2.780 (avg.)	3.004 (avg.)	2.230 (avg.) 2.274 (avg.)	2.229 (avg.)		Trigonal	[13]
$Cu_4[2,4,6\text{-}(CH_3)_3C_6H_2CSS_2]_4 \cdot C_3H_6O$	2.916 (avg.)	3.431 (avg.)	2.206 (avg.) 2.234 (avg.)	2.230 (avg.)		Trigonal	[13]
$Cu_4(C_6H_5CSS_2)_2(C_6H_5CS_2)(Py)_2$	2.724 (avg.)		2.257 (avg.) 2.436 (avg.)	2.200 (avg.)	2.061	Tetrahedral Trigonal	[12]
$Cu_4(p\text{-}CH_3C_6H_4CS_2)_4(PPh_3)_2$	2.664 (avg.)	3.575 (avg.)	2.326 (avg.)	2.252 (avg.)	Cu-P(A) 2.288	Tetrahedral Trigonal	[15b]
$Ag_4(\alpha\text{-}C_{10}H_7CS_2)_4(Py)_4$	2.924 (avg.)	3.563 (avg.)	2.503 (avg.) 2.823 (avg.)	2.470 (avg.)	2.423 (avg.)	Tetrahedral	[6]
$Ag_4(o\text{-}CH_3C_6H_4CS_2)_4(Py)_4$	2.872 (avg.) 3.022 (avg.)	3.194 3.977	2.537 (avg.) 2.721 (avg.)	2.496 (avg.)	2.458 (avg.)	Tetrahedral	[7]
$Ag_6(Ph_3CCS_2)_6$	3.512 (avg.)	2.975 (avg.)	2.485 (avg.) 2.500 (avg.)	2.479 (avg.)		Trigonal	[8]
$[Ag(Ph_3CCS_2)_2]^-$	2.890		2.469 (avg.) 2.764 (avg.)	2.383 (avg.)		Diagonal Tetrahedral	[8] [9]
$[Ag_2(C_6H_5CS_2)_2]_n$							

20.3. Cu(I) and Ag(I) Cluster Complexes with Sterically Hindered Thiolate Ligands

20.3.1. Synthesis of Cu(I) and Ag(I) Cluster Complexes with Thiolate Ligands

Block has developed several syntheses of (triorganosilyl)methanethiols $(RR'R''Si)_nCH_{3-n}SH$, summarized in the literature [18]. Arylthiols are prepared by the reduction of corresponding substituted benzenesulfonyl chloride with $LiAlH_4$ [24].

Cu(I) or Ag(I) thiolate complexes are usually synthesized as follows: The reaction is carried out under argon in degassed solvents. A solution of CuCl or $AgNO_3$ in acetonitrile is added to a stirred solution of the corresponding thiol and an organic base (Et_3N or Me_3N) in organic solvents. The reaction mixture is refluxed for 2–3 h, then the precipitate formed is collected. The crude product is usually recrystallized from hot ethanol, chloroform, or benzene–acetonitrile, and so forth [18–22]. For example:

$$AgNO_3 + (Me_3Si)_3CSH + (C_2H_5)_3N \xrightarrow[\text{(recrystallized)}]{CH_3CN \quad \text{hot ethanol}}$$

$$[AgSC(SiMe_3)_3]_4 \quad [18] \quad (20.8)$$

$$AgNO_3 + 2,4,6\text{-}Pr^i_3C_6H_2SH + (C_2H_5)_3N \xrightarrow[\text{(recrystallized)}]{CH_3CN \quad CHCl_3-EtOH}$$

$$[(AgSC_6H_2Pr^i_3)_4 \cdot HCCl_3]_n \quad [22] \quad (20.9)$$

$$CuCl + (Me_3Si)_2CHSH + (C_2H_5)_3N \xrightarrow[EtOH \quad \text{(recrystallized)}]{CH_3CN \quad \text{hot EtOH}}$$

$$[CuSCH(SiMe_3)_2]_4 \quad [19] \quad (20.10)$$

20.3.2. Structures

20.3.2a. Structural Patterns. Relevant structural parameters of Cu(I) and Ag(I) thiolate complexes are summarized in Table 20.2.

Trinuclear Complexes. The trinuclear silver complex $[AgSC(SiMe_2Ph)_3]_3$ [18] and copper complex $[CuSC(SiMe_3)_3]_3$ [20] have similar structures: a nonplanar six-membered cycle of alternating metal and sulfur atoms M_3S_3 (see Figures 20.9 and 20.10). The angles S—Cu—S (149–164°) and Cu—S—Cu (74.7–76.1°) in Cu_3S_3 are close to those of S—Ag—S (151–163°) and Ag—S—Ag (75.7–83.6°) in Ag_3S_3. Although the volume of the ligand $(Me_3Si)_3CS^-$ is smaller than that of $(Me_2PhSi)_3CS^-$, the similar structure might be a consequence of the smaller atomic radius of the copper atom. The ring size of Cu_3S_3 is smaller than that of Ag_3S_3. The smaller-volume $(Me_3Si)_3CS^-$ ligands project steric cones about the Cu_3S_3 ring that might be similar to the steric cones that the bigger $(PhMe_2Si)_3CS^-$ ligands project about the Ag_3S_3 ring.

Tetranuclear Complexes. Both structures of the tetranuclear silver complex $[AgSC(SiMe_3)_3]_4$ [18] (Figure 20.11) and the copper complex with less-hindered thiolate $[CuSCH(SiMe_3)_2]_4$ [19] (Figure 20.12) consist of a discrete eight-membered cycle of alternating M and S atoms, M_4S_4. The four metal atoms exist in an approximately square-planar arrangement. The angles M—M—M and S—M—S are

Table 20.2 ▪ Comparison of Relevant Parameters for Cu(I) and Ag(I) Thiolate Complexes

Compound	M—M (Å) (bridged)	M—S (Å) (bridged)	S—C (Å)	S—M—S (degrees)	M—S—M (degrees)	M—M—M (degrees)	Reference
$[CuSC(SiMe_3)_3]_3$	2.666 (avg.)	2.175		149.7 / 163.6	74.7–76.1	60 (avg.)	[20]
$[CuSCH(SiMe_3)_2]_4$	2.690 (avg.)	2.17	1.85	175.7	76.3 / 77.2	89.8 (avg.)	[19]
$[CuSC_6H_2Pr^i_3\text{-}2,4,6]_8$	3.146 (unbridged) / 3.107, 2.705 (bridged)	2.16		177.5	92.5		[21]
$[AgSC(SiMe_2Ph)_3]_3$	3.096 (avg.)	2.481	1.90	151.8 / 163.1	75.7 / 83.6	60 (avg.)	[18]
$[AgSC(SiMe_3)_3]_4$	3.313 (avg.)	2.378	1.87	177.5	88.3	89.98 (avg.)	[18]
$[AgSCH(SiMe_3)_2]_8$	2.994 (avg.) / 3.270 (avg.) (secondary Ag⋯S)	2.40 (avg.)	1.86	175.6	77.2	89.98 (avg.)	[18]
$[Ag_4\{SCH_2(SiMe_3)_3\}_3]_n(OMe)_n$	3.143 (avg.)	2.41 (dig) / 2.53 (trig)	1.87	165.5 (dig) / 117.5 (trig)			[18]
$[(AgSC_6H_2Pr^i_3\text{-}2,4,6)_4 \cdot HCCl_3]_n$	2.978 / 3.353	2.358	1.807	171.3	112.9		[22]
$[Ag(SC_6H_2Me_3\text{-}2,4,6)_2]^- Et_4N^+$		2.346	1.772	178.8			[23]

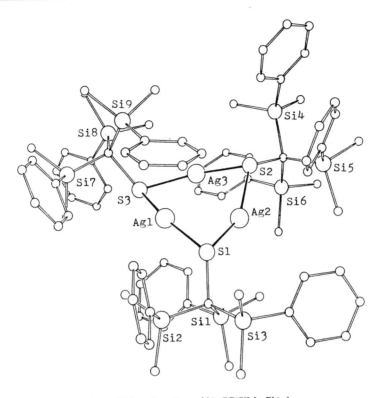

Figure 20.9 ▪ Structure of [AgSC(SiMe$_2$Ph)$_3$]$_3$.

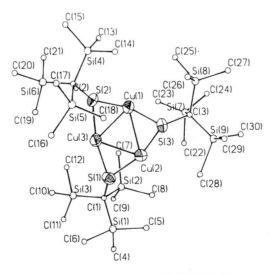

Figure 20.10 ▪ Structure of [CuSC(SiMe$_3$)$_3$]$_3$.

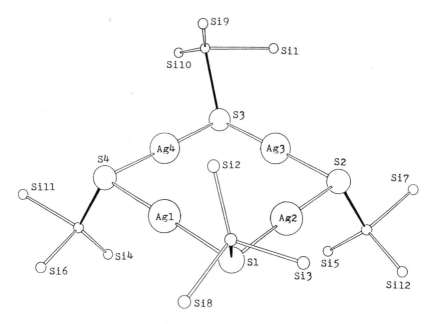

Figure 20.11 ■ Structure of [AgSC(SiMe₃)₃]₄. Methyl groups have been omitted for clarity.

similar in both complexes (see Table 20.2), but the Ag—S—Ag angles in the former (avg. 88.3°) are bigger than the Cu—S—Cu angles in the latter (avg. 76.3°). The eight-membered cycle Ag_4S_4 exists as a shallow crown with a maximum deviation of 0.1 Å from the least-squares plane through the eight atoms whereas the Cu_4S_4 cycle takes on a butterfly configuration. This may be due to the fact that the smaller number of silyl groups in the ligand $(Me_3Si)_2CHS^-$) provides smaller steric congestion to an

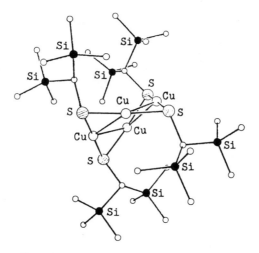

Figure 20.12 ■ Structure of [CuSCH(SiMe₃)₂]₄.

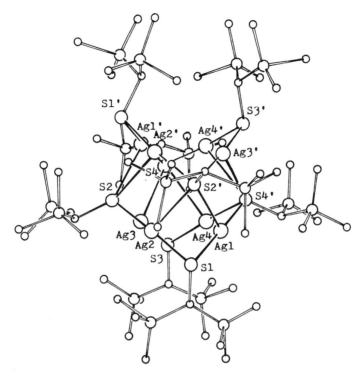

Figure 20.13 ▪ Perspective view of the structure of [{AgSCH(SiMe$_3$)$_2$}$_4$]$_2$, showing secondary Ag···S interactions.

unconstrained M$_4$S$_4$ ring, resulting in different ring configurations. The rings in both complexes are discrete molecular units, exhibiting no secondary M···S interacitons.

Octanuclear Complexes. The less-hindered thiolate (Me$_3$Si)$_2$CHS$^-$ allows the formation of secondary Ag···S interactions to produce the biscyclic structure [AgSCH(SiMe$_3$)$_2$]$_8$, illustrated in Figure 20.13. The species may be formulated as [{AgSCH(SiMe$_3$)$_2$}$_4$]$_2$ because there are clearly two tetranuclear Ag cycles, which face each other and are connected by eight weak secondary Ag···S interacitons. The geometry about silver atoms is distinctly T-shaped, rather than trigonal planar. The [AgSCH(SiMe$_3$)$_2$]$_4$ unit is distinctly nonplanar, similar to [CuSCH(SiMe$_3$)$_2$]$_4$. The M—M—M, S—M—S, and M—S—M angles (see Table 20.2) are very similar in both M$_4$S$_4$ cycles.

The octanuclear copper cluster complex with the sterically hindered arylthiolate ligand (TIPT) [Cu(SC$_6$H$_2$Pr$^i_3$-2,4,6)]$_8$ [21] contains a discrete twisted 16-membered cycle of alternating copper and sulfur atoms, which are at the eight corners. The structure of the Cu$_8$S$_8$ core in crystalline [Cu(SC$_6$H$_2$Pr$^i_3$-2,4,6)]$_8$ is shown in Figure 20.14. The twisted center of the 16-membered cycle is located between Cu-4 and Cu-8. The distance separating the two unbridged copper atoms Cu-4—Cu-8 is 3.146 Å, which is considered rather short in unbridged Cu—Cu distances. All the arylthiolate ligands are doubly bridging and each copper atom exhibits linear 2-coordination.

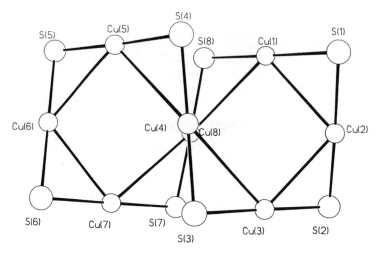

Figure 20.14 ▪ Structure of the Cu_8S_8 core of $[Cu(SC_6H_2Pr^i_3\text{-}2,4,6)]_8$.

Polymeric Complexes. Polymeric complex $[Ag_4\{SCH_2(SiMe_3)\}_3]_n(OCH_3)_n$ was formed by further reduction of the ligand bulk of polysilylated thiolates [18]. The complex polymeric structure may be described in terms of one-dimensional kinked chains of fused Ag_4S_4 cycles I, II, and III, cross-linked through Ag_4S_4 monocycle IV. All sulfur donors are triply bridging, to satisfy the coordination requirements of two trigonal Ag centers and six digonal Ag centers of the asymmetric unit $[Ag_8\{SCH_2(SiMe_3)\}_6]^{2+}$. The charge requirements of $[Ag_4\{SCH_2(SiMe_3)\}_3]_n(OCH_3)_n$ are satisfied by the presence of methoxy groups in the crystal lattice. To our knowledge, this is the first example of a cationic silver-thiolate complex.

The reaction of silver nitrate with the sterically hindered arylthiol $2,4,6\text{-}Pr^i_3C_6H_2SH$ results in another polymeric complex $[(AgSC_6H_2Pr^i_3\text{-}2,4,6)_4 \cdot HCCl_3]_n$ [22]. Its structure consists of one-dimensional double $(Ag\text{—}SR)_n$ strand belts, shown in Figure

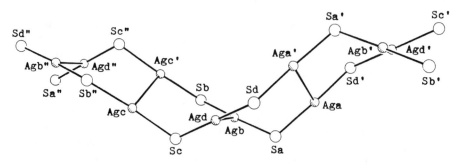

Figure 20.15 ▪ The polymeric structure of $[Ag(SC_6H_2Pr^i_3\text{-}2,4,6)]_n$. Aryl groups have been omitted for clarity.

20.15. The $(Ag—SR)_n$ strands are formed by alternating silver and sulfur atoms of the thiolate ligands. Each belt contains two strands of $(Ag—SR)_n$ connected by Ag—Ag links between the Ag atoms on the two strands. This structure is somewhat similar to that of $(AgSMeEt_2)_n$ containing a chain of two intertwined but totally unconnected $(AgSR)_n$ strands [26]. All S—Ag—S segments are almost linear. However, there are no Ag$\cdots$S interactions between two linked strands. The hindrance of the arylthiolate ligand with bulky substituents $2,4,6\text{-}Pr^i_3C_6H_2S^-$ prevents the neighboring belts from close approach and precludes concomitant Ag$\cdots$Ag interactions to further polymerization leading to the formation of layers.

The use of the less-bulky ligand $2,4,6\text{-}Me_3C_6H_2SH$ gives an insoluble white solid [23]. However, when the ligand $2,4,6\text{-}Me_3C_6H_2SH$ is present in excess, an anionic complex $[Ag(SC_6H_2Me_3\text{-}2,4,6)_2]^-$ is obtained [23]. In this structure, the bond angle S—Ag—S is 178.8°, which is similar to the angle S—Cu—S (178.6°) in the copper anionic complex $[Cu(SC_6HMe_4\text{-}2,3,5,6)_2]^-$ [28].

20.3.2b. Coordination Numbers. A survey of the known structures of copper(I) and silver(I) complexes with halide ligands and various types of sulfur donor ligands reveal a predominance of tetrahedral and trigonal stereochemistries [29]. Copper(I) and silver(I) complexes with 1,1-dithiolate tend to possess coordination numbers of 3 or 4, as discussed in Section 20.2.2b. Linear coordination of copper(I) and silver(I) with analogous ligands is known only in $CuCl_2^-$, the nonmolecular $(Ag_3S)_n^{n+}$ ion [29], and $[Ag(Ph_3CCS_2)_2]^-$ [8], and so on.

The organothiolate anion (RS^-) is a fundamental ligand type. The thiolate anion may be classified as a pseudohalide, comparable to a ligand with Cl^-, Br^-, and I^-. However, alkylthiolate and arylthiolate ligands generally produce lower coordination numbers than other halide, pseudohalide, and uncharged sulfur ligands. This may be interpreted in terms of a greater polarizability of the thiolate ligands. The planar-trigonal and linear-digonal coordination by bridging thiolate are both feasible for both metals. The co-occurrence of two different coordination stereochemistries was found in some copper(I) and silver(I) clusters with thiolate ligands. For example, there are three digonal and two trigonal coordinations in the cluster $[Cu_5(SBu^t)_6]^-$ [30], in which the average coordination number is 2.4. In both clusters, $[Cu_5(SPh)_7]^{2-}$ and $[Ag_5(SPh)_7]^{2-}$ [29], there are four trigonal and one digonal coordinations, in which the average coordination number is 2.8. In the complex $(AgSC_6H_{11})_{12}$ [25], some silver atoms possess approximately linear coordination, whereas others possess severely distorted trigonal planar stereochemistry.

In the Cu(I) and Ag(I) cluster complexes with sterically hindered thiolate ligands investigated by us, with the exception of the polymeric complex $[Ag_4\{SCH_2(SiMe_3)\}_3]_n(OMe)_n$ [18] and the biscyclic structure $[AgSCH(SiMe_3)_2]_8$ [18], all metal centers are linearly coordinated to two bridging sulfur donors, which seldom occurs in other Cu(I) and Ag(I) complexes with thiolate ligands. The reduced coordination at bridged metal centers is attributed to the steric and electronic effects of the sterically hindered substituents of the ligands. Compared with $[Cu_5(SBu^t)_6]^-$ [30] and $[Cu_5(SPh)_7]^{2-}$ [29], respectively, the complexes $[CuSC(SiMe_3)_3]_3$ and $[CuSC_6H_2Pr^i_3\text{-}2,4,6]_8$ with bulky substituents, which have stronger electron-releasing ability, possess a smaller coordination number.

In general, Cu_{trig}—S and Ag_{trig}—S distances are slightly longer than Cu_{dig}—S and Ag_{dig}—S, respectively [29]. In our case (see Tables 20.1 and 20.2), the elongation

Table 20.3 ▪ Ag(I) and Cu(I) Complexes with Polysilylated Thiolates (in Order of Increasing Ligand Volume)

$[Ag_4\{SCH_2(SiMe_3)\}_3]_n(OMe)_n$ [18]	$[CuSCH_2(SiMe_3)]_n$ [23]
$[AgSCH(SiMe_3)_2]_8$ [18]	$[CuSCH(SiMe_3)_2]_4$ [19]
$[AgSC(SiMe_3)_3]_4$ [18]	$[CuSC(SiMe_3)_3]_3$ [20]
$[AgSC(SiMe_2Ph)_3]_3$ [18]	$[CuSC(SiMe_2Ph)_3]_x$

of M—S distances are also observed when the coordination numbers of copper and silver are increased from 2 to 3.

20.3.2c. Structural Influences of the Ligand Volume. In the investigation of the structures of Ag(I) and Cu(I) complexes with sterically hindered polysilylated thiolate and arylthiolate ligands, it has been observed that the variation in the ligand volume remarkably influences the degree of association of both metal thiolates, as summarized in Tables 20.3 and 20.4.

Silver(I) and copper(I) alkylthiolates with unbranched chains are high polymeric compounds, because the metal centers are unprotected from the approach of additional potentially bridging thiolate groups. Simple branched-chain alkylthiolate ligands afford a degree of steric protection to the silver centers, as demonstrated by the formation of discrete octameric and dodecameric cycles in solution. However, the steric cones associated with these ligands are insufficient to shield the ring unit effectively enough to prevent the formation of secondary interactions and consequent aggregation as the species passes from solution to the more dense crystalline phase. The bulky tris(tri-organosilyl)methane substituents of the ligand $(R_3Si)_3CSH$ provide sufficient steric congestion to accomplish both the prevention of secondary interactions between rings and the limitation of ring size. By using the thiolate ligands with different size of substituents $(PhMe_2Si)_3CS^-$ and $(Me_3Si)_3CS^-$, the discrete six-membered cycle and eight-membered cycle structure complexes $[AgSC(SiMe_2Ph)_3]_3$ and $[AgSC(SiMe_3)_3]_4$ are obtained, respectively. In contrast, the less-hindered thiolate $(Me_3Si)_2CHS^-$ is ineffective in preventing the approach of discrete $[AgSR]_n$ monocyclic units. This allows the formation of secondary Ag···S interactions to produce the biscyclic structure $[AgSCH(SiMe_3)_2]_8$. The consequences of further reduction of the ligand bulk are illustrated by the structure of nonmolecular species $[Ag_4\{SCH_2(SiMe_3)\}_3]_n(OMe)_n$. A gradual progression from the discrete molecular structure to the associated biscyclic structure to the complex polymer is observed as a consequence of the increased efficiency of secondary Ag···S bonding. That is, the more bulky the group R in $[AgSR]_n$, the less the degree of the association n. In a sense, this structure rule is analogous to that of the series of copper and silver thiolates with heteroligand

Table 20.4 ▪ Ag(I) and Cu(I) Complexes with Arylthiolates (in Order of Increasing Ligand Volume)

$[AgSC_6H_2Me_3]_n$ [23]	$[CuSC_6H_2Me_3]_n$ [23]
$[(AgSC_6H_2Pr^i_3)_4 \cdot HCCl_3]_n$ [22]	$[CuSC_6H_2Pr^i_3]_8$ [21]

phosphines $[(MSR)_x(PR_3)_y]$ investigated by Dance [4]. In the latter case, the heteroligand usually plays a structure-terminating role to prevent aggregation through secondary interactions. The structure-terminating ligand phosphines lead to a reduction in the size of metal thiolate aggregate according to their proportional presence. In $[(MSR)_x(PR_3)_y]$, the number of M atoms in the aggregate, x, is inversely related to the proportion y/x of phoshpine ligands. That is, the larger the proportion of phosphine ligands, the smaller the number of M atoms in the aggregate.

The same rule as in the silver thiolate complexes is observed in the copper complexes with sterically hindered polysilylated thiolates. The Cu(I) complex with unhindered thiolate $[(Me_3Si)CH_2SCu]_n$ is a polymeric solid that is insoluble in inert solvents, whereas the Cu(I) complexes with bulky substituent thiolates $(Me_3Si)_2CHS^-$ and $(Me_3Si)_3CS^-$ are discrete eight-membered and six-membered cycle structures $[CuSCH(SiMe_3)_2]_4$ and $[CuSC(SiMe_3)_3]_3$, respectively, exhibiting no secondary $Cu\cdots S$ interactions between cycle units.

However, it is remarkable in Tables 20.3 and 20.4 that the copper-thiolate complexes with the same ligand have lower degree of aggregation than the silver-thiolate complexes. For instance, $[AgSCH(SiMe_3)_2]_8$ consists of two $[AgSCH(SiMe_3)_2]_4$ cycles connected by secondary $Ag\cdots S$ interactions, whereas $[CuSCH(SiMe_3)_2]_4$ is a discrete cycle unperturbed by secondary $Cu\cdots S$ interactions. This may be due to the fact that the copper atom has a smaller atomic radius ($r_{Cu} = 1.40$ Å) than silver ($r_{Ag} = 1.70$ Å), so that the less-hindered thiolate $(Me_3Si)_2CHS^-$ is sufficient to shield the ring unit to prevent secondary interactions. On the other hand, if the copper and silver cluster complexes have the same degree of association, the larger-volume ligands fit around the polyhedron of silver atoms. This is consistent with what we obtained from our studies on the silver and copper complexes with dithiocarboxylate ligands (see Section 20.2.2c).

In the copper(I) complexes with sterically hindered arylthiolate ligands, the hindrance of the arylthiolate ligand with bulky substituent $2,4,6$-$Pr^i_3C_6H_2S^-$ is sufficient to prevent secondary interactions, resulting in a discrete 16-membered cycle $[CuSC_6H_2Pr^i_3$-$2,4,6]_8$ [21]; but it is ineffective in preventing the secondary interactions between strands in the silver complex, yielding a one-dimensional double $(Ag-SR)_n$ strand belt complex connected by $Ag-Ag$ links $[AgSC_6H_2Pr^i_3$-$2,4,6]_n$ [22]. Because the substituent is not bulky enough, Cu(I) and Ag(I) trimethylthiophenolate (TMTH) are both insoluble solids. The less bulky the ligand is, the more tightly the strands are linked by secondary interactions. Solvent molecules then would penetrate with difficulty into the solid phase of polymeric complexes. That might account for the fact that $(AgSR)_n$ complexes are usually insoluble when the group R is small enough.

In conclusion, in our studies on the copper(I) and silver(I) complexes with sulfur-containing ligands, such as dithiocarboxylates or perthiocarboxylates and alkylthiolates or arylthiolates, the ligand volume has a profound influence on the degree of association of both metal cluster complexes. However, although there is a suggestion that the steric bulk of the substituent may be tailored to create a particular degree of aggregation, this view may be overly simplistic. The structure rule may be suitable for certain cases, for example, in the case of Ag(I) and Cu(I), but not universally applicable. Thus, mercury(II) 2-methyl-2-propanethiolate is polymeric [31] whereas mercury(II) ethanethiolate exhibits an essentially molecular structure [32]. Likewise, reaction conditions have been demonstrated to have a profound influence on the nature of metal-thiolate complexes. Steric effects may predominate in certain cases, but other factors (charge, stoichiometry, solvent) may also play crucial roles in determining the degree of association of metal-thiolate species.

20.4. Insertion Reaction of CS_2 into Ag—S Bonds

20.4.1. Introduction

The molecular structure of CS_2, which has π bonds, is similar to that of CO_2. Uncoordinated CS_2 is linear with a C—S bond length of 1.554 Å. The activation of CS_2 by transition-metal complexes has been widely studied. Carbon disulfide can coordinate to transition metals to form four distinctly different types of complex [33a]. The insertion of a CS_2 molecule into an M—X bond (M = transition metal, X = H, N, O, S, halogen, alkyl, aryl, amine, alkoxide, etc.) could lead to a variety of known or new dithio or trithio compounds.

In 1931, Duncan, Ott, and Reid [33b] reported that CS_2 reacts with copper mercaptides (CuSR) to give trithiocarbonates (CuSSCSR), but their structures had not been determined. Since 1965, some product structures of the insertion of CS_2 into Cd—S, Mo—S, and Ni—S bonds have been reported. For some copper-thiolate complexes with the copper atom coordinated by triphenylphosphine (PPh_3), the CS_3 molecule could be readily inserted into their Cu—S bonds to give copper alkyl (or phenyl) trithiocarbonate complexes, such as $Cu(PPh_3)_2(S_2CSPh)$ [34] and $Cu(PPh_3)_2(S_2CSEt)$ [35]. In 1986 Chadha, Kumar, and Tuck reported an unusual copper–sulfur cage (Cu_8S_{12}) complex $Cu_8(SC_5H_{11})_4(S_2CSC_5H_{11})_4$ [36], in which only half-units of $(CuSCMe_2Et)_8$ were inserted by CS_2 into Cu—S bonds. However, very few reports on the insertion of CS_2 into Ag—S bonds have been found in the literature.

We have found silver thiolate complexes $(AgSR)_n$ with less-bulky substituents, such as $(AgSBu^t)_n$, $(AgSC_6H_5)_n$, or $(AgSC_6H_2Me_3)_n$, difficult to react with CS_2. However, the complexes react readily with CS_2 to give insertion products when the Ag atom is coordinated with PPh_3 [37]. By changing the ligand to increase its bulk, for example, in the case of the complex $[AgSC_6H_2Pr^i_3\text{-}2,4,6]_n$ reactions with CS_2 occur more readily. A peculiar product $[(AgS_2CSC_6H_2Pr^i_3\text{-}2,4,6)_2(AgSC_6H_2Pr^i_3\text{-}2,4,6)_6] \cdot 8HCCl_3$ with insertion of CS_2 into a quarter of the Ag—S bonds of $[AgSC_6H_2Pr^i_3\text{-}2,4,6]_n$ is formed [22]. A novel structural type of the insertion reaction of CS_2 into M—S bonds has been found.

20.4.2. Insertion of CS_2 into Ag—S Bonds in Ag(I) Alkylthiolate Complexes

Silver(I) alkylthiolates with an unbranched-chain or a simple branched-chain structure are difficult to react with CS_2. However, they could react with CS_2 in the presence of PPh_3. Thus, the white solids of $(Bu^tSAg)_n$ and PPh_3 were dissolved in CS_2, yielding an orange solution, from which orange crystals formed. The structure of the crystal was $(PPh_3)_2Ag(S_2CSBu^t)$, determined by the single-crystal X-ray diffraction method [37]. The molecular configuration is shown in Figure 20.16. The Ag atom is pseudo-tetrahedrally coordinated by two S atoms of thioxanthate ligand ($Bu^tSCS_2^-$) and two P atoms of PPh_3 ligands. The two Ag—S distances average 2.675 Å, whereas the two Ag—P distances average 2.456 Å. The coordination situation is similar to that in $(PPh_3)_2Cu(S_2CSPh)$ [36].

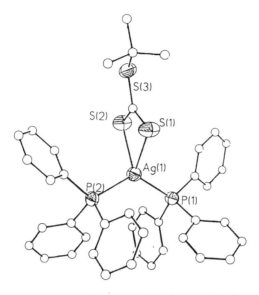

Figure 20.16 ▪ Structure of $(PPh_3)_2Ag(S_2CSBu^t)$.

The trithiocarbonate CS_3 group formed by the insertion of CS_2 into the Ag—S bond is approximately planar. The mean C—S distance in the CS_3 group is 1.702 Å, longer than that of uncoordinated CS_2 (1.554 Å), shorter than that of the single-bond C—S (1.812 Å [16]), and similar to the mean distance of C—S bonds (1.703 Å) in CS_3 of $(PPh_3)_2Cu(S_2CSPh)$ [34].

The orange crystals of $(PPh_3)_2Ag(S_2CSBu^t)$ melted with effervescence at 120–130 °C (lost CS_2) to a colorless liquid, which resolidified to colorless microcrystals that subsequently decomposed at 225 °C. The colorless microcrystals can be dissolved in CS_2 and from the solution form orange crystals again, suggesting that the insertion reaction of CS_2 is reversible. The insertion product is orange due to the CS_3 group. The change of color may be regarded as a characteristic of the insertion of CS_2.

20.4.3. Insertion of CS_2 into Ag—S bonds in Ag(I) Arylthiolate Complexes

The polymeric complex with the sterically hindered arylthiolate ligand $[AgSC_6H_4Pr^i_3$-2,4,6]$_n$ reacts readily with a small amount of CS_2 in chloroform to give the pale-yellow crystals $[AgS_2CSC_6H_2Pr^i_3$-2,4,6]$_2[AgSC_6H_2Pr^i_3$-2,4,6]$_6 \cdot 8HCCl_3$ [22]. The molecular structure consists of eight coplanar silver atoms linked through arylthiolate (RS) and aryltrithiocarbonate $(RSCS_2)$ ligands. Figure 20.17 gives an outline of the $(AgS_2CS)_2(AgS)_6$ core. There are a number of interesting features to this unusual structure.

The eight silver atoms are all in one plane [plane (1)]. The maximal deviation of silver atoms from plane (1) is only 0.01 Å. There is a symmetry center at the midpoint of Ag-2—Ag-2a. The structure can be divided into two parts: One consists of Ag-2 and Ag-2a atoms in the middle of the core that are held together by two aryltrithiocarbonate ligands formed by the insertion of CS_2 into Ag—S bonds. The second part is a

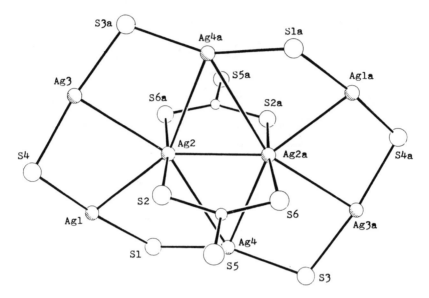

Figure 20.17 ▪ View of the outline of $(AgS_2CS)_2(AgS)_6$ core for $(AgS_2CSC_6H_2Pr^i_3)_2$-$(AgSC_6H_2Pr^i_3)_6$.

12-membered ring of alternating silver and sulfur atoms of the arylthiolate ligand around the core.

In the first part, Ag-2, Ag-2a, and the atoms of two CS_3 segments in the aryltrithio-carbonate ligands are all in the other plane [plane (2), which is approximately perpendicular to plane (1)—the angle between the two planes is 94.0°]. The distance between Ag-2 and Ag-2a is only 2.782 Å, even shorter than the Ag—Ag distance (2.886 Å) in metallic silver, which is rarely seen in silver complexes [38]. The two atoms Ag-2 and Ag-2a are surrounded by six silver atoms doubly bridged by six sulfur atoms of arylthiolate ligands. The distances between Ag-2 and surrounding silver atoms are all longer than the interatomic distance (2.886 Å) in metallic silver and shorter than the van der Waals diameter (3.40 Å) [27]. However, there is no secondary Ag···S interaction between the two parts of the structure. The Ag—Ag interactions make the structure stable.

Another interesting feature of the complex is that only a quarter of AgSR units of the starting complex react with CS_2, forming a novel type of structure of the insertion with CS_2 into M—S bonds.

$$S-C\overset{\displaystyle S-M-S}{\underset{\displaystyle S-M-C}{\Big\langle}}C-S$$

The C—S distance of the resulting CS_3 group averages 1.691 Å, which is close to the distance of C—S bond 1.702 Å in CS_3 of $(PPh_3)_2AgS_2CSBu^t$ [37].

20.4.4. The Factors Influencing the Rate of the Insertion Reaction

The rate of the insertion reaction of CS_2 into Ag—S bonds is markedly affected by a change of the substituents in ligands. We have found that reactions of $(AgSC_6H_5)_n$ or $(AgSC_6H_2Me_3-2,4,6)_n$ with CS_2 are difficult, whereas the reaction of $(AgSC_6H_2Pr^i_3-2,4,6)_n$ with CS_2 is particularly easy, even with only limited CS_2 in the reaction solution. When the ligand $2,4,6-Pr^i_3C_6H_2SH$ (TIPT) is compared with C_6H_5SH or $2,4,6-Me_3C_6H_2SH$ (TMTH), it is found that the substituents of TIPT have the greater steric bulk and the stronger electron-releasing ability. The silver complexes with less bulky ligands are a high polymeric solid $(AgSAr)_n$. Thus, CS_2 molecules penetrate with difficulty into the solid phase, preventing reaction. On the other hand, it seems that the stronger the electron-releasing ability the ligand has, the higher the electron density on the silver atom. When binding to the silver atom, the CS_2 molecule donates a pair of electrons to the silver atom to form a σ bond, then a pair of electrons on the orbit of the silver atom back-donates to the π^* orbit of the C=S bond to form a back-donated π bond. The higher the electron density on the silver atom, the stronger the back-donated π bond. The more tightly the silver atom binds with CS_2, the lower the activation energy the reaction has, then the more readily the insertion reaction takes place. Therefore, the steric and electronic effects of ligands have a profound influence on the rate of the insertion reaction.

As mentioned before, silver thiolate complexes with less bulky substituents are difficult to react with CS_2, but such reactions proceed easily when ligands with a high σ-donor strength, such as phosphine, are coordinated to the silver atoms. That is, the ligands donate the higher electron density to the silver atom and accelerate the insertion reaction.

Silver complexes $[AgSBu^t]_n$ or $[AgSCMeEt_2]_n$ react with CS_2 very slowly. However, when the solvent ethanol is added to the reaction mixture, the insertion reaction is accelerated. This polar solvent might be helpful to the insertion reaction.

In short, there are many factors influencing the rate of the insertion reaction of CS_2 into Ag—S bonds. Once the mechanism of the insertion reaction of CS_2 is thoroughly understood, these factors will be fully realized. We believe that in the next few years the study of the CS_2 insertion reaction will prove to be a fruitful area of research in transition-metal chemistry.

References

1. Karlin, K. D.; Zubieta, J. *Inorg. Persp. Bio. Med.*, **1979**, *2*, 127.
2. Coucouvanis, D.: (a) *Prog. Inorg. Chem.*, **1970**, *11*, 260; (b) *Prog. Inorg. Chem.*, **1979**, *26*, 438 and references therein.
3. Yamaguchi, H.; Kido, A.; Uechi, T.; Yasukouchi, K. *Bull. Chem. Soc. Japan*, **1976**, *49*, 1271.
4. Dance, I. G. *Polyhedron*, **1986**, *5*, 1037 and reference within.
5. (a) Roberie, T.; Hoberman, A. E.; Selbin, J. *J. Coord. Chem.*, **1979**, *9*, 79. (b) Shinzi Kato; Masateru Mizuta, *Bull. Chem. Soc. Japan*, **1972**, *45*, 3492.
6. Tang, K.; Jin, X.; Xie, Y.; Tang, Y. *Scientia Sinica (Section B)*, **1986**, 485.
7. Tang, K.; Jin, X.; Xiao, Q.; Sun, J.; Tang, Y. *J. Struct. Chem.*, **1988**, *7*, 247.
8. Tang, K.; Jin, X.; Hu, Y.; Tang, Y., *Beijing Daxue Xuebao, Ziran Kexueban*, **1990**, *26*, 173.
9. Tang, K.; Jin, X.; Xiao, Q.; Tang, Y. *J. Struct. Chem.*, **1988**, *7*, 159.

10. Tang, K.; Gan, H.; Xu, X-J.; Zhou, G.-D.; Tang, Y.: (a) *Scientia Sinica (Section B)*, **1984**, *27*, 456; (b) *Proceedings of a Symposium for Celebrating the 50th Anniversary of Chinese Chemical Society*, Nanjing, 1982, p. 31.

11. Jin, X.; Jin, Y.; Tang, Y. *Acta Chim. Sinica*, **1986**, *44*, 522.

12. Jin, X.; Jin, Y.; Tang, K.; Tang, Y. *Scientia Sinica (Section B)*, **1987**, 225.

13. Schuerman, J. A.; Fronczek, F. R.; Selbin, J. *Inorg. Chim. Acta*, **1988**, *148*, 177.

14. (a) Maria, A.; Lanfredi, M.; Tiripicchio, A.; Marsich, N.; Camus, A. *Inorg. Chim. Acta*, **1988**, *142*, 269. (b) Camus, A.; Marsich, N.; Jin, X.; Han, Y.; Tang, K., unpublished.

15. (a) Camus, A.; Marsich, N.; Maria, A.; Lanfredi, M.; Ugozzoli, F. *Inorg. Chim. Acta*, in press. (b) Maria, A.; Lanfredi, M.; Tiripicchio, A.; Camus, A.; Marsich, N. *J. Chem. Soc., Dalton Trans.*, in press.

16. (a) Pauling, L. *The Nature of the Chemical Bond*, Cornell University Press, Ithaca, NY, 1960. (b) Wells, A. F. *Structural Inorganic Chemistry*, 4th ed., Oxford University Press, London, 1975, p. 1015.

17. Dietrich, H.; Storck, W.; Manecke, G. *J. Chem. Soc., Chem. Commun.*, **1982**, 1036.

18. Tang, K.; Aslam, M.; Block, E.; Nicholson, T.; Zubieta, J. *Inorg. Chem.*, **1987**, *26*, 1488 and references therein.

19. Tang, K.; Yang, Q.; Liao, H.; Tang, Y. *Kexue Tongbao*, **1988**, *33(6)*, 429.

20. Tang, K.; Yang, Q.; Yang, J.; Tang, Y. *Beijing Daxue Xuebao, Ziran Kexieban*, **1988**, *24(4)*, 399.

21. Yang, Q.; Tang, K.; Liao, H.; Han, Y.; Chen, Z.; Tang, Y. *J. Chem. Soc., Chem. Commun.*, **1987**, 1076.

22. Tang, K.; Yang, J.; Yang, Q.; Tang, Y. *J. Chem. Soc., Dalton Trans.*, **1989**, 2297.

23. Tang, K.; Yang, Y., unpublished results.

24. Truce, W. E.; Guy, M. M. *J. Org. Chem.*, **1961**, *26*, 4331.

25. Dance, I. G. *Inorg. Chim. Acta*, **1977**, *25*, L17.

26. Dance, I. G.; Fitzpatrick, L. J.; Rae, A. D.; Scudder, M. L. *Inorg. Chem.*, **1983**, *22*, 3785.

27. Bondi, A. *J. Phys. Chem.*, **1964**, *68*, 441.

28. Koch, S. A.; Fikar, R.; Millar, M.; O'Sullivan, T. *Inorg. Chem.*, **1984**, *23*, 121.

29. Dance, I. G. *Aust. J. Chem.*, **1978**, *31*, 2195.

30. Dance, I. G. *J. Chem. Soc., Chem. Commun.*, **1976**, 68.

31. Kunchar, N. R. *Nature (London)*, **1964**, *204*, 468.

32. Bradley, D. C.; Kunchar, N. R., unpublished results.

33. (a) Butler, I. S.; Frenster, A. E. *J. Organomet. Chem.*, **1974**, *66*, 161 and references therein. (b) Duncan, W. F.; Ott, E.; Reid, E. E. *Ind. Eng. Chem.*, **1931**, *23*, 381.

34. Eller, P. G.; Ryan, R. R. *Acta Cryst.*, **1977**, *B33*, 619.

35. Avdeff, A.; Fackler, J. P., Jr. *J. Coord. Chem.*, **1975**, *4*, 211.

36. Chadha, R.; Kumar, R.; Tuck, D. G. *J. Chem. Soc., Chem. Commun.*, **1986**, 188.

37. Tang, K.; Xing, Q.; Yang, J.; Yang, Q.; Tang, Y. *Acta Chim. Sinica*, in press.

38. Jansen, M. *Angew. Chem., Int. Ed. Engl.*, **1987**, *26*, 1098.

Index